# Lecture Notes in Mathematics

A collection of informal reports and seminars
Edited by A. Dold, Heidelberg and B. Eckmann, Zürich

Series: Department of Mathematics, University of Maryland,
College Park
Adviser: J. K. Goldhaber

216




Hans Maaß

Universität Heidelberg, Heidelberg/Deutschland


# Siegel's Modular Forms and Dirichlet Series

Course Given at the University of Maryland, 1969–1970


Springer-Verlag
Berlin · Heidelberg · New York 1971

AMS Subject Classifications (1970): 10D20, 10H10

ISBN 3-540-05563-0 Springer-Verlag Berlin · Heidelberg · New York
ISBN 0-387-05563-0 Springer-Verlag New York · Heidelberg · Berlin



Offsetdruck: Julius Beltz, Hemsbach/Bergstr.

It is the large generalization,
limited by a happy particularity,
which is the fruitful conception.

Alfred North Whitehead

Dedicated to

the last great representative

of a passing epoch

CARL LUDWIG SIEGEL

on the occasion of his

seventy-fifth birthday

## PREFACE

These notes present the content of a course I delivered at the University of Maryland, College Park, between September 1969 and April 1970. The choice of the subject was mainly determined by my intention to show how Atle Selberg makes fascinating use of differential operators in order to prove certain functional equations. Of course one has to be somewhat familiar with his theory of weakly symmetric Riemannian spaces, but - as Selberg himself pointed out to me - the main idea can be found already in Riemann's work. Since Selberg never published his idea, it might be of some value for the mathematical community to make available to a wider public the methods which were originally conceived by Selberg a long time ago.

In November 1970 Mrs. Audrey A. Terras sent me two preprints [33, 34] showing that she obtained independently some of the results of §17 by similar methods.

The audience of my course, strongly interested in the subject, had an influence on these notes through many valuable comments; Professor John Horváth undertook the editorial task of preparing them for publication, whereas the typing was done with supreme accuracy by Mrs. John Vanderslice. I enjoyed in gratitude the stimulating atmosphere of the Department of Mathematics of the University of Maryland, and felt encouraged to make more effort.

Heidelberg, May 1971

H. Maaß

CONTENTS

CONTENTS

## §1. Preliminary remarks on topological groups

We shall be concerned in these lectures with the general linear group $GL(n,K)$ and some of its closed subgroups, where $K$ stands either for the field $\mathbb{R}$ of real numbers or the field $\mathbb{C}$ of complex numbers.

We shall employ the following notations. If $U = (u_{\mu\nu})$ is any matrix, $\bar{U} = (\bar{u}_{\mu\nu})$ will be the conjugate complex matrix, $U' = (u_{\nu\mu})$ will be the transposed matrix, $dU$ will be the matrix of differentials $(du_{\mu\nu})$, and if $U = U^{(n)}$ is a square matrix of order $n$, then $\sigma(U) = \sum_{\nu=1}^{n} u_{\nu\nu}$ will be its trace. We shall indicate by $W = W^{(m,n)}$ that $W$ is a matrix with $m$ rows and $n$ columns. Hence in particular $W^{(n,n)} = W^{(n)}$.

On $GL(n,K)$ we define a Riemannian metric with the help of the fundamental form

$$ds^2 = \sigma(U^{-1}\, dU\, d\bar{U}'\, \bar{U}'^{-1}).$$

This $ds^2$ is invariant under the transformation $U \mapsto VU$, where $V$ is any fixed element of $GL(n,K)$. Also $ds^2$ is a positive definite Hermitian form, and therefore $GL(n,K)$ and any Lie subgroup $\Omega$ of $GL(n,K)$ can be considered as a Riemannian manifold. In particular the Riemannian geometry yields an invariant volume element on $\Omega$, and we do not need the concept of a Haar measure.

Let $\Omega$ be a Lie subgroup of $GL(n,K)$ (or, more generally, any Hausdorff topological group), and let $R$ be a topological space with a Hausdorff topology. A *continuous representation of $\Omega$ as a transitive transformation group of $R$* is defined if:

1. there exists a continuous map $(\omega, r) \mapsto \omega r$ from $\Omega \times R$ into $R$ such that

2. $\Omega r = R$ for every $r \in R$, and

3. $\omega_2(\omega_1 r) = (\omega_2\omega_1)r$ for every $\omega_1, \omega_2 \in \Omega$ and $r \in R$.

We call $R$ a representation space of $\Omega$. Two representation spaces $R$ and $R^*$ are said to be *equivalent* if there exists a homeomorphism $\psi\colon r \mapsto r\psi$ from $R$ onto $R^*$ such that $\omega(r\psi) = (\omega r)\psi$ for any $\omega \in \Omega$ and $r \in R$.

If $\Delta$ is a closed subgroup of $\Omega$, then the set of cosets

$$\Omega/\Delta = \{\rho\Delta \mid \rho \in \Omega\}$$

becomes a Hausdorff space when we equip it with the quotient topology. If we define a map from $\Omega \times (\Omega/\Delta)$ into $\Omega/\Delta$ by $(\omega, \rho\Delta) \mapsto \omega\rho\Delta$, then conditions 1-3 listed above are satisfied, i.e., $\Omega/\Delta$ can be considered as a representation space of $\Omega$.

The proof of the following lemma can be found in Pontrjagin's Topological Groups:

*Lemma* 1. *Let $\Omega$ be a locally compact topological group and $R$ a locally compact representation space of $\Omega$. Assume that $R$ is the union of denumerably many compact subsets. Let $r_o$ be an arbitrary point of $R$ and $\Delta$ the closed subgroup $\{\omega \mid \omega \in \Omega,\ \omega r_o = r_o\}$ of $\Omega$ leaving $r_o$ fixed. Then the representation of $\Omega$ as a transitive transformation group of $R$ is equivalent to the representation as a transitive transformation group of the space $\Omega/\Delta$.*

Observe that if $r_1$ is another point of $R$ and $\Delta_1 = \{\omega \mid \omega \in \Omega,\ \omega r_1 = r_1\}$, then the representation spaces $\Omega/\Delta$ and $\Omega/\Delta_1$ are of course equivalent. If $r_1 = \phi r_o$, with $\phi \in \Omega$, then $\Delta_1 = \phi\Delta\phi^{-1}$ and the homeomorphism $\Omega/\Delta \to \Omega/\Delta_1$ defining the equivalence is given by $\rho\Delta \mapsto \rho\Delta\phi^{-1} = \rho\phi^{-1}\Delta_1$.

The assumptions of Lemma 1 will be fulfilled in the special cases of subgroups of $GL(n,K)$ we shall consider, and therefore it will be sufficient to consider only representation spaces of the form $\Omega/\Delta$.

We shall say that a subgroup $\Gamma$ of $\Omega$ is *discontinuous* on the representation space $\Omega/\Delta$, if for any sequence $\{\gamma_n\}$ of elements of $\Gamma$ such that $\gamma_m \neq \gamma_n$ whenever $m \neq n$, and for any element $\rho \in \Omega$, the sequence $\{\gamma_n \rho\Delta\}$ does not converge in $\Omega/\Delta$.

Denote by $\varepsilon$ the unit element of $\Omega$. The group itself $\Omega = \Omega/\{\varepsilon\}$ is a representation space of $\Omega$. We shall say that $\Gamma \subset \Omega$ is *discrete* if it is discontinuous on $\Omega$. If $\varepsilon$ has a denumerable basis of neighborhoods, then $\Gamma$ is discrete if and only if there exists a neighborhood $U(\varepsilon)$ of $\varepsilon$ in $\Omega$ such that $U(\varepsilon) \cap \Gamma = \{\varepsilon\}$.

Let $\Gamma$ be a subgroup of $\Omega$ which is discontinuous on $\Omega/\Delta$. Then for any $\rho \in \Omega$ and any sequence $\{\gamma_n\}$ of elements of $\Gamma$ such that $\gamma_m \neq \gamma_n$ for $m \neq n$, the sequence $\{\gamma_n \rho\Delta\}$ does not converge in $\Omega/\Delta$. But then $\{\gamma_n \rho\}$ cannot have a limit in $\Omega$, hence $\Gamma$ is discrete.

Conversely, assume that $\varepsilon$ has a denumerable basis of neighborhoods in $\Omega$, that $\Delta$ is a compact subgroup of $\Omega$, and that $\Gamma$ is a discrete subgroup of $\Omega$. Let us show that $\Gamma$ is discontinuous on $\Omega/\Delta$.

Assuming the contrary, let $\{\gamma_n\}$ be a sequence of elements of $\Gamma$ whose terms are all different, and $\rho$ an element of $\Omega$ such that $\{\gamma_n \rho\Delta\}$ converges to $\beta\Delta$ in $\Omega/\Delta$. There exists a sequence $\{\delta_n\}$ in $\Delta$ such that $\{\gamma_n \rho \delta_n\}$ converges to $\beta$. Since $\Delta$ is compact, there exists a subsequence $\{\delta_{\nu_n}\}$ of $\{\delta_n\}$ which converges to some element $\delta \in \Delta$. Then $\{\delta_{\nu_n}^{-1}\}$ converges to $\delta^{-1}$ and $\gamma_{\nu_n} = \gamma_{\nu_n} \rho \delta_{\nu_n} \delta_{\nu_n}^{-1} \rho^{-1} \to \beta\delta^{-1}\rho^{-1} = \tau$ as $n \to \infty$. Choose the neighborhoods $U = U(\varepsilon)$ of $\varepsilon$ and $V = V(\tau)$ of $\tau$ such that $U(\varepsilon) \cap \Gamma = \{\varepsilon\}$ and $VV^{-1} \subset U$. For all sufficiently large $m$ and $n$ we have $\gamma_{\nu_m} \in V$ and $\gamma_{\nu_n} \in V$, hence $\gamma_{\nu_m}\gamma_{\nu_n}^{-1} \in VV^{-1} \subset U$. Since $\gamma_{\nu_m}\gamma_{\nu_n}^{-1} \in \Gamma$, we have $\gamma_{\nu_m}\gamma_{\nu_n}^{-1} = \varepsilon$, i.e., $\gamma_{\nu_m} = \gamma_{\nu_n}$, in contradiction with the assumption that $\gamma_m \neq \gamma_n$ for $m \neq n$. We have thus proved

*Lemma* 2. *Let* $\Omega$ *be a Hausdorff topological group with a denumerable basis of neighborhoods of the unit element, and let* $\Delta$ *be a compact subgroup of* $\Omega$. *Then a subgroup* $\Gamma$ *of* $\Omega$ *is discrete if and only if it is discontinuous on* $\Omega/\Delta$.

If $\Omega$ is a Lie subgroup of $GL(n,K)$, and $\Delta$ a closed subgroup of $\Omega$, then $\dim \Omega/\Delta + \dim \Delta = \dim \Omega$. We see therefore that to ask for all representation spaces $\Omega/\Delta$ of minimal dimension, on which the discreteness of $\Gamma \subset \Omega$ is equivalent to the discontinuity of $\Gamma$ on $\Omega/\Delta$, means in general to ask for all compact subgroups $\Delta$ of maximal dimension of $\Omega$. We shall determine therefore the maximal compact subgroups $\Delta$ of some special Lie subgroups $\Omega$ of $GL(n,K)$.

A square matrix $S$ is the matrix of a positive definite quadratic form if and only if $\bar{S} = S = S' > 0$. A square matrix $H$ is the matrix of a positive definite Hermitian form if and only if $\bar{H}' = H > 0$. If $X$ is any matrix with as many rows as the order of $S$ or $H$, we shall write

$$S[X] = X'SX \quad \text{and} \quad H\{X\} = \bar{X}'HX.$$

*Lemma* 3. *Let* $\Omega$ *be a closed subgroup of* $GL(n,K)$. *Any maximal compact subgroup* $\Delta$ *of* $\Omega$ *can be represented in the form*

$$\Delta = \{U \mid H\{U\} = H, \quad U \in \Omega\}$$

*with some positive Hermitian matrix* $H = \bar{H}' > 0$. *In the case of* $K = \mathbb{R}$ *we have* $H\{U\} = H[U]$.

*Proof*. Choose a measurable subset $\mathfrak{F}$ of $\Omega$ which has a finite volume $> 0$ with respect to the invariant volume element $dv$ on $\Omega$.

Since $\Delta$ is compact, the set $\mathcal{L} = \bigcup_{V \in \Delta} V\mathcal{L}$ is also measurable and has finite positive volume. Furthermore $V\mathcal{L} = \mathcal{L}$ for every $V \in \Delta$. The matrix

$$H_0 = \int_{U \in \mathcal{L}} U\bar{U}' \, dv$$

satisfies $\bar{H}_0' = H_0 > 0$ and $VH_0\bar{V}' = H_0$ for $V \in \Delta$. Define $H = H_0^{-1}$. Then $\bar{H}' = H > 0$, and replacing $V$ by $V^{-1}$ we see that $H\{V\} = V$ for every $V \in \Delta$. Thus $\Delta$ is contained in the subgroup $\{U \mid H\{U\} = H,\ U \in \Omega\}$ of $\Omega$. Since this last subgroup is compact and $\Delta$ is maximal, the two must be identical.

*Examples.* 1. If $\Omega = GL(n,K)$, then all maximal compact subgroups are conjugate. Indeed, if $\Delta = \{U \mid H\{U\} = H,\ U \in \Omega\}$, then there exists $Q \in GL(n,K)$ such that $H = \bar{Q}'Q$, and the relation $H\{U\} = H$ means that $\bar{Q}^{-1}\bar{U}'\bar{Q}' = QU^{-1}Q^{-1}$, i.e., for each $U \in \Delta$, the matrix $V = QUQ^{-1}$ is orthogonal ($V' = V^{-1}$) if $K = \mathbb{R}$ or unitary ($\bar{V}' = V^{-1}$) if $K = \mathbb{C}$. Thus we have $\Delta = Q^{-1}O(n)Q$ for $K = \mathbb{R}$ and $\Delta = Q^{-1}U(n)Q$ for $K = \mathbb{C}$, where $O(n)$ denotes the real orthogonal group, and $U(n)$ the unitary group.

As we have observed above, conjugate subgroups yield equivalent representation spaces, hence we have to consider only the subgroups $\Delta = O(n)$ or $\Delta = U(n)$, according as $K = \mathbb{R}$ or $K = \mathbb{C}$.

2. Let $\Omega$ be the special linear group $SL(n,\mathbb{C})$ of all matrices in $GL(n,\mathbb{C})$ whose determinant is equal to 1, and let $\Delta = SU(n) = \{U \mid U \in U(n),\ |U| = 1\}$, where $|U|$ shall always denote the determinant of the square matrix $U$. If with every coset $P\Delta$ we associate the Hermitian matrix $Y = P\bar{P}' > 0$ with $|Y| = 1$, then we obtain a one-to-one correspondence between the representation space $\Omega/\Delta$ and the set $\{Y \mid Y = \bar{Y}' > 0,\ |Y| = 1\}$, and this mapping is a homeomorphism.

In the particular case $n = 2$, we can choose in each coset $P\Delta$ a special representative

$$P = \begin{pmatrix} 1 & z \\ 0 & 1 \end{pmatrix} \begin{pmatrix} \sqrt{t} & 0 \\ 0 & \frac{1}{\sqrt{t}} \end{pmatrix} = \begin{pmatrix} \sqrt{t} & \frac{z}{\sqrt{t}} \\ 0 & \frac{1}{\sqrt{t}} \end{pmatrix},$$

where $z = x + iy \in \mathbb{C}$ is arbitrary, and $t > 0$. We then have

$$Y = P\overline{P}' = \begin{pmatrix} (z\bar{z} + t^2)t^{-1} & zt^{-1} \\ \bar{z}t & t^{-1} \end{pmatrix},$$

and we associate with $Y$ the quaternion

$$\zeta = z + jt = x + iy + jt, \qquad t > 0,$$

where $1, i, j, k = ij$ are the quaternion units. Let us determine what is the effect of

$$U = \begin{pmatrix} a & b \\ c & d \end{pmatrix} \in \Omega$$

on $\zeta$. On $\Omega/\Delta$ the matrix $U$ transforms $P\Delta$ into $UP\Delta$, the corresponding transformation changes $Y = P\overline{P}'$ into $UY\overline{U}'$ and $\zeta$ into a quaternion $\zeta^* = x^* + iy^* + jt^*$ we want to calculate. Now $P^*\Delta = UP\Delta$ means that there exists a matrix $V$ with $V\overline{V}' = E$ (= the unit matrix) and $|V| = 1$, such that $P^*V = UP$. The matrix $V$ has the form

$$V = \begin{pmatrix} u & v \\ -\bar{v} & \bar{u} \end{pmatrix}, \qquad \text{where } u\bar{u} + v\bar{v} = 1,$$

and so

$$V\begin{pmatrix} j \\ 1 \end{pmatrix} = \begin{pmatrix} uj + v \\ -\bar{v}j + \bar{u} \end{pmatrix} = \begin{pmatrix} j \\ 1 \end{pmatrix}(\bar{u} - \bar{v}j)$$

since $ji = -ij$. Also

$$P\begin{pmatrix} j \\ 1 \end{pmatrix} = \begin{pmatrix} jt + z \\ 1 \end{pmatrix}\frac{1}{\sqrt{t}} = \begin{pmatrix} \zeta \\ 1 \end{pmatrix}\frac{1}{\sqrt{t}},$$

hence

$$UP\begin{pmatrix} j \\ 1 \end{pmatrix} = U\begin{pmatrix} \zeta \\ 1 \end{pmatrix}\frac{1}{\sqrt{t}} = P^*V\begin{pmatrix} j \\ 1 \end{pmatrix} = P^*\begin{pmatrix} j \\ 1 \end{pmatrix}(\bar{u} - \bar{v}j)$$

$$= \begin{pmatrix} \zeta^* \\ 1 \end{pmatrix}\frac{1}{\sqrt{t^*}}(\bar{u} - \bar{v}j)$$

and therefore, taking into account the expression of $U$,

$$\begin{pmatrix} a\zeta + b \\ c\zeta + d \end{pmatrix}\frac{1}{\sqrt{t}} = \begin{pmatrix} \zeta^* \\ 1 \end{pmatrix}\frac{1}{\sqrt{t^*}}(\bar{u} - \bar{v}j).$$

Comparing the two sides we get

$$\frac{1}{\sqrt{t}}(a\zeta + b) = \zeta^* \frac{1}{\sqrt{t^*}}(\bar{u} - \bar{v}j),$$

$$\frac{1}{\sqrt{t}}(c\zeta + d) = \frac{1}{\sqrt{t^*}}(\bar{u} - \bar{v}j),$$

which yields finally the required transformation formulas

$$\zeta^* = (a\zeta + b)(c\zeta + d)^{-1}$$

and

$$t^* = \frac{t}{|c\zeta + d|^2}.$$

Let us observe that these mappings $\zeta \mapsto \zeta^*$ are the orientation preserving isometries of the three-dimensional hyperbolic space equipped with the metric

$$ds^2 = \frac{dx^2 + dy^2 + dt^2}{t^2},$$

where $t > 0$.

## §2. *Automorphism groups of bilinear forms*

Let $B = B^{(m)}$ be a real square matrix of order $m$. We define the automorphism group $\Omega(B)$ of $B$ by

$$\Omega(B) = \{U \mid U \in GL(m, \mathbb{R}), \quad B[U] = B\},$$

where we use the notation $B[U] = U'BU$ introduced in §1. The matrix $B$ can be decomposed uniquely into $B = S + A$, where $S' = S$ is symmetric and $A' = -A$ is skew-symmetric. Furthermore $B[U] = B$ if and only if $S[U] = S$ and $A[U] = A$, hence

$$\Omega(B) = \Omega(S) \cap \Omega(A).$$

Therefore it is sufficient to consider the special cases when either $B' = B$ or $B' = -B$. It is easy to see that

$$\Omega(B[T]) = T^{-1}\Omega(B)T.$$

Now we can find an invertible matrix $T$ such that $B[T]$ is of the form

$$\begin{pmatrix} B_1 & 0 \\ 0 & 0 \end{pmatrix},$$

where $B_1 = B_1^{(r)}$ is an $r$-rowed square matrix whose determinant $|B_1|$ is different from zero, and therefore we may assume that $B$ itself is of the form

$$B = \begin{pmatrix} B_1 & 0 \\ 0 & 0 \end{pmatrix}, \qquad B_1 = B_1^{(r)}, \qquad |B_1| \neq 0.$$

If we decompose $U \in GL(m, \mathbb{R})$ into boxes corresponding to the form of $B$:

$$U = \begin{pmatrix} U_1 & U_2 \\ U_3 & U_4 \end{pmatrix}, \qquad U_1 = U_1^{(r)},$$

then $B[U] = B$ is equivalent to $B_1[U_1] = B_1$, $U_1'B_1U_2 = 0$. The second condition implies that $U_2 = 0$. So $U \in \Omega(B)$ must be of the form

$$U = \begin{pmatrix} U_1 & 0 \\ U_3 & U_4 \end{pmatrix},$$

where $B_1[U_1] = B_1$, $U_3$ is arbitrary and $U_4$ non-singular. Let $N$ be the normal subgroup of $\Omega(B)$ generated by the elements

$$\begin{pmatrix} E^{(r)} & 0 \\ W & E \end{pmatrix} \quad \text{and} \quad \begin{pmatrix} E^{(r)} & 0 \\ 0 & U_4 \end{pmatrix},$$

where $W$ is arbitrary and $|U_4| \neq 0$. Then $\Omega(B)/N$ is isomorphic to $\Omega(B_1)$, and we are reduced to the study of the automorphism group of a non-singular matrix.

We shall therefore consider an $m$-rowed square matrix $B$ with $|B| \neq 0$, such that either $B' = B$ or $B' = -B$. Replacing, if necessary, $B$ by $B[T]$, we may finally assume that $B$ is of one of the special types

$$B_{pq} = \begin{pmatrix} E^{(p)} & 0 \\ 0 & -E^{(q)} \end{pmatrix} \qquad \text{with} \quad m = p+q,$$

or

$$B_n = \begin{pmatrix} 0 & E^{(n)} \\ -E^{(n)} & 0 \end{pmatrix} \qquad \text{with} \quad m = 2n.$$

*Lemma* 1. *Let* $S = S^{(m)} = \bar{S} = S' > 0$ *and let* $B$ *be either* $B_{pq}$ *or* $B_n$. *Then there exists* $T \in \Omega(B)$ *such that* $S[T]$ *is equal to a positive diagonal matrix* $D = (\delta_{\mu\nu} d_\nu)$. *In the case* $B = B_n$ *we may even obtain* $d_{n+\nu} = d_\nu$ $(\nu = 1,\ldots,n)$.

*Proof*. Since $S > 0$, it has a uniquely determined positive, real, symmetric square root $\sqrt{S} > 0$. If $S$ is written in the form $S = D[V]$, where $D$ is diagonal and $V'V = E$, then $\sqrt{S} = \sqrt{D}\,[V]$.

*Case* 1: $B' = B = B_{pq}$. Let $U$ be an orthogonal matrix, $U'U = E$, such that

$$B[\sqrt{S}\,U] = \begin{pmatrix} D_1 & 0 \\ 0 & -D_2 \end{pmatrix},$$

where $D_1$ and $D_2$ are positive diagonal matrices. By Sylvester's law of inertia $D_1 = D_1^{(p)}$ and $D_2 = D_2^{(q)}$. Applying the automorphism $X \mapsto X'^{-1}$ and remembering that $B = B' = B^{-1}$, $\sqrt{S}' = \sqrt{S}$, $U'^{-1} = U$, we get

$$B[\sqrt{S}^{-1}\,U] = \begin{pmatrix} D_1^{-1} & 0 \\ 0 & -D_2^{-1} \end{pmatrix}$$

and therefore

$$B[\sqrt{S}^{-1}\,U \begin{pmatrix} \sqrt{D}_1 & 0 \\ 0 & \sqrt{D}_2 \end{pmatrix}] = \begin{pmatrix} E^{(p)} & 0 \\ 0 & -E^{(q)} \end{pmatrix} = B.$$

Thus choosing

$$T = \sqrt{S}^{-1}\,U \begin{pmatrix} \sqrt{D}_1 & 0 \\ 0 & \sqrt{D}_2 \end{pmatrix}$$

we have $B[T] = B$ and $S[T] = \begin{pmatrix} D_1 & 0 \\ 0 & D_2 \end{pmatrix}$, q.e.d.

*Case* 2: $B' = -B = B^{-1}$, $B = B_n$. The matrix $H = iB[\sqrt{S}]$ is Hermitian, $\bar{H}' = H$, hence all the eigenvalues of $H$ are real. Also $H' = -H$, hence $H$ and $-H$ have the same eigenvalues. If $d_1, d_2, \ldots, d_n$ are the positive eigenvalues of $H$, then $d_1, d_2, \ldots, d_n, -d_1, -d_2, \ldots, -d_n$ are all the eigenvalues of $H$. Let

$$D = \begin{pmatrix} d_1 & & \\ & \ddots & \\ & & d_n \end{pmatrix} = (\delta_{\mu\nu} d_\nu).$$

We can determine column vectors $\mathfrak{z}_\nu$ such that $H\mathfrak{z}_\nu = \mathfrak{z}_\nu d_\nu$ for $\nu = 1, \ldots, n$ and such that if $Z = Z^{(m,n)} = (\mathfrak{z}_1, \mathfrak{z}_2, \ldots, \mathfrak{z}_n)$, then $\bar{Z}'Z = E$, i.e., $\bar{\mathfrak{z}}_\mu' \mathfrak{z}_\nu = \delta_{\mu\nu}$. In other words, the $\mathfrak{z}_\nu$ are unitary orthonormal eigenvectors of $H$. Since $\bar{H} = -H$, we have $H\bar{\mathfrak{z}}_\nu = -\bar{\mathfrak{z}}_\nu d_\nu$, and therefore $\bar{\mathfrak{z}}_\mu$ must be unitary orthogonal to all $\mathfrak{z}_\nu$, i.e., $Z'Z = 0$. The $m$-rowed square matrix $U = (Z, \bar{Z})$ is unitary, $\bar{U}'U = E$. We have the relations $HZ = ZD$ and $H\bar{Z} = -\bar{Z}D$ which taken together can be written as

$$\bar{U}'HU = H\{U\} = \begin{pmatrix} D & 0 \\ 0 & -D \end{pmatrix}.$$

Introduce the matrix

$$U_0 = \frac{1}{\sqrt{2}} \begin{pmatrix} iE & iE \\ E & -E \end{pmatrix}.$$

One checks easily that $\bar{U}_0'U_0 = E$ and

$$iB\{U_0\} = \begin{pmatrix} E & 0 \\ 0 & -E \end{pmatrix},$$

and thus we find

$$iB\{\sqrt{S}\,U\} = H\{U\} = \begin{pmatrix} D & 0 \\ 0 & -D \end{pmatrix} =$$

$$= iB\left\{U_o\begin{pmatrix} \sqrt{D} & 0 \\ 0 & \sqrt{D} \end{pmatrix}\right\} = iB\left\{\begin{pmatrix} \sqrt{D} & 0 \\ 0 & \sqrt{D} \end{pmatrix}U_o\right\} .$$

Applying the automorphism $X \mapsto \overline{X}'^{-1}$ we obtain

$$iB\{\sqrt{S}^{-1}U\} = iB\left\{\begin{pmatrix} \sqrt{D}^{-1} & 0 \\ 0 & \sqrt{D}^{-1} \end{pmatrix}U_o\right\} .$$

Define

$$T = i\sqrt{S}^{-1}U\,U_o^{-1}\begin{pmatrix} \sqrt{D} & 0 \\ 0 & \sqrt{D} \end{pmatrix} .$$

This matrix is real since

$$iUU_o^{-1} = \frac{i}{\sqrt{2}}\,(Z,\overline{Z})\begin{pmatrix} -iE & E \\ -iE & -E \end{pmatrix} =$$

$$= \frac{1}{\sqrt{2}}\,(Z+\overline{Z}\,,\,i(Z-\overline{Z}))$$

is real. Furthermore we have $B\{T\} = B[T] = B$ and

$$S[T] = S\{T\} = E\left\{UU_o^{-1}\begin{pmatrix} \sqrt{D} & 0 \\ 0 & \sqrt{D} \end{pmatrix}\right\} = \begin{pmatrix} D & 0 \\ 0 & D \end{pmatrix} .$$

Thus the lemma is proved.

In the case when $\Omega = \Omega(B)$, Lemma 3 of §1 can be strengthened as follows:

*Lemma 2. Let $B \in GL(m,\mathbb{R})$ with either $B' = B$ or $B' = -B$. Then all maximal compact subgroups $\Delta$ of $\Omega(B)$ are conjugate. If $B = B_{pq}$ or $B = B_n$, then*

$$\Delta = T(\Omega(B) \cap \Omega(E))T^{-1}$$

*for some* $T \in \Omega(B)$.

*Proof.* By the transformation property of $\Omega(B)$ we may assume that $B = B_{pq}$ or $B = B_n$. In either case it follows from Lemma 3 of §1 that

$$\Delta = \Omega(B) \cap \Omega(S)$$

for some $S > 0$. In accordance with Lemma 1 determine $T \in \Omega(B)$ such that $S[T]$ is the positive diagonal matrix $D$, i.e., $S = D[T^{-1}]$. We have obviously

$$\Delta = \Omega(B) \cap \Omega(D[T^{-1}]) = \Omega(B) \cap T\Omega(D)T^{-1}$$

$$= T(\Omega(B) \cap \Omega(D))T^{-1}.$$

We want to show that $\Omega(B) \cap \Omega(D)$ is contained in the orthogonal group $\Omega(E) = O(m)$. First of all, observe that in the case $B = B_n$ we may assume that $D$ is of the form $\begin{pmatrix} D_1 & 0 \\ 0 & D_1 \end{pmatrix}$. Let $U \in \Omega(B) \cap \Omega(D)$, i.e., $U'BU = B$ and $U'DU = D$. Then $U^{-1}B^{-1}U'^{-1} = B^{-1}$, and so $U^{-1}BU'^{-1} = B$, i.e., $BU'^{-1} = UB$. It follows that $BDU = BU'^{-1}D = UBD$ and $(BD)^2U = U(BD)^2$. Now one checks directly that $(BD)^2 = D^2$ if $B = B_{pq}$ and $(BD)^2 = -D^2$ if $B = B_n$. Therefore in either case we have $D^2U = UD^2$. Set $D = (\delta_{\mu\nu}d_\nu)$, $U = (u_{\mu\nu})$. Then we have $(d_\mu^2 - d_\nu^2)u_{\mu\nu} = 0$, and since $d_\mu + d_\nu > 0$, also $(d_\mu - d_\nu)u_{\mu\nu} = 0$ for all $\mu$, $\nu$, i.e., $DU = UD$. From $U'DU = D$ it follows now that $U'UD = D$, i.e., $U'U = E$, and so $U$ is indeed orthogonal.

Thus we see that $\Omega(B) \cap \Omega(D) \subset \Omega(E)$, and therefore $\Delta \subset T(\Omega(B) \cap \Omega(E))T^{-1}$. Now $\Delta$ is a maximal compact subgroup of $\Omega(B)$ and $T(\Omega(B) \cap \Omega(E))T^{-1}$ is a compact subgroup of $\Omega(B)$, hence we must have $\Delta = T(\Omega(B) \cap \Omega(E))T^{-1}$.

The group $\Omega(B)$ is itself compact if and only if $\Omega(B) \subset \Omega(E)$. This happens exactly in the cases $B = B_{om}$ and $B = B_{mo}$, which we shall exclude from now on.

Let $B = B_{pq}$ or $B = B_n$ and assume that the automorphism group $\Omega(B)$ is not compact. Let $\Delta$ be the maximal compact subgroup $\Omega(B) \cap \Omega(E)$ of $\Omega(B)$ and consider the representation space $\Omega(B)/\Delta$ of $\Omega(B)$. We want to prove that the map

$$U\Delta \mapsto P = UU', \qquad U \in \Omega(B),$$

establishes a homeomorphism from $\Omega(B)/\Delta$ onto

$$\mathcal{P}(B) = \{P \mid P = \overline{P} = P' > 0, \quad PBP = B\}.$$

*Proof*. 1. The map is well-defined. Indeed, if $U_1\Delta = U_2\Delta$ with $U_1, U_2 \in \Omega(B)$, then $U_2 = U_1V$ with $V \in \Delta \subset \Omega(E)$ and therefore $U_2U_2' = U_1VV'U_1' = U_1U_1'$. 2. The map is one-to-one. Let $U_1U_1' = U_2U_2'$, then $U_1^{-1}U_2 \in \Omega(B) \cap \Omega(E) = \Delta$ and therefore $U_1\Delta = U_2\Delta$. 3. Let $P \in \mathcal{P}(B)$. We want to show that there exists $U \in \Omega(B)$ such that $UU' = P$. By Lemma 1 we can choose $V \in \Omega(B)$ such that $P[V]$ is equal to a positive diagonal matrix $D$. Setting $U' = V^{-1} \in \Omega(B)$, we have $D[U'] = P$. Also $U \in \Omega(B)$ since $UBU' = B$ implies $U'^{-1}B^{-1}U^{-1} = B^{-1}$ or $U'BU = B$. We have $UDBDU' = UDU'BUDU' = PBP = B$, and so $DBD = U^{-1}BU'^{-1} = B$, i.e., $(DB)^2 = \pm E$. Recalling that in the case $B = B_n$ we can choose $D = \begin{pmatrix} D_1 & 0 \\ 0 & D_1 \end{pmatrix}$, we obtain as in the proof of Lemma 2 that $(DB)^2 = \pm D^2$ and therefore $D^2 = \pm E$. Since $D$ is positive, we have $D = E$ and so $P = UU'$. Finally, it is obvious that the maps are continuous.

Next we want to give a special parametric representation of $\mathcal{P}(B)$. We shall treat the two cases $B = B_{pq}$ and $B = B_n$ separately.

| (I) $\quad B = B_{pq} \quad (pq > 0)$ |
|---|

A matrix $U = \begin{pmatrix} U_1 & U_2 \\ U_3 & U_4 \end{pmatrix}$ with $U_1 = U_1^{(p)}$ belongs to $\Omega(B)$ if and only if

$$U_1'U_1 - U_3'U_3 = E^{(p)}, \quad U_4'U_4 - U_2'U_2 = E^{(q)}, \quad U_1'U_2 = U_3'U_4.$$

Clearly $U_3'U_3$ is the matrix of a semi-definite quadratic form; we shall say simply that $U_3'U_3$ is a semi-definite matrix and write $U_3U_3' \geqq 0$. Similarly of course $U_2'U_2 \geqq 0$, from where it follows that $U_1'U_1 = E + U_3'U_3 > 0$, $U_4'U_4 = E + U_4'U_4 > 0$, and in particular $|U_1| \neq 0$, $|U_4| \neq 0$. Let $X = X^{(p,q)}$ be the matrix $U_2U_4^{-1}$. Then $U_2 = XU_4$, $U_2' = U_4'X'$, $U_3' = U_1'X$, $U_3 = X'U_1$, and therefore $U_1'(E - XX')U_1 = E$, $U_4'(E - X'X)U_4 = E$. From these relations we obtain that $E - XX' > 0$ and $E - X'X > 0$, and these last two conditions are equivalent since

$$\begin{pmatrix} E - XX' & 0 \\ 0 & E \end{pmatrix} = \begin{pmatrix} E & 0 \\ 0 & E - X'X \end{pmatrix} \begin{bmatrix} E - XX' & -X \\ X' & E \end{bmatrix} .$$

If we set $W_1 = \sqrt{E - XX'} \cdot U_1$ and $W_2 = \sqrt{E - X'X} \cdot U_4$, then these matrices are orthogonal, $W_1'W_1 = E$, $W_2'W_2 = E$, and we have $U_1 = (E - XX')^{-\frac{1}{2}}W_1$, $U_4 = (E - X'X)^{-\frac{1}{2}}W_2$. Thus we obtain the representation

$$U = \begin{pmatrix} (E - XX')^{-\frac{1}{2}} & X(E - X'X)^{-\frac{1}{2}} \\ X'(E - XX')^{-\frac{1}{2}} & (E - X'X)^{-\frac{1}{2}} \end{pmatrix} \begin{pmatrix} W_1 & 0 \\ 0 & W_2 \end{pmatrix} ,$$

where the first factor is a symmetric matrix, the second factor is an orthogonal matrix, and $X$, $W_1$, $W_2$ are uniquely determined by $U$.

We see furthermore that $U$ belongs to the maximal compact subgroup

$\Delta = \Omega(B) \cap \Omega(E)$ if and only if $U_2 = 0$ and $U_3 = 0$, i.e., if $X = 0$. Thus each coset in $\Omega(B)/\Delta$ can be written uniquely in the form

$$U\Delta = \begin{pmatrix} (E - XX')^{-\frac{1}{2}} & X(E - X'X)^{-\frac{1}{2}} \\ X'(E - XX')^{-\frac{1}{2}} & (E - X'X)^{-\frac{1}{2}} \end{pmatrix} \Delta$$

and the corresponding element in $\mathcal{P}(B)$ is

$$P = UU' = \begin{pmatrix} \frac{E + XX'}{E - XX'} & \frac{2X}{E - X'X} \\ \frac{2X'}{E - XX'} & \frac{E + X'X}{E - X'X} \end{pmatrix}.$$

This is the required parametric representation of $\mathcal{P}(B)$. The parameter is the matrix $X = X^{(p,q)}$ which satisfies the equivalent conditions $E^{(p)} - XX' > 0$ and $E^{(q)} - X'X > 0$, and in particular the dimension of $\mathcal{P}(B)$ is $p \cdot q$. Also $\mathcal{P}(B)$ is a bounded domain, since if $X = (\mathfrak{r}_1, \ldots, \mathfrak{r}_q)$, where the $\mathfrak{r}_i$ are column vectors of order $p$, then the condition on $X$ implies $1 > \mathfrak{r}_i' \mathfrak{r}_i$.

Let us determine how an element $M = \begin{pmatrix} A & B \\ C & D \end{pmatrix} \in \Omega(B)$ acts on $X$. If to $U\Delta$ there corresponds the matrix $X = U_2 U_4^{-1}$, then to $MU\Delta$ there corresponds the matrix $X^* = M\langle X\rangle = U_2^* U_4^{*-1}$, where

$$MU = \begin{pmatrix} U_1^* & U_2^* \\ U_3^* & U_4^* \end{pmatrix}.$$

Now $U_2^* = AU_2 + BU_4$, $U_4^* = CU_2 + DU_4$ and so we obtain the formula

$$M\langle X\rangle = (AX + B)(CX + D)^{-1}.$$

(II) $B = B_n$ $(m = 2n)$

Write the matrix $P \in \mathfrak{P}(B)$ in the form

$$P = \begin{pmatrix} P_1 & P_2 \\ P_2' & P_3 \end{pmatrix},$$

where the $P_j$ are square matrices of order $n$, $P_1' = P_1$, $P_3' = P_3$. The conditions $PBP = B$ and $P > 0$ are equivalent to

$$P_1P_3 - P_2^2 = E, \quad P_1P_2' = P_2P_1, \quad P_2'P_3 = P_3P_2, \quad P_1 > 0, \quad P_3 > 0.$$

There exists a matrix $Y > 0$ such that $P_3 = Y^{-1}$, and define $X = P_2P_3^{-1} = P_2Y$. Then $X = P_2P_3^{-1} = P_3^{-1}P_2' = (P_2P_3^{-1})' = X'$, i.e., $X$ is a symmetric matrix. Furthermore $P_1 = (E + P_2^2)P_3^{-1} = Y + XY^{-1}X$, and so we can write

$$P = \begin{pmatrix} Y + XY^{-1}X & XY^{-1} \\ Y^{-1}X & Y^{-1} \end{pmatrix} = \begin{pmatrix} Y & 0 \\ 0 & Y^{-1} \end{pmatrix} \begin{bmatrix} E & 0 \\ X & E \end{bmatrix}.$$

Conversely, if $Y > 0$ and $X = X'$, then the matrix $P$ defined by this equation belongs to $\mathfrak{P}(B)$. Thus we have established a one-to-one correspondence between the representation space $\mathfrak{P}(B)$ and the generalized upper half-plane, or Siegel's upper half-plane,

$$\mathfrak{H}_n = \{Z \mid Z = X + iY = Z', \ X, Y \text{ real}, \ Y > 0\}.$$

If $Y > 0$, then we can find a unique upper triangular matrix $R = (r_{\mu\nu})$, with $r_{\nu\nu} > 0$ $(\nu = 1,\ldots,n)$ and $r_{\mu\nu} = 0$ for $1 \leqq \nu < \mu \leqq n$, such that $Y = RR'$ and the corresponding matrix $P$ can then be written as

$$P = E\begin{bmatrix} R' & 0 \\ R^{-1}X & R^{-1} \end{bmatrix} .$$

We have associated with each coset $U\Delta$ in $\Omega(B)/\Delta$ the element $P = UU' \in \mathcal{P}(B)$. We see now that if $P$ is determined by $X$ and $R$, then

$$U = \begin{pmatrix} R & XR'^{-1} \\ 0 & R'^{-1} \end{pmatrix} K ,$$

where $K = \begin{pmatrix} A_o & B_o \\ C_o & D_o \end{pmatrix}$ is an arbitrary element in $\Delta = \Omega(B) \cap \Omega(E)$.

We have $K' = K^{-1}$ and $K'BK = B$, i.e., $K^{-1} = -BK'B$ since $B^2 = -E$. But

$$K' = \begin{pmatrix} A_o' & C_o' \\ B_o' & D_o' \end{pmatrix}, \qquad -BK'B = \begin{pmatrix} D_o' & -B_o' \\ -C_o' & A_o' \end{pmatrix},$$

hence $A_o = D_o$, $B_o = -C_o$, and we can write $K$ in the form

$$K = \begin{pmatrix} D_o & -C_o \\ C_o & D_o \end{pmatrix},$$

where the condition $K \in \Omega(B)$ is expressed by the relations $C_oC_o' + D_oD_o' = E$, $C_oD_o' = C_oD_o'$. If we set $W = iC_o + D_o$, then $\overline{W}W' = E$, i.e., $W$ is a unitary matrix of order $n$, and to every unitary matrix $W$ there corresponds exactly one matrix $K \in \Delta$. Thus we have established the one-to-one correspondences

$$U \leftrightarrow X,R,K \leftrightarrow X,Y,W \leftrightarrow Z,W$$

$U \in \Omega(B_n)$, $Z \in \mathfrak{H}_n$, $W \in U(n)$.

Finally let us determine how an element $M \in \Omega(B_n)$ operates on $\mathfrak{H}_n$. If we write

$$M = \begin{pmatrix} A & B \\ C & D \end{pmatrix},$$

where $A$, $B$, $C$, $D$ are square matrices of order $n$, then the condition $M \in \Omega(B_n)$ is expressed by the equations

$$A'D - C'B = E, \quad A'C = C'A, \quad B'D = D'B.$$

If $Z \in \mathfrak{H}_n$ corresponds to the coset $U\Delta$, then $Z^* = M\langle Z\rangle$ corresponds to the coset $MU\Delta$, i.e., there exists a matrix $K = \begin{pmatrix} D_0 & -C_0 \\ C_0 & D_0 \end{pmatrix}$, with $W = iC_0 + D_0$ unitary, such that

$$\begin{pmatrix} R^* & X^*R^{*'-1} \\ 0 & R^{*'-1} \end{pmatrix} K = M \begin{pmatrix} R & XR^{'-1} \\ 0 & R^{'-1} \end{pmatrix},$$

where, of course, $Y = RR'$, $Z = X + iY$, $Y^* = R^*R^{*'}$ and $Z^* = X^* + iY^*$. Using the relations

$$K \begin{pmatrix} iE \\ E \end{pmatrix} = \begin{pmatrix} D_0 & -C_0 \\ C_0 & D_0 \end{pmatrix} \begin{pmatrix} iE \\ E \end{pmatrix} = \begin{pmatrix} iE \\ E \end{pmatrix} W$$

and

$$\begin{pmatrix} R & XR^{'-1} \\ 0 & R^{'-1} \end{pmatrix} \begin{pmatrix} iE \\ E \end{pmatrix} = \begin{pmatrix} Z \\ E \end{pmatrix} R^{'-1},$$

we have

$$\begin{pmatrix} R^* & X^*R^{*'-1} \\ 0 & R^{*'-1} \end{pmatrix} K \begin{pmatrix} iE \\ E \end{pmatrix} = \begin{pmatrix} Z^* \\ E \end{pmatrix} R^{*'-1} W,$$

and so multiplying the above equation from the right by $\begin{pmatrix} iE \\ E \end{pmatrix}$ we obtain

$$\begin{pmatrix} Z^* \\ E \end{pmatrix} R^{*'-1}W = \begin{pmatrix} AZ + B \\ CZ + D \end{pmatrix} R^{'-1}.$$

This yields the formulas

$$Z^*R^{*'-1}W = (AZ + B)R^{'-1}$$

$$R^{*'-1}W = (CZ + D)R^{'-1}$$

from where

$$Z^* = M\langle Z\rangle = (AZ + B)(CZ + D)^{-1}.$$

Furthermore

$$Y^{*-1} = R^{*'-1}R^{*-1} = R^{*'-1}W\overline{W}'R^{*-1} =$$

$$= (CZ + D)R^{'-1}R^{-1}(\overline{Z}C' + D') = Y^{-1}\{\overline{Z}C' + D'\},$$

hence

$$Y^* = Y\{(CZ + D)^{-1}\}.$$

In particular for the determinant of $Y$ we have the transformation formula

$$|Y^*| = |Y| \cdot \| CZ + D\|^{-2}.$$

To conclude this section, we mention incidentally that if $B$ is an arbitrary symmetric or skew-symmetric matrix, not necessarily $B_{pq}$ or $B_n$, then the representation space $\mathcal{P}(B)$ has to be defined by

$$\mathcal{P}(B) = \{P \mid P = \overline{P} = P' > 0, \quad PB^{-1}P = \varepsilon P\},$$

where $\varepsilon = +1$ if $B' = B$ and $\varepsilon = -1$ if $B' = -B$. This definition coincides with the one given above in the special cases $B = B_{pq}$ and $B = B_n$. Furthermore $\mathfrak{p}(B)$ has the following invariance property: if $V \in GL(m, \mathbb{R})$ is given and if we set $B^* = B[V]$, $P^* = P[V]$, $U^* = V^{-1}UV$, then

$$P \mapsto P[U] \qquad \text{with} \quad P \in \mathfrak{p}(B), \quad U \in \Omega(B)$$

will be transformed into

$$P^* \mapsto P^*[U^*] \qquad \text{with} \quad P^* \in \mathfrak{p}(B^*), \quad U^* \in \Omega(B^*).$$

## §3. *Geometry in the representation space*

The representation space

$$\mathcal{P}(B) = \{P \mid P = \overline{P} = P' > 0, \quad PBP = B\}$$

of $\Omega(B)$, where $B = B_{pq}$ or $B = B_n$, is a subspace of the set

$$\mathcal{T}_m = \{T \mid T = \overline{T} = T' > 0\}$$

of square matrices $T = T^{(m)}$. This set $\mathcal{T}_m$ is a representation space of the group $\Omega = GL(m, \mathbb{R})$. An element $U \in \Omega$ operates on $\mathcal{T}_m$ according to $T \mapsto UTU'$. The orthogonal group $O(m)$ is a maximal compact subgroup of $\Omega$, and the map $UO(m) \mapsto T = UU'$ establishes an equivalence between the representation spaces $\Omega/O(m)$ and $\mathcal{T}_m$ of $\Omega$.

The map $T \mapsto UTU'$ transforms the matrix $dT = (dt_{\mu\nu})$ according to $dT \mapsto UdT\,U'$. Since $(UTU')^{-1} = U'^{-1}T^{-1}U^{-1}$, the differential form $T^{-1}dT$ on $\mathcal{T}_m$ transforms into $U'^{-1}T^{-1}dT\,U'$ under the action of $U \in \Omega$, and therefore

$$ds^2 = \sigma(T^{-1}dT\ T^{-1}dT) = \sigma(T^{-1}dT)^2$$

defines an invariant metric on $\mathcal{T}_m$. The form $ds^2$ is positive definite and thus $\mathcal{T}_m$ is a Riemannian manifold. Indeed, if we set $W = W' = \sqrt{T}^{-1}dT\sqrt{T}^{-1} = (w_{\mu\nu})$, then $ds^2 = \sigma(WW') = \sum_{\mu,\nu} w_{\mu\nu}^2 \geqq 0$.

We shall often write $\hat{T}$ for $T^{-1}$. The relation $\hat{T}T = E$ gives $d\hat{T}\cdot T + \hat{T}\cdot dT = 0$, i.e., $d\hat{T}\cdot\hat{T}^{-1} = -T^{-1}\cdot dT$. Thus $\sigma(T^{-1}dT)^2 = \sigma(\hat{T}^{-1}d\hat{T})^2$, i.e., the map $T \mapsto \hat{T}$ is an isometry of $\mathcal{T}_m$.

Next we compute the invariant volume element $dv$ on $\mathcal{T}_m$. For $T = (t_{\mu\nu}) \in \mathcal{T}_m$ and $U \in \Omega$ let us write

$$T^* = UTU'.$$

Each $T \in \mathcal{T}_m$ depends on the $\frac{m(m+1)}{2}$ independent parameters $t_{\mu\nu}$ with $\mu \leqq \nu$. The Jacobian determinant

$$f(U) = \frac{\partial(T^*)}{\partial(T)}$$

is a rational function of the entries of $U$, and satisfies $f(UV) = f(U)f(V)$ for $U, V \in \Omega$. If $U$ is the diagonal matrix $D = (\delta_{\mu\nu}d_\nu)$, and if we write $T^* = DTD' = (t^*_{\mu\nu})$, then $t^*_{\mu\nu} = d_\mu d_\nu t_{\mu\nu}$, hence

$$f(D) = \prod_{\mu\leqq\nu} d_\mu d_\nu = (d_1 d_2 \dots d_m)^{m+1} = |D|^{m+1}.$$

Assume now that the characteristic roots of $U$ are all different. Then there exists a nonsingular matrix $T$ such that $U = TDT^{-1}$, and therefore

$$f(U) = f(T)f(D)f(T^{-1}) = f(D)f(E) = |D|^{m+1} = |U|^{m+1}.$$

Since the set of matrices, whose characteristic roots are all different, is everywhere dense in $\Omega$, and since $f(U)$ is rational, we have $f(U) = |U|^{m+1}$ for all $U \in \Omega$. Writing $[dT] = \prod_{\mu\leqq\nu} dt_{\mu\nu}$ and taking into account that

$$|T^*| = |T| \cdot |U|^2,$$

we have

$$[dT^*] = \left|\frac{\partial(T^*)}{\partial(T)}\right| [dT] = \|U\|^{m+1}[dT] = \frac{|T^*|^{\frac{m+1}{2}}}{|T|^{\frac{m+1}{2}}} [dT].$$

Thus we see that

$$dv = |T|^{-\frac{m+1}{2}} [dT]$$

is an invariant volume element on $\mathcal{T}_m$. It is, up to a constant factor, the only invariant volume element on $\mathcal{T}_m$. It is of course also invariant with respect to the isometry $T \mapsto \hat{T}$.

Let $r$ be an integer, $0 < r < m$. We define a one-to-one birational transformation $T \leftrightarrow F = F^{(r)}, G, H$ by setting

$$T = T^{(m)} = \begin{pmatrix} F & 0 \\ 0 & G \end{pmatrix} \begin{bmatrix} E & 0 \\ H & E \end{bmatrix} .$$

Writing this in the form

$$T = \begin{pmatrix} F + G[H] & H'G \\ GH & G \end{pmatrix}$$

we see that given $T \in \mathcal{T}_m$ we can first determine $G$, whose entries are the $t_{\mu\nu}$ with $r+1 \leqq \mu,\nu \leqq m$, then we determine $H$ and finally $F$. The condition $T > 0$ is equivalent to the conditions $F > 0$, $G > 0$ and $H$ arbitrary. Let us express $ds^2$ in terms of $F$, $G$, $H$. In the first place we find

$$T^{-1} = \begin{pmatrix} F^{-1} & 0 \\ 0 & G^{-1} \end{pmatrix} \begin{bmatrix} E & -H' \\ 0 & E \end{bmatrix} = \begin{pmatrix} F^{-1} & -F^{-1}H' \\ -HF^{-1} & G^{-1} + F^{-1}[H'] \end{pmatrix}$$

and

$$dT = \begin{pmatrix} dF + dG[H] + dH'\cdot GH + H'G\cdot dH & dH'\cdot G + H'\cdot dG \\ dG\cdot H + G\cdot dH & dG \end{pmatrix}.$$

With the abbreviation

$$dT\cdot T^{-1} = \begin{pmatrix} L_0 & L_1 \\ L_2 & L_3 \end{pmatrix}$$

we have

$$ds^2 = \sigma(dT\cdot T^{-1}\cdot dT\cdot T^{-1}) = \sigma(L_0^2 + L_1L_2) + \sigma(L_2L_1 + L_3^2).$$

A straightforward calculation yields

$$L_0 = dF\cdot F^{-1} + H'G\cdot dH\cdot F^{-1},$$

$$L_1 = -dF\cdot F^{-1}H' - H'G\cdot dH\cdot F^{-1}H' + dH' + H'\cdot dG\cdot G^{-1},$$

$$L_2 = G\cdot dH\cdot F^{-1},$$

$$L_3 = dG\cdot G^{-1} - G\cdot dH\cdot F^{-1}H',$$

and therefore

$$ds^2 = \sigma(F^{-1}dF)^2 + \sigma(G^{-1}dG)^2 + 2\sigma(F^{-1}\cdot dH'\cdot G\cdot dH).$$

We see that the first two quadratic forms in the sum are again positive definite. Let $A$ and $B$ be two square matrices such that $F = A'A$, $G = B'B$, and set $W = BdHA^{-1}$. Then

$$\sigma(F^{-1}dH'GdH) = \sigma(W'W),$$

which shows that also the third quadratic form is positive definite.

Next we compute $[dT]$ in terms of $F$, $G$, $H$. Let us decompose

$$T = \begin{pmatrix} T_1 & T_2 \\ T_2' & T_3 \end{pmatrix}$$

with $T_1 = T_1^{(r)}$. Observe that $T_1$ and $F$ depend on $\frac{r(r+1)}{2}$ parameters, $T_2$ and $H$ on $(m-r)r$ parameters, $T_3$ and $G$ on $\frac{(m-r)(m-r+1)}{2}$ parameters, which adds up to the correct total of $\frac{m(m+1)}{2}$. From the expression of $dT$ we find the value of the Jacobian determinant

$$\frac{\partial(T_1,T_2,T_3)}{\partial(F,H,G)} = \left|\begin{array}{c:c:c} E^{(\frac{1}{2}r(r+1))} & * & * \\ \hdashline 0 & \begin{matrix} G & & & 0 \\ & G & & \\ & & \ddots & \\ 0 & & & G \end{matrix}\ (r \text{ times}) & * \\ \hdashline 0 & 0 & E \end{array}\right| = |G|^r.$$

If for an arbitrary $H = (h_{\mu\nu})$ we define $[dH] = \prod_{\mu\nu} dh_{\mu\nu}$, we obtain

$$[dT] = [dT_1][dT_2][dT_3] = \frac{\partial(T_1,T_2,T_3)}{\partial(F,H,G)}[dF][dH][dG],$$

and finally

$$[dT] = |G|^r[dF][dH][dG].$$

Introducing the notations $\hat{T} = T^{-1}$, $\hat{F} = F^{-1}$, $\hat{G} = G^{-1}$, $\hat{H} = -H'$ we obtain from the above expression of $T^{-1}$ that

$$\hat{T} = \begin{pmatrix} \hat{F} & 0 \\ 0 & \hat{G} \end{pmatrix} \begin{bmatrix} E & \hat{H} \\ 0 & E \end{bmatrix}.$$

Taking into account

$$|\hat{T}|^{-\frac{m+1}{2}}[d\hat{T}] = |T|^{-\frac{m+1}{2}}[dT],$$

the analogous formulas for $F$ and $G$, and the relation $|T| = |F|\cdot|G|$, we get

$$\begin{aligned} [d\hat{T}] &= |T|^{-m-1}[dT] = |T|^{-m-1}|G|^r[dF][dH][dG] \\ &= |T|^{-m-1}|G|^r|F|^{r+1}[d\hat{F}][d\hat{H}]|G|^{m-r+1}[d\hat{G}] \\ &= |\hat{F}|^{m-r}[d\hat{F}][d\hat{H}][d\hat{G}]. \end{aligned}$$

Since $T \mapsto \hat{T}$ is an isometry of $\mathcal{F}_m$, we see that if $T \in \mathcal{F}_m$ is written in the form

$$T = \begin{pmatrix} F & 0 \\ 0 & G \end{pmatrix} \begin{bmatrix} E & H \\ 0 & E \end{bmatrix}$$

with $F = F^{(r)}$, then

$$[dT] = |F|^{m-r}[dF][dH][dG].$$

*Theorem.* 1. *Let two arbitrary points $T_0$ and $T_1$ of $\mathcal{F}_m$ be given. There exists a unique curve of shortest length joining $T_0$ and $T_1$. The length of this curve is given by*

$$\left(\sum_{\nu=1}^{m} \log^2 t_\nu\right)^{\frac{1}{2}},$$

*where $t_1,\ldots,t_m$ denote the zeros of $|tT_0 - T_1|$.*

2. *All geodesic curves through $T_0 = E$ have a representation $T(t) = (\delta_{\mu\nu} e^{\lambda_\nu t})[V]$, where $(\lambda_1,\ldots,\lambda_m) \neq (0,\ldots,0)$ are real numbers, $V$ is a constant orthogonal matrix, and $t$ is a real variable.*

3. *In the cases $B = B_{pq}$ or $B = B_n$ the representation space $\mathfrak{P}(B)$ is a geodesic submanifold of $\mathcal{F}_m$, i.e., together with any two points $T_0, T_1 \in \mathfrak{P}(B)$, the geodesic curve joining $T_0$ and $T_1$ belongs to $\mathfrak{P}(B)$.*

*Proof.* 1. Determine $R = R^{(m)}$ so that $T_0[R] = E$ and $T_1[R]$ is the diagonal matrix $D = (\delta_{\mu\nu} t_\nu)$. Then

$$|tT_0 - T_1| = |T_0|\cdot|tE - D| = |T_0|\cdot\prod_{\nu=1}^{m}(t-t_m),$$

and so we may assume without loss of generality that $T_0 = E$, $T_1 = D$.

Assume that $T = T(t)$ $(0 \leqq t \leqq 1)$ is a continuously differentiable geodesic curve joining $T_0$ and $T_1$. Let $r$ be an integer, $0 < r < m$. Then using the birational transformation introduced above, we have a one-to-one correspondence between $T(t)$ and $F(t)$, $G(t)$, $H(t)$, where the order of the square matrix $F(t)$ is $r$, and we have the boundary conditions $F(0) = E^{(r)}$, $G(0) = E^{(m-r)}$, $H(0) = 0$,

$$F(1) = \begin{pmatrix} t_1 & & 0 \\ & \ddots & \\ 0 & & t_r \end{pmatrix}, \qquad G(1) = \begin{pmatrix} t_{r+1} & & 0 \\ & \ddots & \\ 0 & & t_m \end{pmatrix}, \qquad H(1) = 0.$$

Denoting by $s(T_0,T_1)$ the length of the curve, using the expression obtained above for $ds^2$, and the abbreviation $\frac{d}{dt}(\ ) = (\dot{\ })$, we get

$$s(T_0,T_1) = \int_0^1 \{\sigma(F^{-1}\dot{F})^2 + \sigma(G^{-1}\dot{G})^2 + 2\sigma(F^{-1}\dot{H}'G\dot{H}\}^{\frac{1}{2}}dt$$

$$\geqq \int_0^1 \{\sigma(F^{-1}\dot{F})^2 + \sigma(G^{-1}\dot{G})^2\}^{\frac{1}{2}}dt,$$

because the terms are positive definite quadratic forms. In the last inequality the equal sign must hold since otherwise the curve given by

$$T(t) = \begin{pmatrix} F(t) & 0 \\ 0 & G(t) \end{pmatrix} \qquad (0 \leqq t \leqq 1)$$

would be shorter than the given geodesic. Thus we have $\sigma(F^{-1}\dot{H}'G\dot{H}) \equiv 0$ and so $\dot{H} \equiv 0$ because the quadratic form in $\dot{H}$ is positive definite. It follows that $H$ is a constant and because of the boundary conditions $H(t) \equiv 0$.

For every $r$ with $0 < r < m$ we get therefore

$$T(t) = \begin{pmatrix} F(t) & 0 \\ 0 & G(t) \end{pmatrix} \qquad (0 \leqq t \leqq 1),$$

which is possible only if $T(t)$ is a diagonal matrix. Let us set

$$T(t) = (\delta_{\mu\nu}e^{g_\nu(t)}),$$

where the functions $g_\nu$ $(1 \leqq \nu \leqq m)$ are continuously differentiable for $0 \leqq t \leqq 1$, $g_\nu(0) = 0$, $g_\nu(1) = \log t_\nu$. We have

$$\dot{T}(t) = (\delta_{\mu\nu} e^{g_\nu(t)} \dot{g}_\nu(t)),$$

hence

$$T^{-1}\dot{T} = (\delta_{\mu\nu}\dot{g}_\nu),$$

and therefore

$$s(T_0,T_1) = \int_0^1 \{\sigma(T^{-1}\dot{T})^2\}^{\frac{1}{2}} dt$$

$$= \int_0^1 \{\sum_{\nu=1}^m \dot{g}_\nu^2\}^{\frac{1}{2}} dt.$$

The minimal value of $s(T_0,T_1)$ is obtained if the curve is the straight line $g_\nu(t) = t \log t_\nu$, $1 \leqq \nu \leqq m$ in the $(g_1,\ldots,g_m)$-space. Thus $T(t) = (\delta_{\mu\nu} e^{\lambda_\nu t})$, with $\lambda_\nu = \log t_\nu$, is the unique curve of shortest length joining $E$ and $D = (\delta_{\mu\nu} t_\nu)$ and its length is

$$s(T_0,T_1) = \int_0^1 \{\sum_{\nu=1}^m \log^2 t_\nu\}^{\frac{1}{2}} dt = (\sum_{\nu=1}^m \log^2 t_\nu)^{\frac{1}{2}}.$$

2. Any matrix $T_1 \in \mathcal{P}_m$ can be written in the form $T_1 = (\delta_{\mu\nu} e^{\lambda_\nu})[V]$, where $V$ is some orthogonal matrix. By the first part of the proof $(\delta_{\mu\nu} e^{\lambda_\nu t})$ is the geodesic joining $E$ and $(\delta_{\mu\nu} e^{\lambda_\nu})$, hence $(\delta_{\mu\nu} e^{\lambda_\nu t})[V]$ is the geodesic joining $E$ and $T_1$.

3. Since $\Omega(B)$ acts transitively on $\mathcal{P}(B)$, we may assume that $T_0 = E$. Consider the geodesic curve $T(t) = (\delta_{\mu\nu} e^{\lambda_\nu t})[V]$, $V'V = E$, $0 \leqq t \leqq 1$, and suppose that $T_1 = T(1)$ belongs to $\mathcal{P}(B)$. Then $T_1 B T_1 = B$, i.e.,

$$V'(\delta_{\mu\nu} e^{\lambda_\nu}) VBV'(\delta_{\mu\nu} e^{\lambda_\nu}) V = B,$$

or writing $VBV' = (a_{\mu\nu})$,

$$e^{\lambda_\mu + \lambda_\nu} a_{\mu\nu} = a_{\mu\nu}.$$

Therefore we have either $a_{\mu\nu} = 0$ or $a_{\mu\nu} \neq 0$ and $\lambda_\mu + \lambda_\nu = 0$. In either case we have

$$e^{(\lambda_\mu + \lambda_\nu)t} a_{\mu\nu} = a_{\mu\nu}$$

for $0 \leqq t \leqq 1$, which implies

$$V'(\delta_{\mu\nu}e^{\lambda_\nu t})VBV'(\delta_{\mu\nu}e^{\lambda_\nu t})V = B,$$

i.e., $T(t) \in \mathfrak{P}(B)$ for $0 \leqq t \leqq 1$.

## §4. *Symplectic geometry*

We shall denote by $I_n$ or simply by $I$ the square matrix of order $2n$

$$\begin{pmatrix} 0 & E \\ -E & 0 \end{pmatrix},$$

where $E = E^{(n)}$. This is the matrix we denoted by $B_n$ in §2. The automorphism group

$$\Omega = \Omega(I) = \{M \mid M \in GL(m, \mathbb{R}), \quad I[M] = I\}$$

of $I$ is known as the *symplectic group*. We have seen in §2 that if we write

$$M = \begin{pmatrix} A & B \\ C & D \end{pmatrix},$$

where $A$, $B$, $C$, $D$ are square matrices of order $n$, then the condition $I[M] = M'IM = I$ can be written out as

$$A'D - C'B = E, \quad A'C = C'A, \quad B'D = D'B.$$

Taking inverses, and using the fact that $M^{-1} = I'M'I$, we see that the condition $I[M] = I$ is equivalent to $I[M'] = MIM' = I$, i.e., to

$$AD' - BC' = E, \quad AB' = BA', \quad CD' = DC'.$$

If $M \in \Omega$, i.e., $M'IM = I$, then $|M|^2 = 1$ and so $|M| = \pm 1$. We shall see that $|M| = 1$. We know from §2 that

$$M = \begin{pmatrix} A & B \\ C & D \end{pmatrix} \in \Omega$$

operates on the generalized upper half-plane

$$\mathfrak{H} = \{Z \mid Z = Z' = X + iY, \quad Y > 0\}$$

according to

$$Z \mapsto Z^* = M\langle Z\rangle = (AZ + B)(CZ + D)^{-1}.$$

Writing $\overline{Z} = X - iY$ and $\overline{Z}^* = M\langle\overline{Z}\rangle$, we have

$$\begin{pmatrix} A & B \\ C & D \end{pmatrix}\begin{pmatrix} Z & \overline{Z} \\ E & E \end{pmatrix} = \begin{pmatrix} AZ + B & A\overline{Z} + B \\ CZ + D & C\overline{Z} + D \end{pmatrix}$$

$$= \begin{pmatrix} Z^* & \overline{Z}^* \\ E & E \end{pmatrix}\begin{pmatrix} CZ + D & 0 \\ 0 & C\overline{Z} + D \end{pmatrix},$$

and therefore

$$|M|\,|Z - \overline{Z}| = |Z^* - \overline{Z}^*|\,|CZ + D|\,|C\overline{Z} + D|.$$

Taking into account that $Z - \overline{Z} = 2iY$ and $Z^* - \overline{Z}^* = 2iY^*$, we obtain

$$|M|\,|Y| = |Y^*|\,|CZ + D|^2.$$

Now $Y > 0$, $Y^* > 0$ and so $|M| > 0$. This proves that $|M| = 1$.

If $Z = M\langle Z\rangle$, i.e., $Z(CZ + D) = AZ + B$ for every $Z \in \mathfrak{H}$, then $B = C = 0$ and $A = D$. But $ZA = AZ$ for every $Z \in \mathfrak{H}$ implies $A = \lambda E$ for some $\lambda \in \mathbb{R}$. Finally $|M| = 1$ yields $\lambda^2 = 1$, i.e., $\lambda = \pm 1$ and therefore $M = \pm E^{(2n)}$. In particular, the group of transformations of $\mathfrak{H}$ corresponding to the action of $\Omega$ is isomorphic to $\Omega/\{\pm E\}$.

Let $Z_1$ and $Z_2$ be two points of $\mathfrak{H}$ and write $Z_\nu^* = M\langle Z_\nu\rangle$ for $\nu = 1,2$. Then

$$Z_2^* - Z_1^* = (Z_2C' + D')^{-1}(Z_2A' + B') - (AZ_1 + B)(CZ_1 + D)^{-1}$$

$$= (Z_2C' + D')^{-1}\{(Z_2A' + B')(CZ_1 + D) - (Z_2C' + D')(AZ_1 + B)\}(CZ_1 + D)^{-1}$$

$$= (Z_2C' + D')^{-1}\{Z_2(A'C - C'A)Z_1 + Z_2(A'D - C'B) + (B'C - D'A)Z_1$$

$$+ B'D - D'B\}(CZ_1 + D)^{-1}$$

$$= (Z_2C' + D')^{-1}(Z_2 - Z_1)(CZ_1 + D)^{-1},$$

where we have used the formulas which express that $M \in \Omega$. Let us observe that this formula holds even if $Z_\nu = Z_\nu'$ does not belong to $Z_\nu$, and let us apply it to the case $Z_2 = Z = X + iY \in \mathfrak{H}$, $Z_1 = \overline{Z} = X - iY \notin \mathfrak{H}$. Setting $Z^* = X^* + iY^*$ we obtain

$$Y^* = Y\{(C\overline{Z} + D)^{-1}\} = Y\{(CZ + D)^{-1}\}.$$

If we substitute $Z_1 = Z$, $Z_2 = Z + dZ$ in the above formula and write $Z_2^* - Z_1^* = dZ^*$, we obtain

$$dZ^* = dZ[(CZ + D)^{-1}].$$

We proved already that

$$\frac{\partial(Z^*)}{\partial(Z)} = |CZ + D|^{-n-1}, \qquad \frac{\partial(\overline{Z}^*)}{\partial(\overline{Z})} = |C\overline{Z} + D|^{-n-1},$$

hence

$$\frac{\partial(Z^*, \overline{Z}^*)}{\partial(Z, \overline{Z})} = \|CZ + D\|^{-2(n+1)} = \frac{|Y^*|^{n+1}}{|Y|^{n+1}} .$$

From the relations $dZ = dX + idY$ and $d\overline{Z} = dX - idY$ it follows that

$$\frac{\partial(Z,\overline{Z})}{\partial(X,Y)} = \begin{vmatrix} E & iE \\ E & -iE \end{vmatrix} = \begin{vmatrix} 2E & 0 \\ E & -iE \end{vmatrix} = (-2i)^{\frac{1}{2}n(n+1)}$$

and similarly

$$\frac{\partial(Z^*,\overline{Z}^*)}{\partial(X^*,Y^*)} = (-2i)^{\frac{1}{2}n(n+1)}.$$

Consequently

$$\frac{\partial(X^*,Y^*)}{\partial(X,Y)} = \frac{\partial(X^*,Y^*)}{\partial(Z^*,\overline{Z}^*)}\frac{\partial(Z^*,\overline{Z}^*)}{\partial(Z,\overline{Z})}\frac{\partial(Z,\overline{Z})}{\partial(X,Y)} = \frac{|Y^*|^{n+1}}{|Y|^{n+1}}.$$

Writing, as usual, $[dX] = \prod_{\mu \leq \nu} dX_{\mu\nu}$, and similarly for the other differentials, we obtain

$$[dX^*][dY^*] = \frac{|Y^*|^{n+1}}{|Y|^{n+1}}[dX][dY],$$

which shows that

$$dv = |Y|^{-n-1}[dX][dY]$$

is a *volume element* on $\mathcal{H}$ *invariant under the action of* $\Omega$. It is uniquely determined up to a constant factor.

In §3 we have written any $T \in \mathcal{T}_m$ in the form

$$T = \begin{pmatrix} F & 0 \\ 0 & G \end{pmatrix} \begin{bmatrix} E & 0 \\ H & E \end{bmatrix}.$$

It follows from §2 that if $T = P \in \mathcal{P}(I)$, then we can choose $F = Y$, $G = Y^{-1}$, $H = X$, where $Z = X + iY$ is the point of $\mathcal{H}$ corresponding to $P$. Therefore the formula obtained for $ds^2$ in §3 yields

$$ds^2 = \sigma(Y^{-1}dY)^2 + \sigma(Y^{-1}dY)^2 + 2\sigma(Y^{-1}dX\,Y^{-1}\,dX)$$
$$= 2\sigma(Y^{-1}dY)^2 + 2\sigma(Y^{-1}dX)^2$$
$$= 2\sigma(Y^{-1}dZ\,Y^{-1}\,d\overline{Z}) .$$

*Lemma* 1. *The mapping*

$$Z \mapsto W = (Z - iE)(Z + iE)^{-1}$$

*establishes a birational transformation from* $\mathfrak{H}$ *onto the generalized unit circle*

$$\mathfrak{E} = \{W \mid W = W', \quad E - W\overline{W} > 0\}.$$

*Proof*. 1. If $Z \in \mathfrak{H}$, then $Z + iE \in \mathfrak{H}$, hence $|Z + iE| \neq 0$ and so $W$ is well-defined. It is obvious that $W$ is symmetric. We compute

$$E - W\overline{W} = E - (Z + iE)^{-1}(Z - iE)(\overline{Z} + iE)(\overline{Z} - iE)^{-1}$$
$$= (Z + iE)^{-1}\{(Z + iE)(\overline{Z} - iE) - (Z - iE)(\overline{Z} + iE)\}(\overline{Z} - iE)^{-1}$$
$$= 4(Z + iE)^{-1}Y(\overline{Z} - iE)^{-1} = 4Y\{(\overline{Z} - iE)^{-1}\} > 0$$

since $Y > 0$. We have $|E - W| \neq 0$. Indeed, if $\varphi$ is a column vector such that $(E - W)\varphi = 0$ then $\varphi' = \varphi' W$ and $(E - \overline{W})\overline{\varphi} = 0$, hence $\varphi'(E - W\overline{W})\overline{\varphi} = 0$. Since $E - W\overline{W} > 0$, it follows that $\varphi = 0$.

Thus $(E - W)^{-1}$ exists, and solving the equation $W(Z + iE) = Z - iE$, i.e., $i(E + W) = (E - W)Z$ for $Z$, we obtain

$$Z = i(E + W)(E - W)^{-1}.$$

2. It remains to prove that every $W \in \mathfrak{E}$ is the image of some $Z \in \mathfrak{H}$. Given $W \in \mathfrak{E}$, determine $Z$ according to the last formula. Then

$$2Y = -i(Z - \bar{Z}) = (E - W)^{-1}(E + W) + (E + \bar{W})(E - \bar{W})^{-1}$$

$$= (E - W)^{-1}\{(E + W)(E - \bar{W}) + (E - W)(E + \bar{W})\}(E - \bar{W})^{-1}$$

$$= 2(E - W)^{-1}(E - W\bar{W})(E - \bar{W})^{-1},$$

i.e.,

$$Y = (E - W\bar{W})\{(E - \bar{W})^{-1}\} > 0,$$

and so $Z$ belongs to $\mathcal{G}$. This proves the lemma

With the help of the matrix

$$T = T^{(2n)} = \frac{1}{\sqrt{2}}\begin{pmatrix} E & -iE \\ E & iE \end{pmatrix},$$

which satisfies $\bar{T}'T = E$, we can write the transformation

$$Z \mapsto W = (Z - iE)(Z + iE)^{-1}$$

as $W = T\langle Z\rangle$ and $Z = T^{-1}\langle W\rangle$.

Through the correspondence $Z \leftrightarrow W$ the symplectic group $\Omega$ acts on $\mathfrak{E}$. If $Z^* = M\langle Z\rangle$ for some $M \in \Omega$, then

$$W^* = T\langle Z^*\rangle = TM\langle Z\rangle = TMT^{-1}\langle W\rangle = \hat{M}\langle W\rangle,$$

where $\hat{M} = TMT^{-1}$ belongs to the group $\hat{\Omega} = T\Omega T^{-1}$. A matrix $M$ belongs to $\Omega$ if and only if

$$I[M] = I\{M\} = I,$$

hence the matrices $\hat{M} = TMT^{-1}$ belonging to $\hat{\Omega}$ are characterized by

$$I[T^{-1}\hat{M}] = I[T^{-1}], \quad I\{T^{-1}\hat{M}\} = I\{T^{-1}\}.$$

Now

$$I[T^{-1}] = -iI, \quad I\{T^{-1}\} = i\begin{pmatrix} E & 0 \\ 0 & -E \end{pmatrix}$$

and so the above two conditions can be rewritten in the form

$$I[\hat{M}] = I, \qquad \begin{pmatrix} E & 0 \\ 0 & -E \end{pmatrix}\{\hat{M}\} = \begin{pmatrix} E & 0 \\ 0 & -E \end{pmatrix}.$$

From the first relation we get

$$\hat{M}^{-1} = I^{-1}\hat{M}'I = I'\hat{M}'I,$$

and from the second

$$\hat{M}^{-1} = \begin{pmatrix} E & 0 \\ 0 & -E \end{pmatrix}\bar{\hat{M}}'\begin{pmatrix} E & 0 \\ 0 & -E \end{pmatrix}.$$

Since

$$I\begin{pmatrix} E & 0 \\ 0 & -E \end{pmatrix} = -\begin{pmatrix} 0 & E \\ E & 0 \end{pmatrix},$$

the second condition above can be replaced by

$$\hat{M}\begin{bmatrix} 0 & E \\ E & 0 \end{bmatrix} = \bar{\hat{M}}.$$

Changing notations, we obtained that

$$M = \begin{pmatrix} A & B \\ C & D \end{pmatrix}$$

belongs to $\hat{\Omega}$ if and only if

$$I[M] = I \quad \text{and} \quad M\begin{bmatrix} 0 & E \\ E & 0 \end{bmatrix} = \bar{M}.$$

These conditions are equivalent to

$$\bar{A} = D, \quad \bar{B} = C, \quad A'D - C'B = E, \quad A'C = C'A, \quad B'D = D'B,$$

i.e.,

$$\hat{\Omega} = \left\{ \begin{pmatrix} A & B \\ \bar{B} & \bar{A} \end{pmatrix} \;\middle|\; A'\bar{A} - \bar{B}'B = E, \quad A'\bar{B} = \bar{B}'A \right\}.$$

*Lemma* 2. *For any complex matrix* $W = W' = W^{(n)}$ *there exists a matrix* $U$ *such that* $\bar{U}'U = E$ *and that* $W[U] = U'WU$ *is a diagonal matrix* $(\delta_{\mu\nu}r_\nu)$ *with* $0 \leqq r_1 \leqq r_2 \leqq \ldots \leqq r_n$.

*Proof*. Since $W\bar{W}$ is Hermitian, there exists a unitary matrix $U_1$ such that

$$W\bar{W} = (\delta_{\mu\nu}d_\nu)\{U_1\} = \bar{U}_1'(\delta_{\mu\nu}d_\nu)U_1,$$

where the $d_\nu$ are real numbers. Set $F = W[U_1']$. Then $F = F'$ and $F\bar{F} = (\delta_{\mu\nu}d_\nu)$. If we decompose $F$ into $F_1 + iF_2$ with real $F_1$ and $F_2$, the imaginary part of $F\bar{F}$ is zero, i.e., $F_1F_2 = F_2F_1$. Now two symmetric real matrices which commute can be transformed simultaneously by an orthogonal matrix into diagonal matrices, and so we have

$$F_1[V] = D_1, \quad F_2[V] = D_2, \qquad V'V = E.$$

Thus $F[V] = D$ and $W[U_1'V] = D$, where $D$ is a diagonal matrix with complex elements and $U_1'V$ is unitary. Now $D$ can be written in the form

$$D = \begin{pmatrix} e^{i\rho_1/2} & & \\ & \ddots & \\ & & e^{i\rho_n/2} \end{pmatrix} \begin{pmatrix} r_1 & & \\ & \ddots & \\ & & r_n \end{pmatrix} \begin{pmatrix} e^{i\rho_1/2} & & \\ & \ddots & \\ & & e^{i\rho_n/2} \end{pmatrix},$$

where the $r_j$ are real positive elements. We see therefore that multiplying $U_1'V$ by an appropriate unitary matrix, we obtain the matrix $U$ which satisfies the conditions of the lemma.

*Theorem* 1. 1. *Let* $Z_0$ *and* $Z_1$ *be two points of the generalized upper half-plane* $\mathfrak{H}$. *There exists a symplectic matrix* $M$ *such that*

$$M\langle Z_0\rangle = iE, \quad M\langle Z_1\rangle = (i\delta_{\mu\nu}\lambda_\nu),$$

*where* $1 \leqq \lambda_1 \leqq \lambda_2 \leqq \dots \leqq \lambda_n$.

2. *Set*

$$\rho(Z_1,Z_0) = (Z_1 - Z_0)(Z_1 - \overline{Z}_0)^{-1}(\overline{Z}_1 - \overline{Z}_0)(\overline{Z}_1 - Z_0)^{-1}.$$

*A given pair* $Z_0$, $Z_1$ *of points of* $\mathfrak{H}$ *can be transformed by the same* $M \in \Omega$ *into another pair* $Z_0^*$, $Z_1^*$ *of points of* $\mathfrak{H}$ *if and only if the matrices* $\rho(Z_1,Z_0)$ *and* $\rho(Z_1^*,Z_0^*)$ *have the same characteristic roots.*

3. *Denote the characteristic roots of* $\rho(Z_1,Z_0)$ *by* $r_1,r_2,\dots,r_n$. *Then the symplectic distance of the points* $Z_0$ *and* $Z_1$ *is given by*

$$s(Z_0,Z_1) = \sqrt{2}\left\{\sum_{\nu=1}^{n} \log^2 \frac{1+\sqrt{r_\nu}}{1-\sqrt{r_\nu}}\right\}^{\frac{1}{2}}.$$

*Proof.* 1. Since $\Omega$ is transitive on $\mathfrak{H}$, there exists $M \in \Omega$ such that $M\langle Z_0\rangle = iE$. Thus we may assume from the outset that $Z_0 = iE$.

To $Z_0$ and $Z_1$ there correspond the points $W_0 = 0$ and $W_1$ of the generalized unit circle $\mathfrak{E}$. It suffices to show that there exists $\hat{M} \in \hat{\Omega}$ such that $\hat{M}\langle 0\rangle = 0$, $\hat{M}(W_1) = (\delta_{\mu\nu}d_\nu)$, where $0 \leqq d_1 \leqq d_2 \leqq \dots \leqq d_n < 1$, since then we would have

$$M\langle Z_1\rangle = i\left(\delta_{\mu\nu}\frac{1+d_\nu}{1-d_\nu}\right), \qquad \hat{M} = TMT^{-1}.$$

Now $\hat{M} \in \hat{\Omega}$ is such that $\hat{M}\langle 0\rangle = 0$ if and only if it is of the form

$$\hat{M} = \begin{pmatrix} U & 0 \\ 0 & \bar{U} \end{pmatrix},$$

where $U$ is unitary. Determine, according to Lemma 2, a unitary matrix $U$ such that $W_1[U'] = (\delta_{\mu\nu} d_\nu)$. Then the corresponding $\hat{M}$ satisfies $\hat{M}\langle 0\rangle = 0$ and $\hat{M}\langle W_1\rangle = (\delta_{\mu\nu} d_\nu)$.

2. Assume that $Z_j^* = M\langle Z_j\rangle$ for $j = 0,1$, where

$$M = \begin{pmatrix} A & B \\ C & D \end{pmatrix} \in \Omega .$$

Then, as we have seen at the beginning of this section,

$$Z_1^* - Z_0^* = (Z_1 C' + D')^{-1}(Z_1 - Z_0)(CZ_0 + D)^{-1},$$

and therefore

$$\rho(Z_1^*, Z_0^*) = (Z_0 C' + D')^{-1} \rho(Z_1, Z_0)(Z_0 C' + D'),$$

which shows that $\rho(Z_1, Z_0)$ and $\rho(Z_1^*, Z_0^*)$ have the same characteristic roots.

Let now $Z_j = X_j + iY_j$ $(j = 0,1)$ be two points of $\mathfrak{H}$, $W_j = T\langle Z_j\rangle$ the corresponding points of $\mathfrak{E}$, and

$$P_j = \begin{pmatrix} Y_j & 0 \\ 0 & Y_j^{-1} \end{pmatrix} \begin{bmatrix} E & 0 \\ X_j & E \end{bmatrix}$$

the corresponding points of $\mathcal{P}(I)$. According to part 1 of the Theorem, determine $M \in \Omega$ so that $M\langle Z_0\rangle = iE$ and $M\langle Z_1\rangle = (i\delta_{\mu\nu}\lambda_\nu) = iD$ with $1 \leqq \lambda_1 \leqq \lambda_2 \leqq \dots \leqq \lambda_n$. Then $\hat{M} = TMT^{-1} \in \hat{\Omega}$ satisfies $\hat{M}\langle W_0\rangle = 0$, $\hat{M}\langle W_1\rangle = (\delta_{\mu\nu}\sqrt{r_\nu})$, $0 \leqq r_1 \leqq r_2 \leqq \dots \leqq r_n < 1$, where

$$\sqrt{r_\nu} = \frac{\lambda_\nu - 1}{\lambda_\nu + 1} \qquad (\nu = 1,2,\ldots,n).$$

The symplectic distance of the points $Z_0$ and $Z_1$ is

$$s(Z_0,Z_1) = s(P_0,P_1) = s(MP_0M',MP_1M').$$

But the matrix $MP_0M' \in \mathfrak{P}(I)$ corresponding to $iE \in \mathfrak{H}$ is $E$, and the matrix $MP_1M' \in \mathfrak{P}(I)$ corresponding to $iD \in \mathfrak{H}$ is

$$\begin{pmatrix} D & 0 \\ 0 & D^{-1} \end{pmatrix}.$$

The zeros of the polynomial

$$\left| tE - \begin{pmatrix} D & 0 \\ 0 & D^{-1} \end{pmatrix} \right|$$

are $\lambda_\nu$ and $\lambda_\nu^{-1}$ $(\nu = 1,2,\ldots,n)$, hence by the Theorem of §3 we have

$$s(Z_0,Z_1) = \{ \sum_{\nu=1}^{n} \log^2\lambda_\nu + \sum_{\nu=1}^{n} \log^2 \frac{1}{\lambda_\nu} \}^{\frac{1}{2}}$$

$$= \sqrt{2}\{\sum_{\nu=1}^{n} \log^2\lambda_\nu\}^{\frac{1}{2}}$$

$$= \sqrt{2}\left\{ \sum_{\nu=1}^{n} \log^2 \frac{1+\sqrt{r_\nu}}{1-\sqrt{r_\nu}} \right\}^{\frac{1}{2}}.$$

As we have seen above, the characteristic roots of $\rho(Z_1,Z_0)$ are the same as the characteristic roots of

$$\rho(iD,iE) = (D-E)(D+E)^{-1}(D-E)(D+E)^{-1}$$

$$= \left( \delta_{\mu\nu}\left(\frac{\lambda_\nu - 1}{\lambda_\nu + 1}\right)^2 \right) = (\delta_{\mu\nu}r_\nu),$$

i.e., they are indeed $r_1, r_2, \ldots, r_n$. This proves part 3 of the Theorem.

We have also proved that if $0 \leqq r_1 \leqq r_2 \leqq \ldots \leqq r_n < 1$ is given, if $D = (\delta_{\mu\nu}\lambda_\nu)$ is the corresponding matrix, and if the characteristic roots of $\rho(Z_1, Z_0)$ are $r_1, r_2, \ldots, r_n$, then there exists $M \in \Omega$ such that $M\langle Z_0\rangle = iE$, $M\langle Z_1\rangle = iD$. From here the remaining half of assertion 2 of the Theorem follows.

We say that a topological mapping $W \mapsto W^*$ from $\mathfrak{E}$ onto $\mathfrak{E}$ is an *analytic automorphism* of $\mathfrak{E}$ if the elements $w^*_{\mu\nu}$ of $W^*$ are holomorphic functions of the elements $w_{\alpha\beta}$ of $W$, and if

$$\frac{\partial(W^*)}{\partial(W)} \neq 0$$

everywhere in $\mathfrak{E}$. Under these conditions the $w_{\mu\nu}$ are holomorphic functions of the $w^*_{\alpha\beta}$. The analytic automorphisms of $\mathfrak{E}$ form obviously a group.

*Theorem* 2. *The group of analytic automorphisms of $\mathfrak{E}$ is identical with the group of symplectic transformations.*

*Proof*. It is sufficient to prove that any analytic automorphism is a symplectic transformation. We may assume that the analytic automorphism maps 0 into 0, since otherwise we can compose with a suitable symplectic transformation.

Consider an analytic transformation, let $W_0$ be a general point of $\mathfrak{E}$, and let $W_0^*$ be its image. We set $W = tW_0$, where $t$ is a complex variable, and denote by $r_1, r_2, \ldots, r_n$ the characteristic roots of $W_0'\overline{W}_0 = W_0\overline{W}_0$. There exists a unitary matrix $U$ such that $U'W_0'\overline{W}_0\overline{U} = (\delta_{\mu\nu}r_\nu)$, where we assume furthermore that $0 \leqq r_1 \leqq r_2 \leqq \ldots \leqq r_n$. We have

$$U'(E - W'\overline{W})\overline{U} = E - t\bar{t}(\delta_{\mu\nu}r_\nu) > 0 \quad \text{for} \quad t\bar{t}r_n < 1.$$

It follows from $E - W_0\overline{W}_0 > 0$ that $r_n < 1$ and so $W \in \mathfrak{E}$ for $|t| \leqq 1$.

Denote by $W^*$ the image of $W$ for $|t| \leqq 1$. It is a holomorphic function of $t$ and has a power series expansion

$$W^* = \sum_{k=1}^{\infty} t^k W_k^* \qquad \text{for } t\bar{t}r_n < 1,$$

where the matrices $W_k^*$ depend only on $W_0$. If we write $W_0 = (w_{\mu\nu})$, then $W^*$ is a holomorphic function of the elements $tw_{\mu\nu}$ in a neighborhood of $t = 0$, and therefore can be developed into a uniquely determined power series for small values of $|t|$. It follows from the uniqueness that $t^k W_k^*$ is the sum of all terms of degree $k$ in the $tw_{\mu\nu}$ which figure in the power series development. Taking $t = 1$ we see that

$$W_0^* = \sum_{k=1}^{\infty} W_k^* ,$$

where the series converges for every $W_0 \in \mathfrak{E}$, provided that we do not separate the terms of the same degree.

Since $W^* \in \mathfrak{E}$ for $|t| \leqq 1$, we have $E - W^*\overline{W}^* > 0$ on $t\bar{t} = 1$, and integration yields

$$\frac{1}{2\pi i} \int_{|t|=1} (E - W^*\overline{W}^*) \frac{dt}{t} > 0.$$

If we substitute for $W^*$ its power series expansion, and take into account that

$$\frac{1}{2\pi i} \int_{|t|=1} t^k \bar{t}^l W_k^* \overline{W}_l^* \frac{dt}{t} = \begin{cases} 0 & \text{if } k \neq l, \\ W_k^* \overline{W}_k^* & \text{if } k = l, \end{cases}$$

we obtain

$$E - \sum_{k=1}^{\infty} W_k^* \overline{W}_k^* > 0.$$

We may assume that the absolute value of the Jacobian determinant

$$D = \frac{\partial(W^*)}{\partial(W_0)} = \frac{\partial(W_0^*)}{\partial(W_0)}\bigg|_{W_0=0}$$

is at least 1, since otherwise the inverse automorphism would have this property. Denote by $\mathfrak{E}_1$ the image of $\mathfrak{E}$ under the linear mapping $W_0 \mapsto W_1^*$. We consider the real and imaginary parts of the entries of the matrices as Cartesian coordinates, and denote by $I(\mathfrak{E})$ and $I(\mathfrak{E}_1)$ the Euclidean volume of $\mathfrak{E}$ and $\mathfrak{E}_1$, respectively. Then

$$I(\mathfrak{E}_1) = D\overline{D}I(\mathfrak{E}) \geqq I(\mathfrak{E}),$$

but since $\mathfrak{E}_1 \subset \mathfrak{E}$, we have $|D| = 1$. This shows that $\mathfrak{E}_1 = \mathfrak{E}$ and $W_0 \mapsto W_1^*$ maps $\partial(\mathfrak{E})$ onto $\partial(\mathfrak{E})$.

We now choose $W_0$ of the form $W_0 = U'PU$, where $P = (\delta_{\mu\nu}p_\nu)$, the $p_\nu$ are real and $U$ is unitary, $\overline{U}U' = E$. The relation

$$E - W_0\overline{W}_0 = U'(\delta_{\mu\nu}(1-p_\nu^2))\overline{U}$$

shows that $W_0$ belongs to $\partial(\mathfrak{E})$ if and only if $-1 \leqq p_\nu \leqq 1$ for $\nu = 1,2,\ldots,n$ and at least once an equal sign is valid. For a fixed $U$ the determinant $|E - W_1^*\overline{W}_1^*|$ is a polynomial in the $p_1,p_2,\ldots,p_n$ of degree $\leqq 2n$. It vanishes if and only if $W_1^* \subset \partial(\mathfrak{E})$, i.e., if $W_0 \in \partial(\mathfrak{E})$, which happens if and only if $p_\nu = \pm 1$ for some $\nu$ $(\nu = 1,2,\ldots,n)$. Thus $|E - W_1^*\overline{W}_1^*|$ is divisible by $\prod_{\nu=1}^{n} (p_\nu-1)(p_\nu+1)$, i.e.,

$$|E - W_1^*\overline{W}_1^*| = c\cdot\prod_{\nu=1}^{n}(1-p_\nu^2),$$

where the constant factor $c$ depends on $U$. If we set $p_\nu = 0$ for $\nu = 1,2,\ldots,n$, then $W_0 = 0$, $W_1^* = 0$, hence $c = 1$ and so

$$|E - W_1^*\overline{W}_1^*| = \prod_{\nu=1}^{n} (1-p_\nu^2) = |E - W_0\overline{W}_0|.$$

This identity holds for every complex symmetric matrix $W_0$ because any such matrix can be written in the form $W_0 = U'PU$, where $P$ is a real diagonal matrix and $U$ is unitary. Since $W_1^*$ is a linear function of $W_0$, we have

$$|\lambda E - W_1^*\overline{W}_1^*| = |\lambda E - W_0\overline{W}_0|$$

for $\lambda \in \mathbb{C}$ and every $W_0 = W_0'$.

Determine the unitary matrices $U_1$ and $U_2$ so that

$$W_0[U_1] = (\delta_{\mu\nu}p_\nu) \qquad 0 \leqq p_1 \leqq p_2 \leqq \dots \leqq p_n,$$

$$W_1^*[U_2] = (\delta_{\mu\nu}q_\nu) \qquad 0 \leqq q_1 \leqq q_2 \leqq \dots \leqq q_n.$$

It follows that

$$U_1'W_0\overline{W}_0\overline{U}_1 = (\delta_{\mu\nu}p_\nu^2), \qquad U_2'W_1^*\overline{W}_1^*\,\overline{U}_2 = (\delta_{\mu\nu}q_\nu^2),$$

hence by the equality of the two characteristic polynomials $p_\nu^2 = q_\nu^2$ and $p_\nu = q_\nu$ for $\nu = 1,2,\dots,n$. So we get

$$W_1^* = W_0[U], \quad \text{where } U = U_1U_2^{-1}, \quad \overline{U}U' = E.$$

From the inequality $E - \sum_{k=1}^{\infty} W_k^*\overline{W}_k^* > 0$ proved above, it follows in particular that

$$E - W_1^*\overline{W}_1^* - W_k^*\overline{W}_k^* > 0$$

for $k = 2,3,\dots$ . Choose $W_0 = ue^{iS}$, where $0 < u < 1$ and $S = \overline{S} = S'$. Then

$$W_1^*\overline{W}_1^* = U'W_0\overline{W}_0\overline{U} = u^2U'e^{iS}e^{-iS}\overline{U} = u^2E,$$

hence

$$(1-u^2)E \geqq W_k^* \overline{W}_k^* \quad \text{for} \quad k > 1, \quad W_0 = ue^{iS}.$$

Letting $u$ tend to 1, we see that $W_k^* = 0$ for $k = 2,3,\ldots$ and $W_0 = e^{iS}$, $S = \overline{S} = S'$. Since the elements of $W_k^*$ are holomorphic functions of the elements of $S$, we have $W_k^* = 0$ for $k = 2,3,\ldots$ and $W_0 = e^{iS}$, where $S$ is any complex symmetric matrix $S = S'$. But these matrices $W_0 = e^{iS}$ fill a neighborhood of the unit matrix $E$ in $\mathfrak{E}$, hence $W_k^*$ is identically zero for $k = 2,3,\ldots$ and our analytic transformation $W_0 \mapsto W_0^*$ is the linear map $W_0 \mapsto W_1^*$.

We need only to prove that $W_1^* = W_0[U]$ with a fixed unitary matrix $U$. Set

$$W_1^* = W_1^*(W_0) = \sum_{\mu \leqq \nu} w_{\mu\nu} A_{\mu\nu} ,$$

where the $A_{\mu\nu}$ are constant matrices, and define

$$W_1(W_0) = \sum_{\mu \leqq \nu} w_{\mu\nu} \overline{A}_{\mu\nu} .$$

We have obviously

$$\overline{W}_1^* = W_1(\overline{W}_0).$$

Since $W_1^* = U' W_0 U$, with a unitary matrix $U$ which for the moment might depend on $W_0$, we have

$$W_1^* \overline{W}_1^* = U' W_0 \overline{W}_0 \overline{U} = E \quad \text{whenever} \quad W_0 \overline{W}_0 = E,$$

and therefore in particular

$$W_1^*(W_0) W_1(W_0^{-1}) = E \quad \text{for} \quad W_0 = e^{iS}, \quad S = S' = \overline{S}.$$

Since the elements of $W_1^*$ and of $W_1$ are holomorphic functions of the elements of $S$, we see by the same argument as above that

$$W_1^*(W_0)W_1(W_0^{-1}) = E$$

holds first for every $W_0 = e^{iS}$ with a complex symmetric $S = S'$, and hence identically in $W_0 \in \mathfrak{E}$.

Let us write $w_{\mu\mu} = w_\mu$, $A_{\mu\mu} = A_\mu$, and introduce the matrices

$$W_{00} = \begin{pmatrix} w_1 & & \\ & \ddots & \\ & & w_n \end{pmatrix}, \qquad W_{01} = W_0 - W_{00},$$

i.e., $W_0 = W_{00} + W_{01}$. Considering $W_{00}$ as fixed, with $|W_{00}| \neq 0$, we develop $W_0^{-1}$ into a Taylor series in a neighborhood of $W_{01} = 0$ and obtain

$$W_0^{-1} = (W_{00} + W_{01})^{-1} = W_{00}^{-1}(E + W_{01}W_{00}^{-1})^{-1}$$

$$= W_{00}^{-1} - W_{00}^{-1}W_{01}W_{00}^{-1} + \ldots \; .$$

The identity

$$W_1^*(W_0)W_1(W_0^{-1}) =$$

$$= \{W_1^*(W_{00}) + W_1^*(W_{01})\}\{W_1(W_{00}^{-1}) - W_1(W_{00}^{-1}W_{01}W_{00}^{-1}) + \ldots\} = E$$

yields

$$W_1^*(W_{00})W_1(W_{00}^{-1}) = E,$$

$$W_1^*(W_{00})W_1(W_{00}^{-1}W_{01}W_{00}^{-1}) = W_1^*(W_{01})W_1(W_{00}^{-1}),$$

and in particular

$$\sum_{\mu,\nu=1}^{n} w_\mu w_\nu^{-1} A_\mu \bar{A}_\nu = E.$$

Multiplying by $w_1 w_2 \cdots w_n$ and comparing coefficients we see that

$$A_\mu \bar{A}_\nu = 0 \quad \text{for} \quad \mu \neq \nu.$$

Let us denote by $U_0$ the value of $U$ which corresponds to $W_{01} = 0$, i.e.,

$$W_1^*(W_{00}) = U_0' W_{00} U_0 = \sum_{\nu=1}^{n} w_\nu A_\nu.$$

Choose now for $W_{00} = (\delta_{\mu\nu} w_\nu)$ the matrix $e_{\mu\mu}$, with $w_\mu = 1$ and $w_\nu = 0$ for $\nu \neq \mu$. Then $A_\mu = U_0' e_{\mu\mu} U_0$ and so

$$A_\mu \bar{A}_\mu = U_0' e_{\mu\mu} U_0.$$

There exists a unitary matrix $V$ such that $V' A_\mu V = (\delta_{\alpha\beta} p_\beta)$. Since $V' A_\mu \bar{A}_\mu V = (\delta_{\alpha\beta} p_\beta^2)$, and since the rank of $A_\mu$ is 1, it follows from the above relations that

$$V' A_\mu V = \begin{pmatrix} 1 & & & \\ & 0 & & \\ & & \ddots & \\ & & & 0 \end{pmatrix}$$

for an appropriate unitary matrix $V$.

We may therefore assume without loss of generality that

$$A_1 = \begin{pmatrix} 1 & & & \\ & 0 & & \\ & & \ddots & \\ & & & 0 \end{pmatrix},$$

since otherwise we can replace $W_1^*$ by $W_1^*[V]$. From the relation $A_1 A_\nu = 0$ and $A_\nu = A_\nu'$ $(\nu > 1)$ it follows that $A_\nu$ is of the form

$$A_\nu = \begin{pmatrix} 0 & 0 \\ 0 & B_\nu^{(n-1)} \end{pmatrix},$$

where $B_\nu \bar{B}_\nu$ has characteristic roots 1,0,...,0. By repeating the same argument we see that we can assume

$$A_\nu = e_{\nu\nu} = \begin{pmatrix} 0 & & & & & \\ & \ddots & & & & \\ & & 0 & & & \\ & & & 1 & & \\ & & & & 0 & \\ & & & & & \ddots \\ & & & & & & 0 \end{pmatrix},$$

or

$$W_1^*(W_{00}) = \sum_{\nu=1}^{n} \omega_\nu A_\nu = W_{00},$$

$$W_1(W_{00}) = \sum_{\nu=1}^{n} \omega_\nu \bar{A}_\nu = W_{00},$$

and thus

$$W_1^*(W_{01}) = W_{00} W_1(W_{00}^{-1} W_{01} W_{00}^{-1}) W_{00}.$$

Taking into account that

$$W_1^*(W_{01}) = \sum_{\mu<\nu} \omega_{\mu\nu} A_{\mu\nu}, \quad W_1(W_{01}) = \sum_{\mu<\nu} \omega_{\mu\nu} \bar{A}_{\mu\nu},$$

we obtain

$$\sum_{\mu<\nu} \omega_{\mu\nu} A_{\mu\nu} = \sum_{\mu<\nu} \frac{\omega_{\mu\nu}}{\omega_\mu \omega_\nu} W_{00} \bar{A}_{\mu\nu} W_{00},$$

or

$$\omega_\mu \omega_\nu A_{\mu\nu} = W_{00} \bar{A}_{\mu\nu} W_{00}.$$

If $A_{\mu\nu} = (a_{ik})$, then $\omega_\mu \omega_\nu a_{ik} = \omega_i \omega_k \bar{a}_{ik}$, from where it follows that $a_{ik} = 0$ if $(i,k)$ is different from $(\mu,\nu)$ or from $(\nu,\mu)$ and $a_{\mu\nu} = \bar{a}_{\mu\nu}$. In other words,

$$A_{\mu\nu} = a_{\mu\nu}(e_{\mu\nu} + e_{\nu\mu}),$$

where $a_{\mu\nu}$ is a real number and $e_{\mu\nu}$ is the matrix which has a 1 at the intersection of the $\mu^{th}$ row and the $\nu^{th}$ column, and 0 everywhere else. If we set $a_{\nu\nu} = \frac{1}{2}$, then the preceding relation also holds for $A_{\nu\nu} = A_\nu$. We have thus obtained that

$$W_1^* = \sum_{\mu \leq \nu} a_{\mu\nu}(e_{\mu\nu} + e_{\nu\mu})w_{\mu\nu} = (a_{\mu\nu}^* w_{\mu\nu}),$$

where $a_{\mu\mu}^* = 1$, $a_{\mu\nu}^*$ is real and $a_{\mu\nu}^* = a_{\nu\mu}^*$.

Since $W_1^* = U'W_0U$ with a unitary $U$, the quotient $|W_1^*|\cdot|W_0|^{-1}$ has a constant absolute value. On the other hand, it is a rational function of the elements of $W_0$, therefore it is a constant, i.e.,

$$|W_1^*| = c|W_0|.$$

Now the product of the diagonal elements in the matrices $W_0 = (w_{\mu\nu})$ and $W_1^* = (a_{\mu\nu}^* w_{\mu\nu})$ is the same since $a_{\mu\mu}^* = 1$. Thus $c = 1$.

We may assume that $a_{1\nu}^* = 0$ for $\nu > 1$ since otherwise we could transform $W_0^*$ by a unitary matrix of the form $(\pm\delta_{\mu\nu})$. Let us write $w = w_1 w_2 \cdots w_n$. The term $(w_1 w_\nu)^{-1} w w_{1\nu}^2$ has in $|W_1^*|$ the factor $-a_{1\nu}^{*2}$, and in $|W_0|$ the factor -1, so that $a_{1\nu}^* = 1$ for $\nu > 1$. The term $(w_1 w_\mu w_\nu)^{-1} w w_{1\mu} w_{1\nu} w_{\mu\nu}$ has in $|W_1^*|$ the factor $2a_{\mu\nu}^*$ $(1 < \mu < \nu)$, and in $|W_0|$ the factor 2, therefore $a_{\mu\nu}^* = 1$ for all $\mu$ and $\nu$. Thus we have $W_1^* = W_0$ and this proves the theorem.

## §5. *Weakly symmetric Riemannian spaces*

Let $\mathcal{T}$ be an infinitely differentiable Riemannian space of dimension $n$. This means that the following conditions are satisfied:

1. Every point $x_0 \in \mathcal{T}$ belongs to an open set $\mathfrak{U} \subset \mathcal{T}$ which is mapped topologically into $\mathbb{R}^n$ by a system of local coordinates $x \mapsto (x^1, x^2, \ldots, x^n)$.

2. If $x^1, x^2, \ldots, x^n$ and $y^1, y^2, \ldots, y^n$ are two systems of local coordinates having a common domain of definition $\mathfrak{U} \subset \mathcal{T}$, and if $\mathfrak{V} \subset \mathbb{R}^n$ is the image of $\mathfrak{U}$ with respect to the map $x \mapsto (x^1, x^2, \ldots, x^n)$, then each $y^\nu$ $(1 \leq \nu \leq n)$ is an infinitely differentiable function of $x^1, x^2, \ldots, x^n$ on $\mathfrak{V}$.

3. There exists a metric fundamental form which in local coordinates has the expression

$$ds^2 = \sum_{i,k=1}^{n} g_{ik} dx^i dx^k ,$$

where the functions $g_{ik} = g_{ki}$ are infinitely differentiable on the image under the map $x \mapsto (x^1, x^2, \ldots, x^n)$ of the domain of definition $\mathfrak{U} \subset \mathcal{T}$ of the local coordinate system $x^1, x^2, \ldots, x^n$.

For any open subset $\mathfrak{U}$ of $\mathcal{T}$ we denote by $C^\infty(\mathfrak{U})$ the set of all infinitely differentiable complex-valued functions defined on $\mathfrak{U}$.

Let $G$ be a locally compact group of isometries of the Riemannian space $\mathcal{T}$ which acts transitively on $\mathcal{T}$. Furthermore let $\mu$ be an isometry of $\mathcal{T}$. We say that $\mathcal{T}$ is a *weakly symmetric Riemannian space* with respect to $G$ and $\mu$ if the following two conditions are satisfied:

a) $\mu$ commutes with $G$, i.e., $\mu G \mu^{-1} = G$, and $\mu^2$ belongs to $G$;

b) for any pair of points $x$ and $y$ of $\mathcal{Y}$, there exists an $m \in G$ such that $mx = \mu y$ and $my = \mu x$.

A linear map $L$ from $C^\infty(\mathcal{Y})$ into $C^\infty(\mathcal{Y})$ is called a *differential operator* if in terms of a local coordinate system $x^1, x^2, \dots, x^n$ it can be expressed as a differential operator

$$L_0 = \sum_{\nu_1,\dots,\nu_n=1}^{k} c_{\nu_1\dots\nu_n}(x^1,\dots,x^n)\left(\frac{\partial}{\partial x^1}\right)^{\nu_1}\cdots\left(\frac{\partial}{\partial x^n}\right)^{\nu_n},$$

where the $c_{\nu_1\dots\nu_n}$ are complex-valued, infinitely differentiable functions. More precisely, if $x = \phi(x^1,\dots,x^n)$ is the point of $\mathcal{Y}$ corresponding to the point $(x^1,\dots,x^n) \in \mathbb{R}^n$, then

$$(Lf)(x) = (Lf)(\phi(x^1,\dots,x^n)) = L_0(f\phi)(x^1,\dots,x^n)$$

for every $f \in C^\infty(\mathcal{Y})$.

Let $y^1,\dots,y^n$ be another local coordinate system and $x = \psi(y^1,\dots,y^n) \in \mathcal{Y}$. Suppose that in terms of $y^1,\dots,y^n$ the differential operator $L$ is given by

$$(Lf)(x) = L_0^*(f\psi)(y^1,\dots,y^n),$$

where

$$L_0^* = \sum_{\nu_1,\dots,\nu_n=1}^{k} c^*_{\nu_1\dots\nu_n}(y^1,\dots,y^n)\left(\frac{\partial}{\partial y^1}\right)^{\nu_1}\cdots\left(\frac{\partial}{\partial y^n}\right)^{\nu_n}.$$

Then any of the coordinates $x^\nu$ is an infinitely differentiable function of the $y^1,\dots,y^n$, and vice versa. Moreover

$$L_0(f\phi)(x^1,\dots,x^n) = L_0^*(f\psi)(y^1,\dots,y^n)$$

becomes an identity if we replace, for instance, on the left hand side $x^\nu$ by $x^\nu(y^1,\dots,y^n)$ and $\frac{\partial}{\partial x^\nu}$ by $\sum_{\alpha=1}^{n} \frac{\partial y^\alpha}{\partial x^\nu}\frac{\partial}{\partial y^\alpha}$.

A differential operator $L$ is said to be *invariant* (with respect to $G$) if for any $m \in G$ and $f \in C^\infty(\mathfrak{X})$ the operations $x \mapsto mx$ and $f \mapsto Lf$ commute, i.e., if

$$((Lf)m)(x) = (Lf)(mx) = (Lf)(x)\Big|_{x \mapsto mx}$$

$$= L(f(mx)) = L((fm)(x)) = (L(fm))(x),$$

or succinctly

$$(Lf)m = L(fm).$$

We denote by $\mathbf{L} = \mathbf{L}(\mathfrak{X}, G)$ the set of all invariant differential operators. Obviously $\mathbf{L}$ is an algebra over the field $\mathbb{C}$ of complex numbers.

For $L \in \mathbf{L}$ and $f \in C^\infty(\mathfrak{X})$ we set

$$\tilde{L}f = (L(f\mu^{-1}))\mu,$$

and claim that $\tilde{L}$ is an invariant differential operator. To show that $\tilde{L}$ is a differential operator, let $x \mapsto (x^1, x^2, \ldots, x^n)$ be a local coordinate system with domain of definition $\mathcal{U}$, and set $x = \phi(x^1, \ldots, x^n)$. Then $\mathcal{U}^* = \mu\,\mathcal{U}$ is the domain of definition of the coordinate system defined by $y = \mu x = \psi(x^1, \ldots, x^n)$, where $\psi = \mu\phi$, since $\mu$ is infinitely differentiable. For $f \in C^\infty(\mathfrak{X})$ and $x \in \mathcal{U}$ we have

$$(\tilde{L}f)(x) = (L(f\mu^{-1}))(y) = L_0((f\mu^{-1})\psi)(x^1, \ldots, x^n),$$

where

$$L_0 = \sum_{\nu_1, \ldots, \nu_n = 1}^{k} a_{\nu_1 \ldots \nu_n} \left(\frac{\partial}{\partial x^1}\right)^{\nu_1} \cdots \left(\frac{\partial}{\partial x^n}\right)^{\nu_n}$$

is the expression of $L$ in terms of the local coordinates $y \mapsto (x^1, \ldots, x^n)$

in $\mathfrak{U}^*$. Thus $\tilde{L}$ is indeed a differential operator.

To show that $\tilde{L}$ is invariant, let $m \in G$ and $\mu m = m_1\mu$ with $m_1 \in G$. Then

$$(\tilde{L}f)m = (L(f\mu^{-1}))(m_1\mu) = \{(L(f\mu^{-1}))m_1\}\mu$$

$$= \{L((f\mu^{-1})m_1)\}\mu = \{L(f(\mu^{-1}m_1))\}\mu$$

$$= \{L(f(m\mu^{-1}))\}\mu = \{L((fm)\mu^{-1})\}\mu$$

$$= \tilde{L}(fm).$$

This proves our claim.

Clearly $L \mapsto \tilde{L}$ is a linear map from $L$ into itself. Furthermore it follows easily from $\mu^2 \in G$ and from the invariance of $L$ that $\tilde{\tilde{L}} = L$.

A function $k \in C^\infty(\mathcal{Y} \times \mathcal{Y})$ is said to be *point pair invariant* if

$$k(mx,my) = k(x,y) \quad \text{for} \quad x,y \in \mathcal{Y}, \quad m \in G.$$

We then have

$$k(x,y) = k(\mu y,\mu x) \quad \text{for} \quad x,y \in \mathcal{Y}.$$

Indeed, choose $m \in G$ so that $mx = \mu y$ and $my = \mu x$. Then $k(x,y) = k(mx,my) = k(\mu y,\mu x)$.

Let $k$ be a point pair invariant function and $L$ an invariant differential operator. Then $L_x k(x,y)$ is a point pair invariant function, where the subscript $x$ indicates that the differential operator acts on $k$ as a function of the variable $x$ only. Let us set $k(x,y) = k_y(x)$ and $k'(x,y) = L_x k(x,y)$. Then for any $m \in G$ we have

$$k'(mx,my) = (Lk_y(x))_{\substack{x \mapsto mx \\ y \mapsto my}} = (Lk_{my}(x))_{x \mapsto mx} \ .$$

Since $L$ is invariant, this is equal to

$$L(k_{my}(mx)) = L_x k(mx,my) = L_x k(x,y) = k'(x,y).$$

Thus we have indeed $k'(mx,my) = k'(x,y)$.

Next we prove that if $k$ is point pair invariant, then

$$L_x k(x,y) = \tilde{L}_y k(x,y)$$

for any $L \in \mathcal{L}$. Indeed, since $L_x k(x,y) = k'(x,y)$ is point pair invariant, we have

$$L_x k(x,y) = k'(\mu y,\mu x) = (k'(y,\mu x))_{y \mapsto \mu y}$$

$$= (L_y k(y,\mu x))_{y \mapsto \mu y} = \tilde{L}_y k(\mu y,\mu x) = \tilde{L}_y k(x,y).$$

A point pair invariant function can be constructed from a given function $f \in C^{\infty}(\mathfrak{X})$ as follows. Let $x_o \in \mathfrak{X}$ and let

$$G_{x_o} = \{r \mid r \in G, \quad rx_o = x_o\}$$

be the stabilizer of $x_o$ in $G$. Denote by $dr$ the invariant volume element (Haar measure) on the compact group $G_{x_o}$ normalized so that

$$\int_{G_{x_o}} dr = 1.$$

The function

$$f(x,x_o) = \int_{G_{x_o}} f(rx)\,dr$$

is symmetric with respect to rotations, i.e.,

$$f(rx,x_o) = f(x,x_o) \quad \text{for} \quad r \in G_{x_o}.$$

Define now the function $k$ by

$$k(x,y) = f(mx,x_o),$$

where $m \in G$ is such that $my = x_o$. The element $m$ is not uniquely defined by $y$, but if $my = m'y = x_o$, then $m'm^{-1}x_o = x_o$, i.e., $m' = rm$ for some $r \in G_{x_o}$ and so $f(m'x,x_o) = f(rmx,x_o) = f(mx,x_o)$. Clearly $k \in C^\infty(\mathfrak{X} \times \mathfrak{X})$. Furthermore $k$ is point pair invariant, since if $my = x_o$, then $mm_1^{-1}(m_1y) = x_o$, and so

$$k(m_1x,m_1y) = f(mm_1^{-1}(m_1x),x_o) = f(mx,x_o) = k(x,y)$$

for every $m_1 \in G$.

For $y = x_o$ the element $m$ is the identity of $G$ and thus $k(x,x_o) = f(x,x_o)$. If $L$ belongs to $\mathbb{L}$, we have

$$\{L_x k(x,x_o)\}_{x=x_o} = \{L_x f(x,x_o)\}_{x=x_o}$$

$$= \left\{ \int_{G_{x_o}} (L_x(fr))(x)dr \right\}_{x=x_o} = \left\{ \int_{G_{x_o}} (L_x f)(rx)dr \right\}_{x=x_o}$$

$$= \int_{G_{x_o}} (L_x f)(x_o)dr = (Lf)(x_o).$$

Since $x_o$ is an arbitrary point of $\mathfrak{X}$, we can conclude that *the effect of* $L = L_x \in \mathbb{L}$ *on all point pair invariant functions* $k(x,y)$ *determines* $L$ *uniquely*.

*Theorem. If* $\mathfrak{X}$ *is a weakly symmetric Riemannian space, then the ring* $\mathbb{L}$ *of all invariant differential operators on* $\mathfrak{X}$ *is commutative.*

*Proof.* Let $L^{(1)}, L^{(2)} \in \mathbb{L}$ and let $k(x,y)$ be a point pair invariant function. Then

$$L_x^{(1)} L_x^{(2)} k(x,y) = L_x^{(1)} \tilde{L}_y^{(2)} k(x,y)$$

and this is obviously equal to

$$\tilde{L}_y^{(2)} L_x^{(1)} k(x,y) = L_x^{(2)} L_x^{(1)} k(x,y).$$

Since $L^{(1)}L^{(2)}$ and $L^{(2)}L^{(1)}$ have the same effect on point pair invariant functions, they are identical, q.e.d.

The *degree* $g$ of an invariant differential operator $L \neq 0$ is defined as follows. Choose any system of local coordinates $x^1, x^2, \ldots, x^n$, so that $x = \phi(x^1, \ldots, x^n)$ and $(Lf)(x) = L_0(f\phi)(x^1, \ldots, x^n)$, with

$$L_0 = \sum_{\nu_1,\ldots,\nu_n=1}^{k} c_{\nu_1 \ldots \nu_n} \left(\frac{\partial}{\partial x^1}\right)^{\nu_1} \cdots \left(\frac{\partial}{\partial x^n}\right)^{\nu_n} .$$

Then $g$ is the largest of the numbers $\nu_1 + \nu_2 + \ldots + \nu_n$ such that $c_{\nu_1 \ldots \nu_n}$ is not identically zero. It is independent of the choice of the system of local coordinates.

Given a differential operator $L$, its *adjoint* $\hat{L}$ is defined by requiring that for $f,g \in C^\infty(\mathcal{T})$ we shall have

$$\left(f(x)\,\overline{Lg(x)} - \overline{g(x)}\,\hat{L}f(x)\right) dv = d\omega,$$

where $dv$ is the invariant volume element on $\mathcal{T}$ and $\omega = \omega(f,g)$ is an exterior differential form of degree $n-1$ such that

a) $\omega(f,g)$ is bilinear in $f$ and $g$,

b) if the point $x = \phi(x^1, \ldots, x^n)$ is expressed in terms of the local coordinates $x^1, \ldots, x^n$, then $\omega(f,g)$ is uniquely determined by the partial derivatives of $f\phi(x^1, \ldots, x^n)$ and $g\phi(x^1, \ldots, x^n)$ of orders at most $g-1$, where $g$ is the degree of $L$.

We first prove that $\hat{L}$ is uniquely determined if it exists. Let $x_0$ be a point of $\mathcal{T}$ and in a small neighborhood of $x_0$ consider the geodesic distance

$$s(x,x_0) = \min \int_{x_0}^{x} ds$$

from $x_0$ to $x$. Denote by $\mathcal{B}_\rho$ the ball $s(x,x_0) \leqq \rho$, where $\rho$ is a small positive number. Assume now that $L \neq 0$ has two adjoints $\hat{L}$ and $\hat{L}'$ and set $M = \hat{L} - \hat{L}'$. We then have

$$\overline{g(x)}\ Mf(x)dv = d\omega',$$

where $\omega'$ has the same properties as $\omega$. Choose in particular

$$g(x) = (\rho^2 - s^2(x,x_0))^g\ \overline{Mf(x)},$$

where $g$ is the degree of $L$. We get

$$\int_{\mathcal{B}_\rho} (\rho^2 - s^2(x,x_0))^g |Mf(x)|^2 dv = \int_{\partial\mathcal{B}_\rho} \omega' = 0,$$

since the derivatives of order $\leqq g-1$ of $g(x)$ vanish on $\partial\mathcal{B}_\rho = \{x \mid s(x,x_0) = \rho\}$. It follows that $Mf(x) = 0$ for $x \in \mathcal{B}_\rho$. Since $x_0$ is arbitrary, $M = 0$, i.e., $\hat{L} = \hat{L}'$.

Next we prove the existence of the adjoint. By the preceding uniqueness result, it is sufficient to establish the existence of $\hat{L}$ locally. Let $x = \phi(x^1,\ldots,x^n)$ be the expression of $x \in \mathcal{V}$ in terms of a local coordinate system $x^1,\ldots,x^n$. Since $L \mapsto \hat{L}$ is linear, we may assume that $L$ is a monomial operator of the form

$$Lf(x) = L_0(f\phi)(x^1,\ldots,x^n) = c \frac{\partial}{\partial x^{\nu_1}} \cdots \frac{\partial}{\partial x^{\nu_h}} (f\phi)(x^1,\ldots,x^n),$$

with $c = c(x^1,\ldots,x^n)$. Let us write

$$ds^2 = \sum_{i,k=1}^{n} g_{ik} dx^i dx^k ,$$

$g_0 = |g_{ik}|$, $dv = \sqrt{g_0}\,[dx]$, where $[dx] = dx^1 dx^2 \ldots dx^n$. We then define the exterior differential form $\omega$ by

$$\omega = \sum_{j=1}^{h} (-1)^j \left\{ \frac{\partial}{\partial x^{\nu_1}} \cdots \frac{\partial}{\partial x^{\nu_{j-1}}} \bar{c} \sqrt{g_0}\, \psi \right\}$$

$$\left\{ \frac{\partial}{\partial x^{\nu_{j+1}}} \cdots \frac{\partial}{\partial x^{\nu_h}} \bar{\chi} \right\} (-1)^{\nu_j} dx^1 \ldots \widehat{dx^{\nu_j}} \ldots dx^n,$$

where $\psi$, $\chi$ are arbitrary functions and the hat over $dx^{\nu_j}$ means that this factor should be omitted from $dx^1 dx^2 \ldots dx^n$. We compute

$$d\omega = \sum_{j=1}^{h} (-1)^{j-1} \left\{ \frac{\partial}{\partial x^{\nu_1}} \cdots \frac{\partial}{\partial x^{\nu_j}} \bar{c} \sqrt{g_0}\, \psi \right\} \left\{ \frac{\partial}{\partial x^{\nu_{j+1}}} \cdots \frac{\partial}{\partial x^{\nu_h}} \bar{\chi} \right\} [dx] \; +$$

$$+ \sum_{j=1}^{h} (-1)^{j-1} \left\{ \frac{\partial}{\partial x^{\nu_1}} \cdots \frac{\partial}{\partial x^{\nu_{j-1}}} \bar{c} \sqrt{g_0}\, \psi \right\} \left\{ \frac{\partial}{\partial x^{\nu_j}} \cdots \frac{\partial}{\partial x^{\nu_h}} \bar{\chi} \right\} [dx] \; =$$

$$= \bar{c} \sqrt{g_0}\, \psi \frac{\partial}{\partial x^{\nu_1}} \cdots \frac{\partial}{\partial x^{\nu_h}} \bar{\chi}\, [dx] - (-1)^h \bar{\chi} \frac{\partial}{\partial x^{\nu_1}} \cdots \frac{\partial}{\partial x^{\nu_h}} \bar{c} \sqrt{g_0}\, \psi\, [dx]$$

$$= \left( \psi\, \overline{c \frac{\partial}{\partial x^{\nu_1}} \cdots \frac{\partial}{\partial x^{\nu_h}} \chi} - \bar{\chi} \frac{(-1)^h}{\sqrt{g_0}} \frac{\partial}{\partial x^{\nu_1}} \cdots \frac{\partial}{\partial x^{\nu_h}} \bar{c} \sqrt{g_0}\, \psi \right) dv.$$

Setting $\psi = f\phi$ and $\chi = g\phi$ we see that we can define $\hat{L}$ locally by

$$\hat{L}f(x) = \frac{(-1)^h}{\sqrt{g_0}} \frac{\partial}{\partial x^{\nu_1}} \cdots \frac{\partial}{\partial x^{\nu_h}} \bar{c} \sqrt{g_0}\, (f\phi)(x^1, \ldots, x^n).$$

If we replace $f$ and $g$ by $fm$ and $gm$, where $m \in G$, in the defining equation of $\hat{L}$, we obtain

$$\{f(mx)\, \overline{L(gm)(x)} - \overline{g(mx)}\, \hat{L}(fm)(x)\} dv = d\omega'.$$

If on the other hand we make the substitution $x \mapsto mx$ in the same equation, we obtain

$$\{f(mx)\overline{(Lg)(mx)} - \overline{g(mx)}(\hat{L}f)(mx)\}dv = d\omega''.$$

Let us assume now that $L \in \mathcal{L}$, i.e., that the differential operator $L$ is invariant. Then we obtain from the last two equations by subtraction that

$$\overline{g(mx)}\,\{\hat{L}(fm)(x) - (\hat{L}f)(mx)\}dv = d(\omega' - \omega'').$$

From here it follows similarly as in the uniqueness proof above, that $\hat{L}(fm)(x) = (\hat{L}f)(mx)$, i.e., that $\hat{L}$ also belongs to $\mathcal{L}$. Clearly $L \mapsto \hat{L}$ is a linear map from $\mathcal{L}$ into itself.

It is easy to check that $\hat{\hat{L}} = L$, i.e., $L \mapsto \hat{L}$ is an involution of $\mathcal{L}$.

*Examples of weakly symmetric Riemannian spaces.* 1. For the space $\mathfrak{X}$ we take the set

$$\mathcal{P}(B) = \{P \mid P = \bar{P} = P' > 0, \;\; PBP = B\}$$

introduced in §2, where $B$ is either the matrix denoted by $B_{pq}$ in §2, with $pq > 0$, or the matrix denoted by $I$ in §4. The Riemannian metric on $\mathcal{P}(B)$ is given by $ds^2 = \sigma(P^{-1}dP)^2$. For $G$ we take the group

$$G = \Omega(B) = \{U \mid UBU' = B\}$$

which acts on $\mathfrak{X} = \mathcal{P}(B)$ according to $P \mapsto UPU'$. The action of $G = \Omega(B)$ on $\mathfrak{X} = \mathcal{P}(B)$ is transitive and each transformation by an element of $G$ is an isometry of $\mathfrak{X}$. We know that $\Delta = \Omega(B) \cap \Omega(E)$ is a maximal compact subgroup of $G$ and that $\mathfrak{X}$ is isomorphic to $G/\Delta$ which is a representation space of $G$ (§2).

Next we want to prove that if $P \in \mathfrak{X}$, then $\sqrt{P} \in \mathfrak{X}$.

*Case* a): $B = B_{pq}$. We have seen in §2 that given $U \in \Omega(B)$ there exists a unique real matrix $X = X^{(p,q)}$ such that $E - XX' > 0$ and the symmetric positive matrix

$$\begin{pmatrix} (E - XX')^{-\frac{1}{2}} & X(E - X'X)^{-\frac{1}{2}} \\ X'(E - XX')^{-\frac{1}{2}} & (E - X'X)^{-\frac{1}{2}} \end{pmatrix}$$

$$= \begin{pmatrix} E & 0 \\ 0 & E - X'X \end{pmatrix} \begin{bmatrix} (E - XX')^{-\frac{1}{2}} & X(E - X'X)^{-\frac{1}{2}} \\ 0 & (E - X'X)^{-\frac{1}{2}} \end{bmatrix}$$

represents the coset $U\Delta$. For a given matrix $P \in \mathfrak{P}(B)$ there exists $U \in \Omega(B)$ such that $P = UU'$, and $U$ is determined up to a right factor belonging to $\Delta$. Thus in particular we may choose $U = U' > 0$, then $P = U^2$ and $\sqrt{P} = U \in \mathfrak{P}(B)$.

*Case* b): $B = I$. It follows easily from the considerations of §2 that $\Delta$ is the stabilizer of $iE \in \mathfrak{H}$. For any $Z \in \mathfrak{H}$ there exists $U \in \Delta$ such that $U\langle Z\rangle = iD$, where $D$ is a positive diagonal matrix. The one-to-one correspondence between $\mathfrak{H}$ and $\mathfrak{P}(B)$ associates with $Z = X + iY$ the matrix

$$P = \begin{pmatrix} Y & 0 \\ 0 & Y^{-1} \end{pmatrix} \begin{bmatrix} E & 0 \\ X & E \end{bmatrix} ,$$

and the relations $U\langle Z\rangle = iD$, $U\langle iE\rangle = iE$ are equivalent to

$$UPU' = \begin{pmatrix} D & 0 \\ 0 & D^{-1} \end{pmatrix} , \qquad UU' = E.$$

Hence we also have

$$P = U' \begin{pmatrix} D & 0 \\ 0 & D^{-1} \end{pmatrix} U$$

and therefore

$$\sqrt{P} = U' \begin{pmatrix} \sqrt{D} & 0 \\ 0 & \sqrt{D^{-1}} \end{pmatrix} U$$

is a positive, symmetric, symplectic matrix, i.e., $\sqrt{P} \in \mathcal{P}(B)$.

We are now in the position to prove that the isometry $\mu$ of $\mathcal{H}$ given by

$$\mu(P) = P^{-1}$$

satisfies the requirements of the definition of a weakly symmetric Riemannian space. For this it is sufficient to show that given $P_1$ and $P_2$ in $\mathcal{H}$ there exists $U \in G$ so that

$$UP_1U' = P_2^{-1} \quad \text{and} \quad UP_2U' = P_1^{-1}.$$

The first equation is equivalent with $U'P_2U = P_1^{-1}$, hence it is enough to find $U = U'$ which satisfies $UP_1U = P_2^{-1}$. But this equation can be rewritten as

$$(\sqrt{P_1}\, U \sqrt{P_1}\,)^2 = \sqrt{P_1}\; P_2^{-1}\; \sqrt{P_1}\;,$$

i.e., as

$$U = \sqrt{P_1}^{\,-1} \sqrt{\sqrt{P_1}\; P_2^{-1}\; \sqrt{P_1}}\;\; \sqrt{P_1}^{\,-1}\,,$$

and this matrix belongs to $\mathcal{P}(B)$, hence also to $\Omega(B) = G$.

2. The example where we take for $\mathcal{H}$ the space $\{Y \mid Y = Y^{(n)} = Y' > 0\}$ of all positive matrices, $ds^2 = \sigma(Y^{-1}dY)^2$, for $G$ the general linear group $GL(n, \mathbb{R})$ which operates on $\mathcal{H}$ according to $Y \mapsto UYU'$ and for $\mu$ the isometry $Y \to Y^{-1}$, will be discussed in detail in §6.

## §6. *The Riemannian space of all positive matrices*

In this section we shall discuss the weakly symmetric Riemannian space

$$\mathcal{Y} = \{Y \mid Y = Y^{(n)} = Y' > 0\}$$

of all symmetric positive matrices of order $n$. The Riemannian metric on $\mathcal{Y}$ will be $ds^2 = \sigma(Y^{-1}dY)^2$, the group $G$ of isometries of $\mathcal{Y}$ will be the general linear group $GL(n,\mathbb{R})$ which acts on $\mathcal{Y}$ according to $Y \mapsto UYU'$, and for $\mu$ we take $Y \mapsto Y^{-1}$.

For any $f = f(Y) \in C^\infty(\mathcal{Y})$ the differential $df$ is a linear form in the $dy_{\mu\nu}$ $(\mu \leqq \nu)$, where the $y_{\mu\nu}$ are the elements of $Y$. There exists a uniquely determined symmetric matrix $\frac{\partial}{\partial Y}$ such that

$$df = \sigma(dY \tfrac{\partial}{\partial Y})f.$$

Indeed, set $e_{\mu\nu} = \frac{1}{2}(1 + \delta_{\mu\nu})$ and $\omega_{\mu\nu} = e_{\mu\nu}\frac{\partial}{\partial y_{\mu\nu}}$, i.e.,

$$\omega_{\nu\nu} = \frac{\partial}{\partial y_{\nu\nu}}, \qquad \omega_{\mu\nu} = \frac{1}{2}\frac{\partial}{\partial y_{\mu\nu}} \qquad \text{if } \mu \neq \nu.$$

Writing $\frac{\partial}{\partial Y} = (\omega_{\mu\nu})$ we have then

$$\sigma(dY \tfrac{\partial}{\partial Y})f = \sum_{\mu,\nu=1}^{n} dy_{\mu\nu}\omega_{\mu\nu}f$$

$$= \sum_{\nu=1}^{n} dy_{\nu\nu}\omega_{\nu\nu}f + 2\sum_{\mu<\nu} dy_{\mu\nu}\omega_{\mu\nu}f$$

$$= \sum_{\mu\leqq\nu} dy_{\mu\nu}\frac{\partial}{\partial y_{\mu\nu}} f = df.$$

If $U$ is a fixed element of $G = GL(n,\mathbb{R})$, then for the transformation $Y \mapsto Y^* = UYU'$ we get the following formulae

$$dY^* = U \cdot dY \cdot U',$$

$$df = \sigma(dY^* \frac{\partial}{\partial Y^*})f = \sigma(U \cdot dY \cdot U' \frac{\partial}{\partial Y^*})f = \sigma(dY \cdot U' \frac{\partial}{\partial Y^*} U)f.$$

Hence $\frac{\partial}{\partial Y} = U' \frac{\partial}{\partial Y^*} U$ or

$$Y^* = UYU', \qquad \frac{\partial}{\partial Y^*} = U'^{-1} \frac{\partial}{\partial Y} U^{-1}.$$

By the definition given in §5, a differential operator $L(Y, \frac{\partial}{\partial Y})$ on $\mathcal{T}$ is invariant with respect to $G$ if

$$\{L(Y, \frac{\partial}{\partial Y})f(Y)\}_{Y \mapsto Y^*} = L(Y^*, \frac{\partial}{\partial Y^*})f(Y^*) = L(Y, \frac{\partial}{\partial Y})f(Y^*)$$

for all $f \in C^\infty(\mathcal{T})$ and $U \in G$. This can be written more concisely as

$$L(Y^*, \frac{\partial}{\partial Y^*}) = L(Y, \frac{\partial}{\partial Y}).$$

Since $Y^* \frac{\partial}{\partial Y^*} = UY \frac{\partial}{\partial Y} U^{-1}$ we see that the differential operators $\sigma(Y \frac{\partial}{\partial Y})^h$ are invariant for $h = 1,2,3,\ldots$ .

*Theorem. The differential operators* $\sigma(Y \frac{\partial}{\partial Y})^h$, $h = 1,2,\ldots,n$ *form an algebraically independent basis of the commutative ring* $L = L(\mathcal{T},G)$, *i.e.,* $L$ *is isomorphic with the ring* $\mathbb{C}[x_1,\ldots,x_n]$ *of polynomials in* $n$ *indeterminates.*

The proof that each $L(Y, \frac{\partial}{\partial Y})$ can be expressed as a polynomial in the $\sigma(Y \frac{\partial}{\partial Y})^h$, $h = 1,\ldots,n$, will proceed by induction on the degree of $L(Y,X)$ considered as a polynomial in the entries of the matrix

$X = X'$. To the invariance property

$$L(UYU', U'^{-1} \frac{\partial}{\partial Y} U^{-1}) = L(Y, \frac{\partial}{\partial Y})$$

there corresponds the property

$$L(UYU', U'^{-1}XU^{-1}) = L(Y,X) \quad \text{for } U \in G.$$

Determine $U \in G$ such that $UYU' = E$ and set $X_1 = U'^{-1}XU^{-1}$. Since $X$ is a "general" matrix, we may assume that $X_1$ is a diagonal matrix

$$X_1 = \begin{pmatrix} \xi_1 & & \\ & \ddots & \\ & & \xi_n \end{pmatrix} .$$

For any orthogonal matrix $V$ we have

$$L(Y,X) = L(E,X_1) = L(E,VX_1V')$$

and this is true if we take for $V$ a permutation matrix $P$ such that $X_1 \mapsto PX_1P'$ permutes the diagonal entries of $X_1$. It follows that $L(Y,X)$ is a symmetric polynomial in $\xi_1, \xi_2, \ldots, \xi_n$ and therefore by a well-known theorem $L(Y,X)$ is a polynomial in the

$$\xi_1^h + \xi_2^h + \ldots + \xi_n^h = \sigma(X_1^h) = \sigma(U^{-1}U'^{-1}X)^h = \sigma(YX)^h$$

$h = 1,\ldots,n$. Let us write

$$L(Y,X) = p(\sigma(YX), \sigma(YX)^2, \ldots, \sigma(YX)^n)$$

and replace $X$ by $\frac{\partial}{\partial Y}$. In general the differential operator

$$L_1(Y, \frac{\partial}{\partial Y}) = L(Y, \frac{\partial}{\partial Y}) - p\Big(\sigma(Y \frac{\partial}{\partial Y}), \ldots, \sigma(Y \frac{\partial}{\partial Y})^n\Big)$$

is not zero, but it is an invariant differential operator whose degree is less than that of $L(Y, \frac{\partial}{\partial Y})$ since the homogeneous terms of highest degree of two differential operators commute. By the induction hypothesis $L_1(Y, \frac{\partial}{\partial Y})$ is a polynomial in $\sigma(Y \frac{\partial}{\partial Y})^h$, $h = 1,\ldots,n$, and so the same is true for $L(Y, \frac{\partial}{\partial Y})$. Thus we have proved that the $\sigma(Y \frac{\partial}{\partial Y})^h$, $h = 1,\ldots,n$, generate $L$.

To prove the algebraic independence of the $\sigma(Y \frac{\partial}{\partial Y})^h$, $h = 1,2,\ldots,n$, we proceed as follows. For any polynomial $p(x_1,x_2,\ldots,x_n) \neq 0$ we define its weight as the degree of $p(x_1,x_2^2,\ldots,x_n^n)$. If $k$ denotes this number, then we can write

$$p(x_1,\ldots,x_n) = \sum_{\nu=0}^{k} h_\nu(x_1,\ldots,x_n),$$

where $h_\nu$ is the sum of all monomials of weight $\nu$. In particular $h_k(x_1,\ldots,x_n) \neq 0$.

Assume now that

$$p\left(\sigma(Y \frac{\partial}{\partial Y}),\ldots,\sigma(Y \frac{\partial}{\partial Y})^n\right) = 0.$$

From the relation

$$h_\nu\left(\sigma(Y \frac{\partial}{\partial Y}),\ldots,\sigma(Y \frac{\partial}{\partial Y})^n\right) e^{\sigma(Y)}$$

$$= e^{\sigma(Y)}\{h_\nu(\sigma(Y),\ldots,\sigma(Y)^n) + r_\nu(\sigma(Y),\ldots,\sigma(Y)^n)\},$$

where $r_\nu(x_1,\ldots,x_n)$ is a polynomial of weight less than $\nu$, it follows then that

$$h_k(\sigma(Y),\sigma(Y)^2,\ldots,\sigma(Y)^n) = 0.$$

Since the $\sigma(Y)^h$, $h = 1,2,\ldots,n$, are obviously algebraically independent, we have $h_k(x_1,x_2,\ldots,x_n) = 0$ and therefore $p(x_1,x_2,\ldots,x_n) = 0$. Our theorem is completely proved.

We introduce the following notation for the subdeterminants of any matrix $A = A^{(m,n)} = (a_{ik})$. If

$$1 \leqq \alpha_1 < \dots < \alpha_h \leqq m, \qquad 1 \leqq \beta_1 < \dots < \beta_h \leqq n,$$

we set

$$\begin{pmatrix} \alpha_1 & \dots & \alpha_h \\ \beta_1 & \dots & \beta_h \end{pmatrix}_A = \left| a_{\alpha_i \beta_k} \right|_{i,k=1,2,\dots,h} .$$

If $A = A^{(m,r)}$, $B = B^{(r,n)}$ and $h \leqq m, n, r$, then it is well known that

$$\begin{pmatrix} \alpha_1 & \dots & \alpha_h \\ \beta_1 & & \beta_h \end{pmatrix}_{AB} = \sum_{1 \leqq \gamma_1 < \dots < \gamma_h \leqq r} \begin{pmatrix} \alpha_1 & \dots & \alpha_h \\ \gamma_1 & \dots & \gamma_h \end{pmatrix}_A \begin{pmatrix} \gamma_1 & \dots & \gamma_h \\ \beta_1 & \dots & \beta_h \end{pmatrix}_B .$$

Now it can be seen easily that for $1 \leqq h \leqq n$ the operators

$$M_h = \sum_{\substack{1 \leqq \alpha_1 < \dots < \alpha_h \leqq n \\ 1 \leqq \beta_1 < \dots < \beta_h \leqq n}} \begin{pmatrix} \alpha_1 & \dots & \alpha_h \\ \beta_1 & \dots & \beta_h \end{pmatrix}_Y \begin{pmatrix} \beta_1 & \dots & \beta_h \\ \alpha_1 & \dots & \alpha_h \end{pmatrix}_{\frac{\partial}{\partial Y}}$$

are invariant differential operators which form a basis of the ring $L$. In particular

$$M_n = |Y| \left| \frac{\partial}{\partial Y} \right| .$$

In fact, if we set $M_h = M_h(Y, \frac{\partial}{\partial Y})$, then $M_h(Y,X)$ as a function of the variable matrix $X = X'$ is the $h$-th elementary symmetric function of the characteristic roots of $YX$, and these are algebraically independent.

Let us determine the involution $L \mapsto \tilde{L}$ for a given operator

$L = L(Y \frac{\partial}{\partial Y}) \in L$. If we denote $\mu(Y) = Y^{-1}$ by $\hat{Y}$, then we have by definition (§5)

$$\widetilde{L(Y \frac{\partial}{\partial Y})f(Y)} = \{L(Y \frac{\partial}{\partial Y})f(\hat{Y})\}_{Y \mapsto \hat{Y}} = L(\hat{Y} \frac{\partial}{\partial \hat{Y}})f(Y),$$

or

$$\widetilde{L(Y \frac{\partial}{\partial Y})} = L(\hat{Y} \frac{\partial}{\partial \hat{Y}}), \qquad \hat{Y} = Y^{-1}.$$

Differentiation of $\hat{Y}Y = E$ yields $d\hat{Y}\cdot Y + \hat{Y}\cdot dY = 0$ or $d\hat{Y} = - Y^{-1}\cdot dY\cdot Y^{-1}$. Hence

$$\sigma(d\hat{Y} \frac{\partial}{\partial \hat{Y}}) = - \sigma\left(dY\cdot Y^{-1}(Y^{-1} \frac{\partial}{\partial \hat{Y}})'\right) = \sigma(dY \frac{\partial}{\partial Y}),$$

and $- Y^{-1}(Y^{-1} \frac{\partial}{\partial \hat{Y}})' = \frac{\partial}{\partial Y}$ or

$$\hat{Y} \frac{\partial}{\partial \hat{Y}} = - (Y \frac{\partial}{\partial Y})' .$$

So we obtain

$$\widetilde{L(Y \frac{\partial}{\partial Y})} = L\left(- (Y \frac{\partial}{\partial Y})'\right) .$$

A function $f \in C^{\infty}(\mathcal{Y})$ is called an *eigenfunction* of $L$ if for each $L \in L$ there exists a complex number $\lambda = \lambda(L)$ such that

$$Lf = \lambda f.$$

We shall construct a family of eigenfunctions depending on $n$ complex parameters on the basis of the following observation: The set of all real matrices

$$T = \begin{pmatrix} t_{11} & & & 0 \\ t_{21} & t_{22} & & \\ \cdot & \cdot & \cdot & \cdot \\ t_{n1} & t_{n2} & \cdots & t_{nn} \end{pmatrix}, \qquad t_{\nu\nu} > 0, \quad \nu = 1,2,\ldots,n$$

form a subgroup $G_o$ of $G$ which acts transitively on $\mathcal{Y}$ because any $Y \in \mathcal{Y}$ can be represented in the form $Y = TT'$ with $T \in G_o$. Moreover the map $T \mapsto Y = TT'$ gives a one-to-one mapping from $G_o$ onto $\mathcal{Y}$, and therefore every function on $G_o$ can be considered as a function on $\mathcal{Y}$.

In particular let $s = (s_1, s_2, \ldots, s_n) \in \mathbb{C}^n$ and let $f_s$ be the function on $\mathcal{Y}$ defined by

$$f_s(Y) = \Phi_s(T) = \prod_{\nu=1}^{n} t_{\nu\nu}^{2s_\nu + \nu - \frac{1}{2}(n+1)}, \qquad Y = TT'.$$

It is obvious that

$$\Phi_s(T_1T_2) = \Phi_s(T_1)\Phi_s(T_2)$$

and therefore

$$f_s(TY_oT') = f_s(TT_oT_o'T') = \Phi_s(TT_o)$$

$$= \Phi_s(T)\Phi_s(T_o) = \Phi_s(T)f_s(Y_o).$$

Set $Y^* = TY_oT'$ and take any $L \in \mathcal{L}$. Then

$$L_{Y^*}f_s(Y^*) = L_{Y_o}f_s(TY_oT') = \Phi_s(T)L_{Y_o}f_s(Y_o).$$

The substitution $Y_o \mapsto E$ yields

$$Lf_s(Y) = \lambda f_s(Y) \quad \text{with} \quad \lambda = \lambda(s,L) = (Lf_s)(E),$$

i.e., $f_s$ is an eigenfunction of $\mathcal{L}$.

In order to represent $f_s(Y)$ explicitly as a function of $Y$, we introduce the notation

$$Y_h = Y\begin{bmatrix} E^{(h)} \\ 0 \end{bmatrix}, \qquad h = 1,2,\ldots,n,$$

so that

$$Y = \begin{pmatrix} Y_h & * \\ * & * \end{pmatrix}.$$

If $Y = TT'$ with $T = (t_{\mu\nu}) \in G_o$, then clearly $|Y_h| = t_{11}^2 t_{22}^2 \cdots t_{hh}^2$, hence

$$t_{hh}^2 = \frac{|Y_h|}{|Y_{h-1}|}, \qquad h = 1,2,\ldots,n,$$

with $|Y_o| = 1$ and finally

$$f_s(Y) = \prod_{\nu=1}^{n} |Y_\nu|^{s_\nu - s_{\nu+1} - \frac{1}{2}},$$

where $s_{n+1} = -\frac{n+1}{4}$.

We now develop *a constructive method for the computation of the eigenvalues* $\lambda_h(s) = \lambda_h(s_1, s_2, \ldots, s_n)$ of the operators $\sigma(Y \frac{\partial}{\partial Y})^h$, $h = 1,2,\ldots,n$, which we know form a basis of the ring $L$, with respect to the eigenfunctions $f_s(Y)$, i.e., of the complex numbers $\lambda_h(s)$ defined by

$$\sigma(Y \frac{\partial}{\partial Y})^h f_s(Y) = \lambda_h(s) f_s(Y).$$

We shall prove by induction on $n$ that

$$\lambda_h(s) = p_h(s) + q_h(s), \qquad h = 1,2,\ldots$$

where

$$p_h(s) = s_1^h + s_2^h + \ldots + s_n^h,$$

and $q_h(s)$ is a symmetric polynomial of degree less than $h$, so that $q_h(s)$ can be represented as a polynomial in $p_1, p_2, \ldots, p_{h-1}$. It will then follow that $p_h$ can be written as the sum of $\lambda_h$ and of a polynomial in $\lambda_1,\dots,\lambda_{h-1}$. Thus, together with $p_1(s),p_2(s),\dots,p_n(s)$ also the polynomials $\lambda_1(s),\lambda_2(s),\dots,\lambda_n(s)$ are algebraically independent. Furthermore, as $s = (s_1,s_2,\dots,s_n) \in \mathbb{C}^n$ varies, $\lambda_1(s),\lambda_2(s),\dots,\lambda_n(s)$ ranges over all $n$-tuples of complex numbers and the values of $\lambda_1(s),\lambda_2(s),\dots,\lambda_n(s)$ determine $s_1,s_2,\dots,s_n$ uniquely up to order.

The proof runs as follows: We parametrize $Y \in \mathcal{Y}$ with the help of the matrices

$$V = V^{(r)} > 0, \qquad W = W^{(n-r)} > 0, \qquad X = X^{(r,n-r)}$$

according to

$$Y = \begin{pmatrix} V & 0 \\ 0 & W \end{pmatrix}\begin{bmatrix} E & X \\ 0 & E \end{bmatrix} = \begin{pmatrix} V & VX \\ X'V & W + V[X] \end{pmatrix},$$

where $r$ is a given integer, $1 \leqq r \leqq n-1$. Let

$$Y = \begin{pmatrix} F & H \\ H' & G \end{pmatrix} \quad \text{and} \quad \frac{\partial}{\partial Y} = \begin{pmatrix} \frac{\partial}{\partial F} & \frac{1}{2}\frac{\partial}{\partial H} \\ \frac{1}{2}\frac{\partial}{\partial H'} & \frac{\partial}{\partial G} \end{pmatrix}$$

be the corresponding decompositions, i.e., $F = V$, $G = W + V[X]$, $H = VX$, so that

$$dF = dV,$$

$$dG = dW + (dV)[X] + dX'\cdot VX + X'V\cdot dX,$$

$$dH = dV\cdot X + V\cdot dX.$$

Substituting into

$$\sigma(dY \frac{\partial}{\partial Y}) = \sigma(dF \frac{\partial}{\partial F}) + \sigma(dG \frac{\partial}{\partial G}) + \sigma(dH' \frac{\partial}{\partial H})$$

and comparing with

$$\sigma(dY \frac{\partial}{\partial Y}) = \sigma(dV \frac{\partial}{\partial V}) + \sigma(dW \frac{\partial}{\partial W}) + \sigma(dX' \frac{\partial}{\partial X})$$

we obtain after some computation

$$\frac{\partial}{\partial F} = \frac{\partial}{\partial V} - \frac{1}{2}\{X \frac{\partial}{\partial X'} V^{-1} + V^{-1}(X \frac{\partial}{\partial X'})'\} + X \frac{\partial}{\partial W} X',$$

$$\frac{\partial}{\partial G} = \frac{\partial}{\partial W},$$

$$\frac{\partial}{\partial H} = V^{-1} \frac{\partial}{\partial X} - 2X \frac{\partial}{\partial W}.$$

If we write now

$$Y \frac{\partial}{\partial Y} = \begin{pmatrix} \Omega_1 & \Omega_3 \\ \Omega_4 & \Omega_2 \end{pmatrix},$$

then we have

$$\Omega_1 = V \frac{\partial}{\partial V} - \frac{1}{2}(X \frac{\partial}{\partial X'})',$$

$$\Omega_2 = W \frac{\partial}{\partial W} + \frac{1}{2} X' \frac{\partial}{\partial X},$$

$$\Omega_3 = \frac{1}{2} \frac{\partial}{\partial X},$$

$$\Omega_4 = X' V \frac{\partial}{\partial V} - W \frac{\partial}{\partial W} X' - \frac{1}{2} X' (X \frac{\partial}{\partial X'})' + \frac{1}{2} W \frac{\partial}{\partial X'} V^{-1}.$$

If $\omega_1$ and $\omega_2$ are two differential operators, we define the relation $\omega_1 \simeq \omega_2$ by requiring that $(\omega_1 - \omega_2)\phi(V,W) = 0$ for all functions $\phi$ independent of $X$. We prove by induction on $h$ that

$$(Y \frac{\partial}{\partial Y})^h \simeq \begin{pmatrix} A_h & 0 \\ X'A_h - (W \frac{\partial}{\partial W})^h X' & (W \frac{\partial}{\partial W})^h \end{pmatrix},$$

where $A_h$ is an appropriate differential matrix, independent of $X$

and $\frac{\partial}{\partial X}$. For $h = 1$ the assertion is true with $A_1 = V \frac{\partial}{\partial V}$. For a general $h$ we have obviously

$$(Y \frac{\partial}{\partial Y})^{h+1} = \begin{pmatrix} V \frac{\partial}{\partial V} & \frac{1}{2} \frac{\partial}{\partial X} \\ X' V \frac{\partial}{\partial V} - W \frac{\partial}{\partial W} X' & W \frac{\partial}{\partial W} + \frac{1}{2} X' \frac{\partial}{\partial X} \end{pmatrix} (Y \frac{\partial}{\partial Y})^h .$$

Using the fact that for a matrix $C = C^{(n-r)}$ independent of $X$ we have

$$\frac{\partial}{\partial X} CX' = \sigma(C)E,$$

and the induction hypothesis, we see that we have the desired representation for $(Y \frac{\partial}{\partial Y})^{h+1}$ if we choose

$$A_{h+1} = (V \frac{\partial}{\partial V} + \frac{n-r}{2} E)A_h - \frac{1}{2} \sigma (W \frac{\partial}{\partial W})^h E .$$

The recurrence shows that we can take

$$A_h = (V \frac{\partial}{\partial V} + \frac{n-r}{2} E)^h - \frac{1}{2} \sum_{\nu=0}^{h-1} (V \frac{\partial}{\partial V} + \frac{n-r}{2} E)^{\nu} \sigma (W \frac{\partial}{\partial W})^{h-1-\nu}$$

and

$$\sigma(Y \frac{\partial}{\partial Y})^h = \sigma(A_h) + \sigma(W \frac{\partial}{\partial W})^h$$

$$= \sigma(V \frac{\partial}{\partial V} + \frac{n-r}{2} E)^h + \sigma(W \frac{\partial}{\partial W})^h - \frac{1}{2} \sum_{\nu=0}^{h-1} \sigma(V \frac{\partial}{\partial V} + \frac{n-r}{2} E)^{\nu} \sigma (W \frac{\partial}{\partial W})^{h-1-\nu}.$$

To calculate $\lambda_h(s)$ we may replace $\sigma(Y \frac{\partial}{\partial Y})^h$ by this last operator since $f_s(Y)$ is independent of $X$. Indeed, let us write

$$V = \begin{pmatrix} t_1^2 & & 0 \\ & \ddots & \\ 0 & & t_r^2 \end{pmatrix} \begin{bmatrix} 1 & & * \\ & \ddots & \\ 0 & & 1 \end{bmatrix} , \qquad W = \begin{pmatrix} t_{r+1}^2 & & 0 \\ & \ddots & \\ 0 & & t_n^2 \end{pmatrix} \begin{bmatrix} 1 & & * \\ & \ddots & \\ 0 & & 1 \end{bmatrix}$$

so that

$$Y = \begin{pmatrix} t_1^2 & & 0 \\ & \ddots & \\ 0 & & t_n^2 \end{pmatrix} \begin{bmatrix} 1 & & * \\ & \ddots & \\ 0 & & 1 \end{bmatrix} .$$

Introducing the notation

$$f(s_1,\dots,s_r;V) = \prod_{\nu=1}^{r} t_\nu^{2s_\nu + \nu - \frac{1}{2}(r+1)} ,$$

$$f(s_{r+1},\dots,s_n;W) = \prod_{\nu=1}^{n-r} t_{r+\nu}^{2s_{r+\nu} + \nu - \frac{1}{2}(n-r+1)} ,$$

and

$$f_s(Y) = f(s_1,\dots,s_n;Y) = \prod_{\nu=1}^{n} t_\nu^{2s_\nu + \nu - \frac{1}{2}(n+1)} ,$$

we have

$$f(s_1,\dots,s_n;Y) = |V|^{-\frac{1}{4}(n-r)} |W|^{\frac{1}{4}r} f(s_1,\dots,s_r;V) f(s_{r+1},\dots,s_n;W)$$

and therefore the above relation between operators yields

$$\sigma(Y \tfrac{\partial}{\partial Y})^h f_s(Y)$$

$$= |V|^{-\frac{1}{4}(n-r)} |W|^{\frac{1}{4}r} \{\sigma(V \tfrac{\partial}{\partial V} + \tfrac{n-r}{4} E)^h + \sigma(W \tfrac{\partial}{\partial W} + \tfrac{r}{4} E)^h$$

$$- \frac{1}{2} \sum_{\nu=0}^{h-1} \sigma(V \tfrac{\partial}{\partial V} + \tfrac{n-r}{4} E)^\nu \, \sigma(W \tfrac{\partial}{\partial W} + \tfrac{r}{4} E)^{h-1-\nu} \} .$$

$$\cdot\; f(s_1,\dots,s_r;V) f(s_{r+1},\dots,s_n;W) .$$

First consider the case $\underline{n = 2}$, $\underline{r = 1}$: Writing $V = (v)$ and $W = (w)$ we have $f(s_1;v) = v^{s_1}$, $f(s_2;w) = w^{s_2}$ and $f_s(Y) = v^{s_1 - \frac{1}{4}} w^{s_2 + \frac{1}{4}}$. The above formula then yields directly

$$\lambda_h(s) = (s_1 + \tfrac{1}{4})^h + (s_2 + \tfrac{1}{4})^h - \tfrac{1}{2} \sum_{\nu=0}^{h-1} (s_1 + \tfrac{1}{4})^{\nu}(s_2 + \tfrac{1}{4})^{h-1-\nu} ,$$

which is indeed a symmetric polynomial with homogeneous term of highest degree $s_1^h + s_2^h$.

Now let $\underline{n \geq 3}$ and assume that all assertions are true for $r$ and $n-r$ instead of $n$, $1 \leq r \leq n-1$. Since the $\sigma(V \frac{\partial}{\partial V} + \frac{n-r}{4} E)^h$ are invariant operators on the space $\{V = V^{(r)} > 0\}$, and the $\sigma(W \frac{\partial}{\partial W} + \frac{r}{4} E)^h$ are invariant operators on the space $\{W = W^{(n-r)} > 0\}$, we may set

$$\sigma(V \frac{\partial}{\partial V} + \frac{n-r}{4} E)^h f(s_1,\ldots,s_r;V) = \mu_h(s_1,\ldots,s_r) f(s_1,\ldots,s_r;V),$$

$$\sigma(W \frac{\partial}{\partial W} + \frac{r}{4} E)^h f(s_{r+1},\ldots,s_n;W) = \nu_h(s_{r+1},\ldots,s_n) f(s_{r+1},\ldots,s_n;W),$$

where by the induction hypothesis $\mu_h(s_1,\ldots,s_r)$ and $\nu_h(s_{r+1},\ldots,s_n)$ are symmetric polynomials of the form

$$\mu_h(s_1,\ldots,s_r) = s_1^h + \ldots + s_r^h + \text{terms of lower degree},$$

$$\nu_h(s_{r+1},\ldots,s_n) = s_{r+1}^h + \ldots + s_n^h + \text{terms of lower degree}.$$

From the above formula we now obtain

$$\lambda_h(s) = \mu_h(s_1,\ldots,s_r) + \nu_h(s_{r+1},\ldots,s_n)$$
$$- \frac{1}{2} \sum_{\nu=0}^{h-1} \mu_\nu(s_1,\ldots,s_r)\nu_{h-1-\nu}(s_{r+1},\ldots,s_n)$$
$$= p_h(s) + q_h(s).$$

Here $q_h(s)$ denotes a polynomial of degree $< h$, which is symmetric in both sets $s_1,\ldots,s_r$ and $s_{r+1},\ldots,s_n$. But since $r$ is arbitrary in

$0 < r < n$, the polynomial $q_h(s)$ is also symmetric in all variables $s_1, s_2, \ldots, s_n$, q.e.d.

An immediate consequence of our result is the following:
*The effect of an operator*

$$L = p\left(\sigma(Y \tfrac{\partial}{\partial Y}), \ldots, \sigma(Y \tfrac{\partial}{\partial Y})^n\right) \in L$$

*on the eigenfunctions* $f_s(Y)$ *determines* $L$ *uniquely.* Indeed, $Lf_s(Y) = 0$ implies

$$p(\lambda_1(s), \lambda_2(s), \ldots, \lambda_n(s)) = 0$$

for all $s \in \mathbb{C}^n$. Hence the polynomial $p$ vanishes identically, so that $L = 0$.

We now introduce the integral

$$J_s(X) = \int_{Y>0} e^{-\sigma(YX^{-1})} f_s(Y) dv,$$

where $X > 0$ and $dv$ denotes the invariant volume element

$$dv = |Y|^{-\frac{1}{2}(n+1)} \prod_{\mu \leq \nu} dy_{\mu\nu} .$$

We shall see that the integral converges for $\mathcal{R}e\, s_\nu > \frac{n-1}{4}$, $\nu = 1, 2, \ldots, n$.

Let us prove the formula

$$J_s(X) = \pi^{\frac{1}{4}n(n-1)} \prod_{\nu=1}^{n} \Gamma(s_\nu - \tfrac{n-1}{4}) \cdot f_s(X).$$

In the first place set $X = TT'$ with $T \in G_0$. The substitution $Y \mapsto TYT'$ transforms our integral into

$$J_s(X) = \int_{Y>0} e^{-\sigma(Y)} f_s(TYT')dv$$

$$= J_s(E)\Phi_s(T) = J_s(E)f_s(X),$$

hence the proof of our formula is reduced to the case $X = E$.

In order to compute

$$J_s(E) = \int_{Y>0} e^{-\sigma(Y)} f_s(Y)dv$$

we substitute $Y = TT'$ with $T \in G_0$ and set

$$T = T_n = \begin{pmatrix} T_{n-1} & 0 \\ \mathfrak{t}' & t \end{pmatrix}, \qquad Y = Y_n = \begin{pmatrix} Y_{n-1} & \mathfrak{y} \\ \mathfrak{y}' & y \end{pmatrix}.$$

Then

$$Y_{n-1} = T_{n-1}T'_{n-1}, \qquad \mathfrak{y} = T_{n-1}\mathfrak{t}, \qquad y = t^2 + \mathfrak{t}'\mathfrak{t},$$

and

$$\frac{\partial(Y_n)}{\partial(T_n)} = \frac{\partial(Y_{n-1}, \mathfrak{y}, y)}{\partial(T_{n-1}, \mathfrak{t}, t)} = \begin{vmatrix} \frac{\partial(Y_{n-1})}{\partial(T_{n-1})} & 0 & 0 \\ * & T_{n-1} & 0 \\ 0 & * & 2t \end{vmatrix}$$

$$= 2\frac{\partial(Y_{n-1})}{\partial(T_{n-1})}|T_{n-1}|t = 2\frac{\partial(Y_{n-1})}{\partial(T_{n-1})}|T_n|.$$

By induction on $n$ we get therefore

$$\frac{\partial(Y_n)}{\partial(T_n)} = 2^n \prod_{j=1}^{n}\prod_{\nu=1}^{j} t_{\nu\nu} = 2^n \prod_{\nu=1}^{n} t_{\nu\nu}^{n-\nu+1}.$$

Since $|Y|^{-\frac{1}{2}(n+1)} = \prod_{\nu=1}^{n} t_{\nu\nu}^{-n-1}$, we can express the invariant volume element in the form

$$dv = 2^n \prod_{\nu=1}^{n} t_{\nu\nu}^{-\nu} \prod_{\mu\geq\nu} dt_{\mu\nu}$$

and we obtain by integration over $t_{\nu\nu} > 0$ $(\nu = 1,2,\dots,n)$, $t_{\mu\nu}$ arbitrary $(\mu > \nu)$:

$$J_s(E) = 2^n \int\dots\int e^{-\sum\limits_{\mu\geq\nu} t_{\mu\nu}^2} \prod_{\nu=1}^{n} t_{\nu\nu}^{2s_\nu - \frac{1}{2}(n+1)} \prod_{\mu\geq\nu} dt_{\mu\nu}$$

$$= 2^n \left(\int_{-\infty}^{\infty} e^{-t^2} dt\right)^{\frac{1}{2}n(n-1)} \prod_{\nu=1}^{n} \int_0^{\infty} e^{-t^2}\, t^{2s_\nu - \frac{1}{2}(n+1)} dt$$

$$= \pi^{\frac{1}{4}n(n-1)} \prod_{\nu=1}^{n} \Gamma(s_\nu - \tfrac{n-1}{4}) ,$$

and the integrals converge indeed for $\mathcal{Re}\, s_\nu > \frac{n-1}{4}$ $(\nu = 1,2,\dots,n)$. q.e.d.

Now we can prove that the adjoint $\hat{L}$ of a differential operator $L \in \mathcal{L}$ is given by

$$\hat{L} = \bar{\tilde{L}}.$$

First we observe that $e^{-\sigma(YX^{-1})}$ is a point pair invariant function, since it does not change if $Y$ is replaced by $UYU'$ and $X$ by $UXU'$, where $U \in G = GL(n,\mathbb{R})$. It follows that

$$\tilde{L}f_s(X)J_s(E) = \tilde{L}_X \int_{Y>0} e^{-\sigma(YX^{-1})} f_s(Y)dv$$

$$= \int_{Y>0} f_s(Y)\tilde{L}_X e^{-\sigma(YX^{-1})} dv = \int_{Y>0} f_s(Y) L_Y e^{-\sigma(YX^{-1})} dv$$

$$= \int_{Y>0} e^{-\sigma(YX^{-1})} \hat{\tilde{L}}_Y f_s(Y)\,dv = \hat{\tilde{L}} f_s(X) J_s(E)$$

since $\hat{\tilde{L}}_Y f_s(Y) = \lambda(s) f_s(Y)$ with an appropriate $\lambda(s) \in \mathbb{C}$. This shows that $\tilde{L} = \hat{\tilde{L}}$ or $\hat{L} = \tilde{\tilde{L}} = \bar{\tilde{L}}$, q.e.d.

According to the rule given on p. 59 the adjoint of the operator

$$K = \sigma(Y) \prod_{\mu \leqq \nu} \left( \frac{\partial}{\partial y_{\mu\nu}} \right)^{a_{\mu\nu}}$$

is of the form

$$\hat{K} = (-1)^h |Y|^{\frac{1}{2}(n+1)} \prod_{\mu \leqq \nu} \left( \frac{\partial}{\partial y_{\mu\nu}} \right)^{a_{\mu\nu}} \overline{\sigma(Y)} |Y|^{-\frac{1}{2}(n+1)},$$

where $h = \sum_{\mu \leqq \nu} a_{\mu\nu}$. Applying this to the operator

$$L = |Y|^h \left|\frac{\partial}{\partial Y}\right|^h,$$

which is a finite sum of monomials of the type K, we obtain

$$\hat{L} = (-1)^{nh} |Y|^{\frac{1}{2}(n+1)} \left|\frac{\partial}{\partial Y}\right|^h |Y|^{h-\frac{1}{2}(n+1)}.$$

On the other hand we know that

$$\hat{L} = \tilde{L} = |\hat{Y}|^h \left|\frac{\partial}{\partial \hat{Y}}\right|^h,$$

with $\hat{Y} = Y^{-1}$, and therefore

$$|\hat{Y}|^h \left|\frac{\partial}{\partial \hat{Y}}\right|^h = (-1)^{nh} |Y|^{\frac{1}{2}(n+1)} \left|\frac{\partial}{\partial Y}\right|^h |Y|^{h-\frac{1}{2}(n+1)}.$$

Setting in

$$f_s(Y) = \prod_{\nu=1}^{n} |Y_\nu|^{s_\nu - s_{\nu+1} - \frac{1}{2}}$$

the special value of $s$ given by

$$s_\nu - \frac{n-1}{4} = \begin{cases} t + 1 - \frac{\nu-1}{2} & \text{for } 1 \leqq \nu \leqq h, \\ t - \frac{\nu-1}{2} & \text{for } h < \nu \leqq n, \end{cases}$$

we obtain

$$f_s(Y) = |Y_h| \cdot |Y|^t,$$

where

$$Y = \begin{pmatrix} Y_h & * \\ * & * \end{pmatrix}, \qquad \text{i.e.,} \quad |Y_h| = \begin{pmatrix} 1 & \cdots & h \\ 1 & \cdots & h \end{pmatrix}_Y .$$

Replacing the same value of $s$ in the expression we obtained above for $J_s(E)$ we get

$$\pi^{\frac{1}{4}n(n-1)} \prod_{\nu=1}^{n} \Gamma(s_\nu - \frac{n-1}{4}) = \prod_{\nu=1}^{h} (t - \frac{\nu-1}{2}) \prod_{\nu=1}^{n} \pi^{\frac{1}{2}(\nu-1)} \Gamma(t - \frac{\nu-1}{2})$$

$$= \varepsilon_h(t) \Gamma_n(t),$$

where we used the abbreviations

$$\varepsilon_h(t) = t(t - \frac{1}{2}) \ldots (t - \frac{h-1}{2}),$$

$$\Gamma_n(t) = \prod_{\nu=1}^{n} \pi^{\frac{1}{2}(\nu-1)} \Gamma(t - \frac{\nu-1}{2}).$$

Thus we have proved the formula

$$\int_{Y>0} e^{-\sigma(Y)} |Y_h| . |Y|^t dv = \varepsilon_h(t) \Gamma_n(t).$$

Denote by $(\alpha_1, \alpha_2, \ldots, \alpha_n)$ a permutation of $(1,2,\ldots,n)$ and let $(\delta_{\mu\alpha_\nu})$ be the corresponding permutation matrix. If we substitute

$Y[(\delta_{\mu\alpha_\nu})]$ for $Y$, then

$$|Y_h| = \begin{pmatrix} 1 & \dots & h \\ 1 & \dots & h \end{pmatrix}_Y$$

will be replaced by

$$\begin{pmatrix} 1 & \dots & h \\ 1 & \dots & h \end{pmatrix}_{Y[(\delta_{\mu\alpha_\nu})]} = \begin{pmatrix} \alpha_1 & \dots & \alpha_h \\ \alpha_1 & \dots & \alpha_h \end{pmatrix}_Y$$

and so we get

$$\int_{Y>0} e^{-\sigma(Y)} \begin{pmatrix} \alpha_1 & \dots & \alpha_h \\ \alpha_1 & \dots & \alpha_h \end{pmatrix}_Y |Y|^t dv = \varepsilon_h(t)\Gamma_n(t).$$

Applying the invariant differential operator

$$|Y|^t \mathrm{M}_h(Y)|Y|^{-t},$$

where $\mathrm{M}_h = \mathrm{M}_h(Y)$ is the operator introduced on p. 67, to the function which is identically equal to 1 we obtain a polynomial $\mu_h(t)$, so that

$$\mathrm{M}_h|Y|^{-t} = \mu_h(t)|Y|^{-t}.$$

We give to $s \in \mathbb{C}^n$ a special value by setting

$$s_\nu - \frac{n-1}{4} = t - \frac{\nu-1}{2}, \qquad \nu = 1,2,\dots,n,$$

and writing, as usual, $\hat{X}$ for $X^{-1}$, we calculate $\mu_h(t)$ as follows

$$\mu_h(t)|\hat{X}|^{-t}\Gamma_n(t) = \mathrm{M}_h(\hat{X})|\hat{X}|^{-t}\Gamma_n(t)$$

$$= \mathrm{M}_h(\hat{X})|X|^t\Gamma_n(t) = \mathrm{M}_h(\hat{X})J_s(X)$$

$$= \int_{Y>0} |Y|^t \mathrm{M}_h(\hat{X})e^{-\sigma(Y\hat{X})}dv$$

$$= (-1)^h \int_{Y>0} |Y|^t \sum_{\substack{\alpha_1<\dots<\alpha_h \\ \beta_1<\dots<\beta_h}} \begin{pmatrix} \alpha_1 & \dots & \alpha_h \\ \beta_1 & \dots & \beta_h \end{pmatrix}_{\hat{X}} \begin{pmatrix} \beta_1 & \dots & \beta_h \\ \alpha_1 & \dots & \alpha_h \end{pmatrix}_Y e^{-\sigma(Y\hat{X})} dv$$

$$= (-1)^h \int_{Y>0} |Y|^t \sum_{\alpha_1<\dots<\alpha_h} \begin{pmatrix} \alpha_1 & \dots & \alpha_h \\ \alpha_1 & \dots & \alpha_h \end{pmatrix}_{\hat{X}Y} e^{-\sigma(Y\hat{X})} dv.$$

Setting in particular $X = E$ we obtain

$$\mu_h(t)\Gamma_n(t) = (-1)^h \binom{n}{h} \varepsilon_h(t)\Gamma_n(t)$$

and finally

$$M_h(Y)|Y|^{-t} = (-1)^h \binom{n}{h} \varepsilon_h(t)|Y|^{-t}.$$

Our final computation will give us the eigenvalues of the operator $M_n$ with respect to the eigenfunctions $f_s(Y)$. Write

$$\tilde{U} = (\delta_{n+1-\mu,\nu}).$$

Since

$$\begin{pmatrix} 1 & \dots & h \\ 1 & \dots & h \end{pmatrix}_{Y^{-1}[\tilde{U}]} = \begin{pmatrix} n-h+1,\dots,n \\ n-h+1,\dots,n \end{pmatrix}_{Y^{-1}} =$$

$$= |Y|^{-1} \begin{pmatrix} 1,\dots,n-h \\ 1,\dots,n-h \end{pmatrix}_Y$$

and

$$f_s(Y^{-1}[\tilde{U}]) = \prod_{h=1}^{n} \begin{pmatrix} 1 & \dots & h \\ 1 & \dots & h \end{pmatrix}_{Y^{-1}[\tilde{U}]}^{s_h - s_{h+1} - \frac{1}{2}},$$

with $s_{n+1} = -\frac{n+1}{4}$, we get easily that

$$f_s(Y^{-1}[\tilde{U}]) = f_{\tilde{s}}(Y),$$

where

$$\tilde{s}_h = -s_{n+1-h}, \qquad h = 1,2,\ldots,n.$$

Writing $(s+1)_h = s_h + 1$, it follows therefore that

$$J_s(E)\mathrm{M}_n(\hat{X}[\tilde{U}])f_{\tilde{s}}(\hat{X}[\tilde{U}]) = J_s(E)\mathrm{M}_n(\hat{X})f_s(X)$$

$$= \mathrm{M}_s(\hat{X})J_s(X) = \int\limits_{Y>0} f_s(Y)\mathrm{M}_n(\hat{X})e^{-\sigma(Y\hat{X})}dv$$

$$= \int\limits_{Y>0} f_s(Y)\mathrm{M}_n(Y)e^{-\sigma(Y\hat{X})}dv$$

$$= (-1)^n|\hat{X}| \int\limits_{Y>0} e^{-\sigma(Y\hat{X})}|Y|f_s(Y)dv$$

$$= (-1)^n|\hat{X}| \int\limits_{Y>0} e^{-\sigma(Y\hat{X})}f_{s+1}(Y)dv$$

$$= (-1)^n|X|^{-1}J_{s+1}(E)f_{s+1}(X) = (-1)^n J_{s+1}(E)f_s(X)$$

$$= (-1)^n J_{s+1}(E)f_{\tilde{s}}(\hat{X}[\tilde{U}])$$

$$= (-1)^n \prod_{\nu=1}^{n}(s_\nu - \tfrac{n-1}{4})J_s(E)f_{\tilde{s}}(\hat{X}[\tilde{U}]).$$

Substituting $s$ for $\tilde{s}$ and writing $Y = \hat{X}[\tilde{U}]$, we obtain the required result

$$\mathrm{M}_n f_s(Y) = \prod_{\nu=1}^{n}(s_\nu + \tfrac{n-1}{4})f_s(Y).$$

## §7. *A generalization of* $J_s(X)$

Let $\mathcal{Y}$ be again the Riemannian space of all positive matrices $Y = Y^{(n)} > 0$ and denote by $L$ the ring of differential operators on $\mathcal{Y}$, invariant with respect to $G = GL(n,\mathbb{R})$.

If we define $u(Y)$ by

$$f_s(Y) = \Phi_s(T) = |Y|^{s_0} u(Y), \qquad Y = TT',$$

where $ns_0 = s_1 + \ldots + s_n$, then $u(Y)$ is homogeneous of degree 0 since

$$\sum_{\nu=1}^{n} \nu(s_\nu - s_{\nu+1} - \tfrac{1}{2}) = ns_0,$$

and we have

$$u(Y) = |T|^{-2s_0} \Phi_s(T) = \prod_{\nu=1}^{n} t_{\nu\nu}^{2(s_\nu - s_0) + \nu - \frac{1}{2}(n+1)}.$$

Let us show that for a sufficiently large constant $\kappa$ we have

$$|u(Y)| \leqq (\sigma(Y))^{\kappa} \qquad \text{for } Y > 0, \ |Y| = 1.$$

Without loss of generality we may assume that all $s_\nu$ are real. Choose $\kappa$ so that

$$2(s_\nu - s_0 + \tfrac{\kappa}{n}) + \nu - \tfrac{n+1}{2} > 0 \qquad \text{for } 1 \leqq \nu \leqq n.$$

Then we obtain indeed

$$u(Y)|Y|^{\kappa/n} = \prod_{\nu=1}^{n} t_{\nu\nu}^{2(s_\nu - s_0 + \kappa/n) + \nu - \frac{1}{2}(n+1)} \leqq$$

$$\leqq \prod_{\nu=1}^{n} \left( \sum_{\alpha \geqq \beta} t_{\alpha\beta}^2 \right)^{s_\nu - s_0 + \frac{\kappa}{n} + \frac{\nu}{2} - \frac{n+1}{4}} = (\sigma(Y))^{\kappa}.$$

Using the representation

$$f_s(Y) = \prod_{\nu=1}^{n} |Y_\nu|^{s_\nu - s_{\nu+1} - \frac{1}{2}}$$

we can see easily that the partial derivatives of $u(Y)$ of order $h$ satisfy estimates of the following type

$$\left| \frac{\partial^h u(Y)}{\partial y_{\alpha\beta} \cdots \partial y_{\mu\nu}} \right| \leqq C_h (\sigma(Y))^{\kappa_h}$$

for $Y > 0$, $|Y| = 1$ and $h \geqq 0$, with positive constants $C_h$ and $\kappa_h$. Indeed, the derivatives can be represented in the form

$$\sum_a p_a(Y) f_{s - s_0 - a}(Y),$$

where $a = (a_1, a_2, \ldots, a_n)$, $h \geqq a_1 \geqq a_2 \geqq \ldots \geqq a_n \geqq 0$, the $a_\nu$ are integers, $(s - s_0 - a)_\nu = s_\nu - s_0 - a_\nu$, and the $p_a(Y)$ are polynomials in the entries of $Y$. We mention that the derivatives of order $h$ of $u(Y)$ are homogeneous functions of degree $-h$.

It will thus be reasonable to consider, writing now $s$ instead of $s_0$, where this time $s$ is a single complex variable, integrals of the form

$$J_s(X, u) = \int_{Y>0} e^{-\sigma(YX^{-1})} |Y|^s u(Y) dv,$$

where $X > 0$, and where $u(Y)$ satisfies the following conditions:

1. $u(Y) \in C^\infty(\mathcal{Y})$,

2. $u(Y)$ is an eigenfunction of $L$,

3. $u(Y)$ is a homogeneous function of degree 0, so that

$$\sigma(Y \frac{\partial}{\partial Y})u(Y) = 0,$$

4. for $h \geqq 0$ there exist positive constants $C_h$ and $\kappa_h$ such that

$$\left| \frac{\partial^h u(Y)}{\partial y_{\alpha\beta} \cdots \partial y_{\mu\nu}} \right| \leqq C_h(\sigma(Y))^{\kappa_h}$$

for $Y > 0$, $|Y| = 1$.

Under these assumptions we shall prove that

$$J_s(X,u) = \pi^{\frac{1}{4}n(n-1)} \prod_{\nu=1}^{n} \Gamma(s - \alpha_\nu)|X|^s u(X),$$

if $\mathcal{R}e\, s$ is sufficiently large. The constants $\alpha_1, \alpha_2, \ldots, \alpha_n$ are in general complex numbers. They depend only on the eigenvalues with respect to $u$ and satisfy the relation

$$\alpha_1 + \alpha_2 + \ldots + \alpha_n = \frac{n(n-1)}{4}.$$

We reduce the computation of $J_s(X,u)$ to the case $X = E$ with the help of the substitution $Y \mapsto Y[\sqrt{X}]$ which yields

$$J_s(X,u) = J_s(E,u_1)|X|^s,$$

where

$$u_1(Y) = u(Y[\sqrt{X}])$$

has the same properties as $u(Y)$, and also the eigenvalues of the operators in $L$ are the same with respect to $u(Y)$ and $u_1(Y)$. For properties 1, 2 and 3 this assertion is clear. To prove property 4 note that $\sigma(Y[\sqrt{X}]) = \sigma(YX) = \sigma(X[\sqrt{Y}])$. Therefore, if $c_1$ and $c_2$ denote a lower and an upper bound for the eigenvalues of $X$, respectively, we have $c_1 E < X < c_2 E$, i.e., $c_1 Y < X[\sqrt{Y}] < c_2 Y$ and so

$$c_1 \sigma(Y) < \sigma(Y[\sqrt{X}]) < c_2 \sigma(Y),$$

from where inequalities of the type considered in 4. follow for $u_1(Y)$.

Denoting by $M_n(Y)$ the operator

$$M_n(Y) = |Y| \left| \frac{\partial}{\partial Y} \right|$$

introduced on p. 67, and applying the operator $M_n(\hat{X})$, where $\hat{X} = X^{-1}$, to $J_s(X,u)$ we get on one hand

$$M_n(\hat{X}) J_s(X,u) = |\hat{X}| \int_{Y>0} \left| \frac{\partial}{\partial \hat{X}} \right| e^{-\sigma(Y\hat{X})} |Y|^s u(Y) dv$$

$$= (-1)^n |\hat{X}| J_{s+1}(X,u),$$

and on the other hand, by passing to the adjoint,

$$M_n(\hat{X}) J_s(X,u) = \int_{Y>0} |Y|^s u(Y) M_n(\hat{X}) e^{-\sigma(Y\hat{X})} dv$$

$$= \int_{Y>0} |Y|^s u(Y) M_n(Y) e^{-\sigma(Y\hat{X})} dv$$

$$= \int_{Y>0} e^{-\sigma(Y\hat{X})} M_n(\hat{Y}) |Y|^s u(Y) dv,$$

always provided that $\Re\, s$ is sufficiently large. It was in order to justify the application of Green's formula (shifting and replacing $M_n(Y)$ by the adjoint operator $M_n(\hat{Y})$ ) that we introduced condition 4 for $u(Y)$. Since (p. 79)

$$|Y|^{-s} M_n(\hat{Y}) |Y|^s = (-1)^n |Y|^{\frac{1}{2}(n+1) - s} \left| \frac{\partial}{\partial Y} \right| |Y|^{s - \frac{1}{2}(n-1)}$$

$$= (-1)^n |Y|^{\frac{1}{2}(n-1) - s} M_n(Y) |Y|^{s - \frac{1}{2}(n-1)}$$

and

$$M_n(Y) |Y|^{s - \frac{1}{2}(n-1)} = (-1)^n \varepsilon_n\left(\frac{n-1}{2} - s\right) |Y|^{s - \frac{1}{2}(n-1)}$$

$$= (s^n + \ldots) |Y|^{s - \frac{1}{2}(n-1)},$$

where

$$\varepsilon_n(t) = \prod_{\nu=1}^{n} \left(t - \frac{\nu-1}{2}\right)$$

is the function introduced on p. 80, we obtain the operator identity

$$|Y|^{-s} M_n(\hat{Y}) |Y|^s = (-1)^n \left(s^n + p\left(s, \sigma\left(Y \frac{\partial}{\partial Y}\right), \ldots, \sigma\left(Y \frac{\partial}{\partial Y}\right)^n\right)\right),$$

where $p(x_0, x_1, \ldots, x_n)$ is a polynomial whose degree in $x_0$ is less than $n$. If we set

$$\sigma\left(Y \frac{\partial}{\partial Y}\right)^h u(Y) = \lambda_h u(Y)$$

for $h = 1, 2, \ldots, n$, with $\lambda_1 = 0$ because $u(Y)$ is homogeneous of degree zero, then

$$M_n(\hat{Y}) |Y|^s u(Y) = (-1)^n (s^n + p(s, 0, \lambda_2, \ldots, \lambda_n)) |Y|^s u(Y)$$

$$= (-1)^n f(s) |Y|^s u(Y),$$

where

$$f(s) = f(s, u) = s^n + p(s, \lambda_1, \ldots, \lambda_n),$$

and $p(s, \lambda_1, \ldots, \lambda_n)$ is a polynomial in $s$ of degree less than $n$.

Substituting into the last integral we obtain

$$M_n(\hat{X})J_s(X,u) = (-1)^n f(s) \int_{Y>0} e^{-\sigma(Y\hat{X})}|Y|^s u(Y)\,dv$$

$$= (-1)^n f(s) J_s(X,u).$$

A comparison of the two expressions obtained for $M_n(\hat{X})J_s(X,u)$ yields

$$|\hat{X}|J_{s+1}(X,u) = f(s)J_s(X,u)$$

or, taking into account that $|\hat{X}|J_{s+1}(X,u) = |X|^{-1}J_{s+1}(E,u_1)|X|^{s+1} =$
$= |X|^s J_{s+1}(E,u_1)$,

$$J_{s+1}(E,u_1) = f(s)J_s(E,u_1).$$

Write

$$f(s) = (s-\alpha_1)(s-\alpha_2)\dots(s-\alpha_n),$$

where the roots $\alpha_h = \alpha_h(u)$ depend on the function $u(Y)$. Using the functional equation $\Gamma(s+1) = s\Gamma(s)$ we can rewrite this in the form

$$f(s) = \prod_{\nu=1}^{n} \frac{\Gamma(s+1-\alpha_\nu)}{\Gamma(s-\alpha_\nu)}.$$

The function

$$H(s,u_1) = \frac{J_s(E,u_1)}{\prod_{\nu=1}^{n} \Gamma(s-\alpha_\nu)}$$

is periodic, i.e.,

$$H(s+1,u_1) = H(s,u_1).$$

Since it is holomorphic for large values of $\mathcal{R}e\, s$, it is an entire function. We shall show that $\lim_{\sigma\to\infty} H(\sigma+it,u_1)$ ($\sigma,t$ real) exists. It will follow that $H(s,u_1)$ is a constant and it remains then to compute this constant.

We shall calculate

$$J_s(E,u_1) = \int_{Y>0} e^{-\sigma(Y)}|Y|^s u_1(Y)dv,$$

which depends on $X$ through $u_1(Y) = u(Y[\sqrt{X}])$, by integrating first over the hypersurface $|Y|$ = const., and then with respect to $y = \sqrt[n]{|Y|}$. For this purpose we introduce on $|Y| = c$ a volume element $dv_1$, which is independent of $c$ and invariant with respect to the mappings $Y \mapsto Y[U]$, where $|U| = 1$, by setting

$$dv_1 = \frac{dv}{d\log y} = \frac{y\,dv}{dy},$$

with $y^n = |Y|$, $y > 0$. Setting $Y = yY_1$, with $|Y_1| = 1$, we have

$$J_s(E,u_1) = \int_{\substack{Y_1>0\\|Y_1|=1}} \int_0^\infty e^{-y\sigma(Y_1)}y^{ns}u_1(Y_1)dv_1\frac{dy}{y}$$

$$= \Gamma(ns)I(s,u_1),$$

where

$$I(s,u_1) = \int_{\substack{Y_1>0\\|Y_1|=1}} \frac{u_1(Y_1)}{(\sigma(Y_1))^{ns}}dv_1 .$$

Thus

$$H(s,u_1) = \frac{\Gamma(ns)}{\prod_{\nu=1}^{n}\Gamma(s-\alpha_\nu)} I(s,u_1)$$

and Stirling's formula

$$\Gamma(s-\alpha) \sim \sqrt{2\pi}\, s^{s-\alpha-\frac{1}{2}} e^{-s} \qquad \text{as} \quad s \to \infty$$

yields

$$H(s,u_1) \sim (2\pi)^{-\frac{1}{2}(n-1)} n^{ns-\frac{1}{2}} s^{\alpha_1+\alpha_2+\ldots+\alpha_n+\frac{1}{2}(n-1)} I(s,u_1)$$

as $s \to \infty$.

We first apply this formula to the special case $u = u_1 = 1$, when clearly

$$f(s,1) = (-1)^n \varepsilon_n(\tfrac{n-1}{2} - s) = s(s-\tfrac{1}{2}) \ldots (s-\tfrac{n-1}{2}).$$

Using the result

$$J_s(E,1) = \int_{Y>0} e^{-\sigma(Y)} |Y|^s dv$$

$$= \Gamma_n(s) = \pi^{\frac{1}{4}n(n-1)} \prod_{\nu=1}^{n} \Gamma(s - \tfrac{\nu-1}{2})$$

proved in §6 we obtain

$$H(s,1) = \frac{J_s(E,1)}{\prod_{\nu=1}^{n} \Gamma(s-\frac{\nu-1}{2})} = \pi^{\frac{1}{4}n(n-1)},$$

i.e.,

$$\pi^{\frac{1}{4}n(n-1)} \sim (2\pi)^{-\frac{1}{2}(n-1)} n^{ns-\frac{1}{2}} s^{\frac{1}{4}n(n-1)+\frac{1}{2}(n-1)} I(s,1)$$

as $s \to \infty$.

Dividing the two asymptotic relations we get

$$H(s,u_1) \sim \pi^{\frac{1}{4}n(n-1)} s^{\alpha_1+\ldots+\alpha_n-\frac{1}{4}n(n-1)} \frac{I(s,u_1)}{I(s,1)},$$

as $s \to \infty$.

Next we prove

$$\lim_{s\to\infty} \frac{I(s,u_1)}{I(s,1)} = u_1(E) = u(X).$$

By virtue of the linearity of the left hand side with respect to $u_1$, we may assume without loss of generality, replacing, if necessary, $u_1(Y)$ by $u_1(Y) - u_1(E)$, that $u_1(E) = 0$. The geometric mean of the characteristic roots of $Y_1$ is at most equal to their arithmetic mean, i.e.,

$$\sigma(Y_1) \geqq n \sqrt[n]{|Y_1|} = n,$$

and we have equality only if $Y_1 = E$. Thus $\sigma(Y_1) \to n$ is equivalent to $Y_1 \to E$. Since $u_1(Y_1)$ is continuous and $u_1(E) = 0$, for a given $\varepsilon > 0$ there exists a $\delta = \delta(\varepsilon) > 0$ so that

$$|u_1(Y_1)| < \varepsilon \quad \text{for} \quad \sigma(Y_1) < n(1+\delta).$$

Moreover we have

$$(\sigma(Y_1))^{\kappa_0 - ns} = (\sigma(Y_1))^{\kappa_0 - \frac{1}{2}ns}(\sigma(Y_1))^{-\frac{1}{2}ns}$$

$$\leqq (n(1+\delta))^{\kappa_0 - \frac{1}{2}ns} \cdot (\sigma(Y_1))^{-\frac{1}{2}ns},$$

provided that $\sigma(Y_1) \geqq n(1+\delta)$ and $\kappa_0 - \frac{1}{2}ns < 0$, i.e., $s > \dfrac{2\kappa_0}{n}$. Remembering that

$$|u_1(Y_1)| \leqq C_0(\sigma(Y_1))^{\kappa_0}$$

we obtain

$$\left|\frac{I(s,u_1)}{I(s,1)}\right| \leqq \left|\frac{\displaystyle\int_{\sigma(Y_1)\leqq n(1+\delta)} \frac{u_1(Y_1)}{(\sigma(Y_1))^{ns}}\,dv_1}{I(s,1)}\right| + \left|\frac{\displaystyle\int_{\sigma(Y_1)\geqq n(1+\delta)} \frac{u_1(Y_1)}{(\sigma(Y_1))^{ns}}\,dv_1}{I(s,1)}\right|$$

$$\leqq \varepsilon + C(n(1+\delta))^{-\frac{1}{2}ns} \frac{\displaystyle\int \frac{dv_1}{(\sigma(Y_1))^{\frac{1}{2}ns}}}{I(s,1)}$$

$$= \varepsilon + C(n(1+\delta))^{-\frac{1}{2}ns} \frac{I(\frac{1}{2}s,1)}{I(s,1)} ,$$

for $s > \dfrac{2\kappa_0}{n}$, where $C = C_0(n(1+\delta))^{\kappa_0}$. The asymptotic relation found on p. 91 yields by division

$$1 \sim n^{-\frac{1}{2}ns}\, 2^{-\frac{1}{4}n(n-1) - \frac{1}{2}(n-1)} \frac{I(\frac{1}{2}s,1)}{I(s,1)}$$

as $s \to \infty$, hence

$$\left|\frac{I(s,u_1)}{I(s,1)}\right| \leqq \varepsilon + C'(1+\delta)^{-\frac{1}{2}ns}$$

for large $s$, from where the assertion follows.

Assume now that $X$ is such that $u_1(E) = u(X) \neq 0$. Then

$$|H(s,u_1)| = \left|\frac{\Gamma(ns)I(s,1)}{\prod_{\nu=1}^{n} \Gamma(s-\alpha_\nu)}\right| \left|\frac{I(s,u_1)}{I(s,1)}\right| \geqq \frac{1}{2} \left|\frac{\Gamma(ns)I(s,1)}{\prod_{\nu=1}^{n} \Gamma(s-\alpha_\nu)}\right| |u(X)|$$

for $s \geqq s_1$ and in particular $H(s,u_1) \neq 0$ for large $s$. For any $h \in \mathbb{N}$ we have

$$H(s,u_1) = H(s+h,u_1) \sim \pi^{\frac{1}{4}n(n-1)} (s+h)^{\alpha_1 + \dots + \alpha_n - \frac{1}{4}n(n-1)} u(X)$$

and therefore necessarily

$$\alpha_1 + \alpha_2 + \dots + \alpha_n = \frac{n(n-1)}{4},$$

so that

$$H(s,u_1) \sim \pi^{\frac{1}{4}n(n-1)} u(X)$$

as $s \to \infty$. As we have mentioned earlier, this implies that

$$H(s,u_1) = \pi^{\frac{1}{4}n(n-1)} u(X)$$

for all $s$. Since both sides of this equation are continuous functions of $X$, it holds also for $X$ such that $u(X) = 0$.

We have

$$J_s(E,u_1) = \pi^{\frac{1}{4}n(n-1)} \prod_{\nu=1}^{n} \Gamma(s - \alpha_\nu) u(X),$$

and obtain the final result

$$J_s(X,u) = \int_{Y>0} e^{-\sigma(YX^{-1})} |Y|^s u(Y) dv$$

$$= \pi^{\frac{1}{4}n(n-1)} \prod_{\nu=1}^{n} \Gamma(s - \alpha_\nu) |X|^s u(X),$$

where the numbers $\alpha_\nu = \alpha_\nu(u)$, $\nu = 1,2,\dots,n$, are defined by

$$|Y|^{\frac{1}{2}(n-1) - s} M_n |Y|^{s - \frac{1}{2}(n-1)} u(Y) = f(s)u(Y),$$

$$f(s) = (s - \alpha_1)(s - \alpha_2)\dots(s - \alpha_n).$$

The hypersurface $\mathcal{Y}_1$ defined by the equation $|Y| = 1$ has a special importance in the Riemannian space $\mathcal{Y}$ of all symmetric positive matrices $Y$, because the metric fundamental form on $\mathcal{Y}$ can be written

$$ds^2 = d^2(-\log |Y|).$$

Indeed, since $d(Y^{-1}) = -Y^{-1}\cdot dY\cdot Y^{-1}$, we have

$$ds^2 = \sigma(Y^{-1}dYY^{-1}dY) = -\sigma(d(Y^{-1})dY) = -d(\sigma(Y^{-1}dY)).$$

Now

$$e_{\mu\nu}\frac{\partial}{\partial y_{\mu\nu}}\log |Y| = \frac{1}{|Y|}e_{\mu\nu}\frac{\partial}{\partial y_{\mu\nu}}|Y|,$$

hence

$$\frac{\partial}{\partial Y}\log |Y| = Y^{-1}$$

and therefore

$$d\log |Y| = \sigma(dY\frac{\partial}{\partial Y})\log |Y| = \sigma(dY\cdot Y^{-1}) = \sigma(Y^{-1}\cdot dY).$$

Replacing in the above formula, we obtain the announced expression for $ds^2$.

Note that the operator

$$d^2 = (\sigma(dY\frac{\partial}{\partial Y}))^2 = \sum_{\substack{\mu\leqq\nu\\ \kappa\leqq\lambda}} dy_{\mu\nu}\, dy_{\kappa\lambda}\frac{\partial}{\partial y_{\mu\nu}}\frac{\partial}{\partial y_{\kappa\lambda}}$$

just introduced is not coordinate invariant.

The following general lemma shows that $\mathcal{Y}_1$ is a geodesic submanifold of $\mathcal{Y}$.

*Lemma. Let* $\mathcal{Y}$ *be a non-empty open subset of* $\mathbb{R}^n = \{(y_1,\ldots,y_n)\}$. *Assume that* $\mathcal{Y}$ *is a cone with vertex at* $(0,\ldots,0)$, *i.e., that if* $(y_1,\ldots,y_n) \in \mathcal{Y}$ *then* $(\lambda y_1,\ldots,\lambda y_n) \in \mathcal{Y}$ *for every* $\lambda \geqq 0$. *Assume furthermore that* $\mathcal{Y}$ *is a Riemannian space, whose metric fundamental form has the expression* $ds^2 = d^2\psi$, *with* $\psi = -\log \chi$, *where* $\chi = \chi(y_1,\ldots,y_n)$ *is a positive function, homogeneous of positive degree. Then the surface* $\mathcal{Y}_1$ *defined by*

$$\chi(y_1,\ldots,y_n) = 1, \qquad (y_1,\ldots,y_n) \in \mathcal{Y}$$

*is a geodesic submanifold of* $\mathcal{Y}$, *i.e., any geodesic curve joining two points of* $\mathcal{Y}_1$ *lies completely in* $\mathcal{Y}_1$.

*Proof.* Let

$$y_\mu = h_\mu(x_1,x_2,\ldots,x_{n-1}) \qquad (1 \leqq \mu \leqq n)$$

be a local representation of $\mathcal{Y}_1$ such that

$$\operatorname{rank}\left(\frac{\partial h_\mu}{\partial x_\nu}\right) = n-1.$$

Then the equation

$$\chi(h_1,h_2,\ldots,h_n) = 1$$

is an identity in $x_1,x_2,\ldots,x_{n-1}$.

Introduce the variable $x_n$ by

$$y_\mu = h_\mu(x_1,\ldots,x_{n-1})x_n.$$

Let $k > 0$ be the degree of homogeneity of $\chi$. Then

$$\chi(y_1,\ldots,y_n) = x_n^k\,\chi(h_1,\ldots,h_n) = x_n^k\,,$$

and $x_n > 0$ is uniquely defined for every $(y_1,\ldots,y_n) \in \mathcal{Y}$. We show that $x_1, x_2, \ldots, x_n$ is a local coordinate system on $\mathfrak{F}$. For this we must show that

$$\operatorname{rank}\begin{pmatrix} \frac{\partial y_1}{\partial x_1} & \cdots & \frac{\partial y_1}{\partial x_n} \\ \cdot & \cdot & \cdot \\ \cdot & \cdot & \cdot \\ \frac{\partial y_n}{\partial x_1} & \cdots & \frac{\partial y_n}{\partial x_n} \end{pmatrix} = \operatorname{rank}\begin{pmatrix} x_n \frac{\partial h_1}{\partial x_1} & \cdots & x_n \frac{\partial h_1}{\partial x_{n-1}} & h_1 \\ \cdot & \cdot & \cdot & \cdot \\ \cdot & \cdot & \cdot & \cdot \\ x_n \frac{\partial h_n}{\partial x_1} & \cdots & x_n \frac{\partial h_n}{\partial x_{n-1}} & h_n \end{pmatrix}$$

$$= n,$$

i.e., that the only solution $(\xi_1, \xi_2, \ldots, \xi_n)$ of the system of homogeneous linear equations

$$\sum_{\nu=1}^{n-1} x_n \frac{\partial h_n}{\partial x_\nu} \xi_\nu + h_\mu \xi_n = 0, \qquad 1 \leqq \mu \leqq n,$$

is $\xi_1 = \xi_2 = \ldots = \xi_n = 0$. From $\chi(h_1, \ldots, h_n) = 1$ we get

$$\sum_{\mu=1}^{n} \frac{\partial \chi}{\partial h_\mu} \frac{\partial h_\mu}{\partial x_\nu} = 0, \qquad 1 \leqq \nu \leqq n-1.$$

The homogeneity of $\chi$ yields

$$\sum_{\mu=1}^{n} \frac{\partial \chi}{\partial h_\mu} h_\mu = k\chi > 0.$$

Multiplying these equations by $x_n \xi_\nu$ and $\xi_n$, respectively, and adding, we get

$$0 = \sum_{\mu=1}^{n} \frac{\partial \chi}{\partial h_\mu} \left\{ \sum_{\nu=1}^{n-1} x_n \frac{\partial h_\mu}{\partial x_\nu} \xi_\nu + h_\mu \xi_n \right\} = k\chi\xi_n .$$

From here we deduce that $\xi_n = 0$. But then, by virtue of the assumption on the $h_\mu(x_1, x_2, \ldots, x_{n-1})$, the original system gives $\xi_1 = \ldots = \xi_{n-1} = 0$.

The metric fundamental form can be written

$$ds^2 = d^2\psi = \sum_{\alpha,\beta=1}^{n} \frac{\partial^2\psi}{\partial y_\alpha \partial y_\beta} dy_\alpha dy_\beta = \sum_{\mu,\nu=1}^{n} g_{\mu\nu} dx_\mu dx_\nu ,$$

with

$$g_{\mu\nu} = \sum_{\alpha,\beta=1}^{n} \frac{\partial^2\psi}{\partial y_\alpha \partial y_\beta} \frac{\partial y_\alpha}{\partial x_\mu} \frac{\partial y_\beta}{\partial x_\nu} .$$

We show that $g_{n\nu} = 0$ for $\nu < n$. Since the matrix $(g_{\mu\nu})$ is symmetric, this will imply that also $g_{\nu n} = 0$ for $\nu < n$. From $y_\alpha = h_\alpha(x_1, \ldots, x_{n-1})x_n$ we get

$$\frac{\partial y_\alpha}{\partial x_n} = h_\alpha \quad \text{and} \quad x_n \frac{\partial y_\alpha}{\partial x_n} = y_\alpha ,$$

so that

$$x_n g_{n\nu} = \sum_{\alpha,\beta=1}^{n} \frac{\partial^2\psi}{\partial y_\alpha \partial y_\beta} y_\alpha \frac{\partial y_\beta}{\partial x_\nu}$$

$$= \sum_{\alpha=1}^{n} y_\alpha \sum_{\beta=1}^{n} \frac{\partial}{\partial y_\beta}\left(\frac{\partial\psi}{\partial y_\alpha}\right) \frac{\partial y_\beta}{\partial x_\nu} = \sum_{\alpha=1}^{n} y_\alpha \frac{\partial}{\partial x_\nu}\left(\frac{\partial\psi}{\partial y_\alpha}\right) .$$

Remembering that $\psi = -\log\chi$ and using the homogeneity of $\chi$ we get from Euler's relation

$$\sum_{\alpha=1}^{n} y_\alpha \frac{\partial\psi}{\partial y_\alpha} = -\frac{1}{\chi} \sum_{\alpha=1}^{n} y_\alpha \frac{\partial\chi}{\partial y_\alpha} = -k .$$

Differentiating we obtain

$$\sum_{\alpha=1}^{n} y_\alpha \frac{\partial}{\partial x_\nu}\left(\frac{\partial\psi}{\partial y_\alpha}\right) = -\sum_{\alpha=1}^{n} \frac{\partial y_\alpha}{\partial x_\nu}\frac{\partial\psi}{\partial y_\alpha} = -\frac{\partial\psi}{\partial x_\nu} = 0$$

because $\psi = -\log\chi = -\log\chi(h_1x_n,\ldots,h_nx_n) = -\log x_n^k = -k\log x_n$. Thus $x_ng_{n\nu} = 0$, i.e., $g_{n\nu} = 0$ for $\nu < n$.

Next we compute $g_{nn}$ given by

$$x_n^2g_{nn} = x_n^2\sum_{\alpha,\beta=1}^{n}\frac{\partial^2\psi}{\partial y_\alpha\partial y_\beta}\frac{\partial y_\alpha}{\partial x_n}\frac{\partial y_\beta}{\partial x_n} =$$

$$= x_n^2\sum_{\alpha,\beta=1}^{n}\frac{\partial^2\psi}{\partial y_\alpha\partial y_\beta}h_\alpha h_\beta = \sum_{\alpha,\beta=1}^{n}\frac{\partial^2\psi}{\partial y_\alpha\partial y_\beta}y_\alpha y_\beta\,.$$

Since $\frac{\partial\chi}{\partial y_\alpha}$ is homogeneous of degree $k-1$,

$$\frac{\partial\psi}{\partial y_\alpha} = -\frac{\partial\log\chi}{\partial y_\alpha}$$

is homogeneous of degree -1 and so by Euler's relation

$$\sum_{\nu=1}^{n}\frac{\partial}{\partial y_\nu}\left(\frac{\partial\psi}{\partial y_\alpha}\right)y_\nu = -\frac{\partial\psi}{\partial y_\alpha}\,.$$

It follows that

$$x_n^2g_{nn} = \sum_{\alpha=1}^{n}\left(\sum_{\beta=1}^{n}\frac{\partial^2\psi}{\partial y_\alpha\partial y_\beta}y_\beta\right)y_\alpha$$

$$= -\sum_{\alpha=1}^{n}\frac{\partial\psi}{\partial y_\alpha}y_\alpha = k$$

and so $g_{nn} = kx_n^{-2}$.

Thus the metric fundamental form on $\mathcal{T}$ can be written as

$$ds^2 = ds_1^2 + \frac{k}{x_n^2}\, dx_n^2 ,$$

where

$$ds_1^2 = \sum_{\mu,\nu=1}^{n-1} g_{\mu\nu} dx_\mu dx_\nu$$

is the metric fundamental form on $\mathcal{Y}_1$. Let $y_\nu(t) = h_\nu(x_1(t),\ldots,x_{n-1}(t))\cdot x_n(t)$, $1 \leqq \nu \leqq n$, $0 \leqq t \leqq 1$, be the parametric representation of a curve joining two points of $\mathcal{Y}_1$, i.e., such that $x_n(0) = x_n(1) = 1$. Since the length of this curve is

$$\int_0^1 \sqrt{\left(\frac{ds_1}{dt}\right)^2 + \frac{k}{x_n^2}\left(\frac{dx_n}{dt}\right)^2}\; dt ,$$

and since $kx_n^{-2} > 0$, it can be a geodesic only if $\frac{dx_n}{dt} = 0$, i.e., $x_n(t) = 1$ for $0 \leqq t \leqq 1$, which means that it lies completely in $\mathcal{Y}_1$. This achieves the proof of our lemma.

In order to explain the geometric significance of the inequality

$$|u(Y_1)| < C(\sigma(Y_1))^\kappa ,$$

satisfied by a function $u(Y_1)$ on the geodesic submanifold $\mathcal{Y}_1 = \{Y_1 > 0,\ |Y_1| = 1\}$, we introduce the geodesic distance $\rho = \rho(Y_1,E)$ of $Y_1$ from $E$, measured on $\mathcal{Y}_1$, or equivalently on $\mathcal{Y}$. We know from the Theorem of §3 that

$$\rho = \sqrt{\sum_{\nu=1}^{n} \log^2 \lambda_\nu} ,$$

where $\lambda_1, \lambda_2, \ldots, \lambda_n$ are the characteristic roots of $Y_1$. Since $\lambda_1\lambda_2\cdots\lambda_n = 1$, we can arrange these roots so that

$$\lambda_1 \leqq \ldots \leqq \lambda_h \leqq 1 \leqq \lambda_{h+1} \leqq \ldots \leqq \lambda_n ,$$

where $h$ is an integer, $1 \leqq h < n$ and furthermore $\lambda_1 < 1$, $\lambda_n > 1$ if $Y_1 \neq E$. It follows that

$$\lambda_1 \lambda_2 \cdots \lambda_h \lambda_n^{n-h} \geqq 1,$$

i.e.,

$$(n-h)\log \lambda_n \geqq \sum_{\nu=1}^{h} |\log \lambda_\nu|$$

and therefore

$$\rho^2 \leqq \left( \sum_{\nu=1}^{h} |\log \lambda_\nu| \right)^2 + (n-h)\log^2 \lambda_n$$

$$\leqq (n-h)^2 \log^2 \lambda_n + (n-h)\log^2 \lambda_n$$

$$= (n-h)(n-h+1)\log^2 \lambda_n \leqq n(n-h)\log^2 \lambda_n .$$

Thus we get the inequalities

$$|\log \lambda_\nu| \leqq \rho \leqq \sqrt{n(n-1)}\, \log \lambda_n \qquad (\nu = 1,\ldots,n)$$

and consequently

$$\sigma(Y_1) = \sum_{\nu=1}^{n} \lambda_\nu \leqq \sum_{\nu=1}^{n} e^{|\log \lambda_\nu|} \leqq n e^{\rho} ,$$

$$\sigma(Y_1) \geqq \lambda_n = e^{\log \lambda_n} \geqq e^{\rho/\sqrt{n(n-1)}} .$$

We see that any estimate

$$|u(Y_1)| \leqq C(\sigma(Y_1))^{\kappa}, \qquad Y_1 > 0, \quad |Y_1| = 1,$$

can be replaced by an estimate of the type

$$|u(Y_1)| \leqq C_1 e^{\kappa_1 \rho}, \qquad Y_1 > 0, \quad |Y_1| = 1,$$

where $C_1$ and $\kappa_1$ are positive constants. A similar observation holds, of course, for the analogous estimates which involve the partial derivatives of $u(Y)$ on $|Y| = 1$.

## §8. *The Riemannian space* $\mathfrak{H} \times \mathbb{R}/2\pi\mathbb{Z}$

We denote by $\mathbb{Z}$ the additive group of rational integers, by $\mathfrak{H}$ Siegel's upper half-plane introduced in §2, and let $\Omega$ be the symplectic group, which operates on $\mathfrak{H}$ (see §4). Set $G = \Omega \times H$, the direct product of the multiplicative group $\Omega$ and of the additive group $H = \mathbb{R}/2\pi\mathbb{Z}$. For the elements of $G$ we shall use notation of the type $M_a$, where $M \in \Omega$, $a \in \mathbb{R}$, so that $M_a = M_{a'}$ if $a \equiv a' \pmod{2\pi}$, and

$$M_a \cdot L_b = (ML)_{a+b} .$$

We also introduce the set $\mathfrak{T} = \mathfrak{H} \times H$, whose elements shall be denoted by $(Z,t)$, where $Z = X + iY \in \mathfrak{H}$, $t \in \mathbb{R}$, so that $(Z,t) = (Z,t')$ if $t \equiv t' \pmod{2\pi}$. The action of $G$ as a transformation group of $\mathfrak{T}$ is defined by

$$M_a(Z,t) = (M\langle Z\rangle , t + a + \arg|CZ + D|),$$

where

$$M = \begin{pmatrix} A & B \\ C & D \end{pmatrix}$$

and, as before,

$$M\langle Z\rangle = (AZ + B)(CZ + D)^{-1}.$$

Next we prove that there exists on $\mathfrak{T}$ a Riemannian metric, which is invariant with respect to $G$. For $(Z,t) \in \mathfrak{T}$, $Z = X + iY \in \mathfrak{H}$ let us set

$$(Z^*,t^*) = M_a(Z,t),$$

where

$$M = \begin{pmatrix} A & B \\ C & D \end{pmatrix},$$

so that $Z^* = M\langle Z\rangle$ and

$$t^* = t + a + \arg|CZ + D|.$$

We know from §4 (p. 33) that

$$dZ^* = (ZC' + D')^{-1} dZ (CZ + D)^{-1}$$

and

$$Y^{*-1} = Y^{-1}\{\overline{Z}C' + D'\} = Y^{-1}\{ZC' + D'\}.$$

We have moreover

$$d \arg|CZ + D| = \frac{1}{2i}\{ d \log|CZ + D| - d \log|C\overline{Z} + D|\}$$

$$= \frac{1}{2i}\{\sigma(C\,dZ(CZ + D)^{-1}) - \sigma(C\,d\overline{Z}(C\overline{Z} + D)^{-1})\},$$

hence

$$\sigma(dX^* \cdot Y^{*-1}) = \frac{1}{2}\{\sigma(dZ^* \cdot Y^{*-1}) + \sigma(d\overline{Z}^* \cdot Y^{*-1})\}$$

$$= \frac{1}{2}\sigma((ZC' + D')^{-1} dZ \cdot Y^{-1}(\overline{Z}C' + D')) + \frac{1}{2}\sigma((\overline{Z}C' + D')^{-1} d\overline{Z} \cdot Y^{-1}(ZC' + D'))$$

$$= \frac{1}{2}\sigma(dZ \cdot Y^{-1}) + \frac{1}{2}\sigma(d\overline{Z} \cdot Y^{-1})$$

$$- i\sigma((ZC' + D')^{-1} dZ \cdot C') + i\sigma((\overline{Z}C' + D')^{-1} d\overline{Z} \cdot C')$$

$$= \sigma(dX \cdot Y^{-1}) - i\sigma(C\,dZ(CZ + D)^{-1}) + i\sigma(C\,d\overline{Z}(C\overline{Z} + D)^{-1})$$

$$= \sigma(dX \cdot Y^{-1}) + 2\,d \arg|CZ + D| = \sigma(dX \cdot Y^{-1}) + 2(dt^* - dt),$$

and therefore

$$dt^* - \frac{1}{2}\sigma(dX^*\cdot Y^{*-1}) = dt - \frac{1}{2}\sigma(dX\cdot Y^{-1}).$$

It follows that

$$ds^2 = \sigma(dX\cdot Y^{-1})^2 + \sigma(dY\cdot Y^{-1})^2 + (dt - \frac{1}{2}\sigma(dX\cdot Y^{-1}))^2$$

is invariant with respect to $G$. Since $ds^2 = 0$ implies $dX = 0$, $dY = 0$, $dt = 0$, we see that $ds^2$ is indeed a positive definite invariant Riemannian metric fundamental form.

Next we prove that $\mathcal{T} = \mathcal{H} \times \mathbb{R}/2\pi\mathbb{Z}$ is a weakly symmetric Riemannian space with respect to $G = \Omega \times \mathbb{R}/2\pi\mathbb{Z}$ and to the isometry $\mu$ defined by

$$\mu(Z,t) = (-\overline{Z},-t),$$

i.e.,

$$\mu\colon X \mapsto -X, \quad Y \mapsto Y, \quad t \mapsto -t.$$

We have to show that given two elements $(Z_1,t_1)$ and $(Z_2,t_2)$ of $\mathcal{T}$ there exists a transformation $M_a \in G$ such that

$$M\langle Z_1\rangle = -\overline{Z}_2, \quad t_1 + a + \arg|CZ_1 + D| \equiv -t_2 \pmod{2\pi},$$

$$M\langle Z_2\rangle = -\overline{Z}_1, \quad t_2 + a + \arg|CZ_2 + D| \equiv -t_1 \pmod{2\pi}.$$

By Theorem 1 of §4 there exists a symplectic matrix $M$ which transforms $(Z_1,Z_2)$ into $(-\overline{Z}_2,-\overline{Z}_1)$ if and only if the cross ratios $\rho(Z_1,Z_2)$ and $\rho(-\overline{Z}_2,-\overline{Z}_1)$ have the same characteristic roots. But this condition is satisfied since from

$$\rho(Z_1,Z_2) = (Z_1 - Z_2)(Z_1 - \overline{Z}_2)^{-1}(\overline{Z}_1 - \overline{Z}_2)(\overline{Z}_1 - Z_2)^{-1}$$

we obtain by transposition

$$(\overline{Z}_1 - Z_2)^{-1}(\overline{Z}_1 - \overline{Z}_2)(Z_1 - \overline{Z}_2)^{-1}(Z_1 - Z_2),$$

and here the factors form a cyclic permutation of the factors of

$$\rho(-\overline{Z}_2,-\overline{Z}_1) = (\overline{Z}_1 - \overline{Z}_2)(Z_1 - \overline{Z}_2)^{-1}(Z_1 - Z_2)(\overline{Z}_1 - Z_2)^{-1}.$$

Moreover we have (cf. p. 33)

$$Z_1 - \overline{Z}_2 = M\langle Z_1\rangle - M\langle \overline{Z}_2\rangle$$

$$= (Z_1 C' + D')^{-1}(Z_1 - \overline{Z}_2)(C\overline{Z}_2 + D)^{-1},$$

and taking determinants

$$|CZ_1 + D| \cdot |C\overline{Z}_2 + D| = 1.$$

In particular

$$\arg|CZ_1 + D| \equiv \arg|CZ_2 + D| \qquad (\mathrm{mod}\ 2\pi),$$

hence there exists also an $a \in \mathbb{R}$ which satisfies the two congruences.

Instead of the entries of $X$ and $Y$, we introduce the entries of

$$Z = X + iY \quad \text{and} \quad \overline{Z} = X - iY,$$

and consider them as independent variables. We define the differential operators $\frac{\partial}{\partial Z}, \frac{\partial}{\partial \overline{Z}}$ by requiring that

$$df = \{\sigma(dX \frac{\partial}{\partial X}) + \sigma(dY \frac{\partial}{\partial Y}) + dt \frac{\partial}{\partial t}\}f$$

$$= \{\sigma(dZ \frac{\partial}{\partial Z}) + \sigma(d\overline{Z} \frac{\partial}{\partial \overline{Z}}) + dt \frac{\partial}{\partial t}\}f$$

hold for every $f(X,Y,t) = f(Z,\overline{Z},t) \in C^{\infty}(\mathfrak{D}^{+})$, or explicitly

$$\frac{\partial}{\partial Z} = \frac{1}{2}\left(\frac{\partial}{\partial X} - i\,\frac{\partial}{\partial Y}\right), \qquad \frac{\partial}{\partial \overline{Z}} = \frac{1}{2}\left(\frac{\partial}{\partial X} + i\,\frac{\partial}{\partial Y}\right).$$

Since

$$dZ^{*} = (ZC' + D')^{-1}dZ(CZ + D)^{-1},$$

$$d\overline{Z}^{*} = (\overline{Z}C' + D')^{-1}d\overline{Z}(C\overline{Z} + D)^{-1},$$

$$dt^{*} = dt + \frac{1}{2i}\{\sigma(C\,dZ(CZ + D)^{-1}) - \sigma(C\,d\overline{Z}(C\overline{Z} + D)^{-1})\},$$

we get

$$df = \{\sigma(dZ^{*}\,\frac{\partial}{\partial Z^{*}}) + \sigma(d\overline{Z}^{*}\,\frac{\partial}{\partial \overline{Z}^{*}}) + dt^{*}\,\frac{\partial}{\partial t^{*}}\}\,f$$

$$= \Big\{\sigma((ZC' + D')^{-1}dZ(CZ + D)^{-1}\,\frac{\partial}{\partial Z^{*}}) + \sigma((\overline{Z}C' + D')^{-1}d\overline{Z}(C\overline{Z} + D)^{-1}\,\frac{\partial}{\partial \overline{Z}^{*}})$$

$$+ dt\,\frac{\partial}{\partial t^{*}} + \frac{1}{2i}\{\sigma(C\,dZ(CZ + D)^{-1}) - \sigma(C\,d\overline{Z}(C\overline{Z} + D)^{-1})\}\,\frac{\partial}{\partial t^{*}}\Big\}\,f\,.$$

Comparing with

$$df = \{\sigma(dZ\,\frac{\partial}{\partial Z}) + \sigma(d\overline{Z}\,\frac{\partial}{\partial \overline{Z}}) + dt\,\frac{\partial}{\partial t}\}f,$$

we obtain after some computation

$$\frac{\partial}{\partial Z^{*}} = (CZ + D)((CZ + D)\frac{\partial}{\partial Z})' + \frac{i}{2}(CZ + D)C'\frac{\partial}{\partial t}\,,$$

$$\frac{\partial}{\partial \overline{Z}^{*}} = (C\overline{Z} + D)((C\overline{Z} + D)\frac{\partial}{\partial \overline{Z}})' - \frac{i}{2}(C\overline{Z} + D)C'\frac{\partial}{\partial t}\,,$$

$$\frac{\partial}{\partial t^{*}} = \frac{\partial}{\partial t}\,.$$

Note that the right-hand sides of these equations are actually symmetric matrices.

To obtain nicer formulas, we introduce

$$Q = \frac{1}{4} Y^{-1} \frac{\partial}{\partial t} + \frac{\partial}{\partial Z}, \qquad \overline{Q} = \frac{1}{4} Y^{-1} \frac{\partial}{\partial t} + \frac{\partial}{\partial \overline{Z}},$$

and get

$$Q^* = (CZ + D)((CZ + D)Q)',$$

$$\overline{Q}^* = (C\overline{Z} + D)((C\overline{Z} + D)\overline{Q})',$$

$$\frac{\partial}{\partial t^*} = \frac{\partial}{\partial t}.$$

The invariant differential operators $L \in \mathcal{L}$ on $\mathcal{T}^+$ are polynomials in $\frac{\partial}{\partial t}$ and in the entries of $Q$, $\overline{Q}$, with coefficients which are functions of $Z$, $\overline{Z}$, $t$, and the invariance with respect to $G$ is expressed by the relation

$$L(Z^*,\overline{Z}^*,t^*,Q^*,\overline{Q}^*,\frac{\partial}{\partial t^*}) = L(Z,\overline{Z},t,Q,\overline{Q},\frac{\partial}{\partial t}).$$

In order to obtain a representation for all differential operators $L \in \mathcal{L}$, we first replace $Q$, $\overline{Q}$ by variable symmetric matrices $W$, $\overline{W}$, and $\frac{\partial}{\partial t}$ by a variable $w$, so that all the new independent variables commute with each other. We want to determine the functions $L(Z,\overline{Z},t,W,\overline{W},w)$, which are polynomials in $w$ and in the entries of $W$, $\overline{W}$, and which have the invariance property

$$L(M\langle Z\rangle, M\langle\overline{Z}\rangle, t + a + \arg|CZ + D|, (CZ + D)W(ZC' + D'),$$
$$(C\overline{D} + D)\overline{W}(\overline{Z}C' + D'), w) =$$
$$= L(Z,\overline{Z},t,W,\overline{W},w).$$

Since $G$ operates transitively on $\mathcal{T}$, there exists a transformation $M_a$ such that

$$M\langle Z\rangle = iE, \quad M\langle\overline{Z}\rangle = -iE, \quad t + a + \arg|CZ + D| = 0,$$

and in particular (p. 33)

$$E = (\overline{Z}C' + D')^{-1}Y(CZ + D)^{-1},$$

i.e.,

$$Y = (\overline{Z}C' + D')(CZ + D) = (ZC' + D')(C\overline{Z} + D).$$

It suffices therefore to consider the polynomials

$$L(W_1, \overline{W}_1) = L(iE, -iE, 0, W_1, \overline{W}_1),$$

where

$$W_1 = (CZ + D)W(ZC' + D'), \qquad \overline{W}_1 = (C\overline{Z} + D)\overline{W}(\overline{Z}C' + D').$$

We have obviously

$$L((iC_o + D_o)W_1(iC_o' + D_o'), (-iC_o + D_o)\overline{W}_1(-iC_o' + D_o')) = L(W_1, \overline{W}_1)$$

for every

$$S = \begin{pmatrix} A_o & B_o \\ C_o & D_o \end{pmatrix} \in \Omega$$

which leaves $iE$ fixed. But we know that these matrices are of the form

$$S = \begin{pmatrix} D_o & -C_o \\ C_o & D_o \end{pmatrix},$$

where $U = iC_o + D_o$ is an arbitrary unitary matrix. If $b \in \mathbb{R}$ is such that $b + \arg|U| \equiv 0 \pmod{2\pi}$, or $|U| = e^{-ib}$, then $S_b$ leaves $(iE, 0)$ fixed. In the relation

$$L(UW_1U', \overline{U}\overline{W}_1\overline{U}') = L(W_1, \overline{W}_1), \qquad \overline{U}U' = E,$$

we can choose $U$ so that $\bar{U}\bar{W}_1\bar{U}'$ is a diagonal matrix $D = (\delta_{\mu\nu}t_\nu)$, hence

$$L(W_2,D) = L(W_1,\bar{W}_1),$$

where $W_2 = UW_1U'$. The invariance property now reduces to the condition

$$L(VW_2V',\bar{V}D\bar{V}') = L(W_2,D)$$

for every $V = (\delta_{\mu\nu}e^{i\phi_\nu})$, where $\phi_1,\phi_2,\ldots,\phi_n$ are arbitrary real numbers.

Let us write $W_2 = (w_{\mu\nu})$ and assume that the monomial

$$\prod_{\nu=1}^{n} t_\nu^{a_\nu} \prod_{\alpha,\beta=1}^{n} w_{\alpha\beta}^{e_{\alpha\beta}b_{\alpha\beta}} \qquad (b_{\alpha\beta} = b_{\beta\alpha})$$

actually appears in $L(W_2,D)$. It is multiplied by the factor

$$\prod_{\nu=1}^{n} e^{-2i\phi_\nu a_\nu} \prod_{\alpha,\beta=1}^{n} e^{i(\phi_\alpha + \phi_\beta)e_{\alpha\beta}b_{\alpha\beta}}$$

when we apply $V$. Since this factor must be 1, and since the $\phi_\nu$ are arbitrary real numbers, we have necessarily

$$a_\nu = \sum_{\alpha=1}^{n} e_{\nu\alpha}b_{\nu\alpha}, \qquad (1 \leqq \nu \leqq n),$$

and the monomial can be rewritten as

$$\prod_{\alpha,\beta=1}^{n} (\sqrt{t_\alpha}\, w_{\alpha\beta} \sqrt{t_\beta})^{e_{\alpha\beta}b_{\alpha\beta}}.$$

This proves that

$$L(W_2,D) = L(W_3,E), \qquad \text{where } W_3 = \sqrt{D}\, W_2 \sqrt{D}\,.$$

Finally we have

$$L(VW_3V',E) = L(W_3,E)$$

for an arbitrary orthogonal matrix $V$, and this shows, as in the proof of the Theorem of §6, that $L(W_3,E)$ is a polynomial in the $\sigma(W_3^h)$, $h = 1,2,\ldots,n$:

$$L(W_3,E) = p(\sigma(W_3),\sigma(W_3^2),\ldots,\sigma(W_3^n)).$$

Substituting back we obtain

$$\begin{aligned}\sigma(W_3^h) &= \sigma(\sqrt{D}\ W_2\ \sqrt{D})^h = \sigma(W_2D)^h \\ &= \sigma(UW_1U'\bar{U}\bar{W}_1\bar{U}')^h = \sigma(W_1\bar{W}_1)^h \\ &= \sigma((CZ+D)W(ZC'+D')(C\bar{Z}+D)\bar{W}(\bar{Z}C'+D'))^h \\ &= \sigma(YWY\bar{W})^h.\end{aligned}$$

Thus we get the representation

$$L(Z,\bar{Z},t,W,\bar{W},\omega) = q(\sigma(Y\bar{W}YW),\ldots,\sigma(Y\bar{W}YW)^n,\omega),$$

where $q$ is a polynomial in $n+1$ indeterminates with constant coefficients.

In order to return to differential operators $L \in \mathcal{L}$ we introduce

$$K = 2iYQ = \frac{i}{2}\frac{\partial}{\partial t}E + (Z-\bar{Z})\frac{\partial}{\partial Z},$$

$$\Lambda = 2iY\bar{Q} = \frac{i}{2}\frac{\partial}{\partial t}E + (Z-\bar{Z})\frac{\partial}{\partial \bar{Z}}.$$

Then $\sigma(Y\bar{Q}YQ)^h = (-4)^{-h}\sigma(\Lambda K)^h$, and so, replacing $W$ by $Q$, $\bar{W}$ by $\bar{Q}$ and $\omega$ by $\frac{\partial}{\partial t}$ in the above formula, we see that

$$L(Z,\overline{Z},t,Q,\overline{Q},\frac{\partial}{\partial t}) - q_1(\sigma(\Lambda K),\ldots,\sigma(\Lambda K)^n,\frac{\partial}{\partial t}),$$

where $q_1$ is a polynomial in $n+1$ indeterminates, is a differential operator whose degree is less than that of $L$.

The operators $\sigma(\Lambda K)^h$, $h = 1,2,\ldots,n$, are not invariant. We shall show that there exist invariant operators $H_j$, $j = 1,2,\ldots,n$, such that $H_1,H_2,\ldots,H_n,\frac{\partial}{\partial t}$ are algebraically independent and

$$H_j = \sigma(\Lambda K)^j + \text{terms of lower degree}$$

for $1 \leqq j \leqq n$. Then we can see by induction that every $L \in \mathcal{L}$ can be expressed uniquely as a polynomial in $\frac{\partial}{\partial t},H_1,H_2,\ldots,H_n$, with constant coefficients.

We now develop an explicit construction of suitable operators $H_j$ not depending on $\frac{\partial}{\partial t}$. The images of points and operators with respect to a given transformation

$$M = \begin{pmatrix} A & B \\ C & D \end{pmatrix} \in \Omega$$

will always be denoted by a star, as in

$$Z \mapsto Z^* = M\langle Z\rangle \,, \qquad \overline{Z} \mapsto \overline{Z}^* = M\langle\overline{Z}\rangle \,.$$

We have

$$K^* = (\overline{Z}C' + D')^{-1}((CZ + D)K')',$$

$$\Lambda^* = (ZC' + D')^{-1}((C\overline{Z} + D)\Lambda')'.$$

Because of

$$((C\overline{Z} + D)\Lambda')' = \Lambda(\overline{Z}C' + D') - \frac{n+1}{2}(Z - \overline{Z})C'$$

we get

$$\Lambda^*K^* = (ZC' + D')^{-1}\{\Lambda(\bar{Z}C' + D') - \tfrac{n+1}{2}(ZC' + D') + \tfrac{n+1}{2}(\bar{Z}C' + D')\} \cdot$$

$$\cdot (\bar{Z}C' + D')^{-1}((CZ + D)K')',$$

which can be rewritten as

$$\Lambda^*K^* + \frac{n+1}{2} K^* = (ZC' + D')^{-1}\{(CZ + D)(\Lambda K + \frac{n+1}{2} K)'\}'$$

so that

$$\sigma(\Lambda^*K^* + \frac{n+1}{2} K^*) = \sigma(\Lambda K + \frac{n+1}{2} K),$$

i.e., $\sigma(\Lambda K + \frac{n+1}{2} K)$ is an invariant operator which we shall take for $H_1$. It might be worth while to mention that

$$\Delta = - H_1 + \frac{\partial^2}{\partial t^2} + \frac{n(n+1)}{2} \frac{i}{2} \frac{\partial}{\partial t} =$$

$$= \sigma(Y(Y \frac{\partial}{\partial X})' \frac{\partial}{\partial X}) + \sigma(Y(Y \frac{\partial}{\partial Y})' \frac{\partial}{\partial Y}) + \sigma(Y \frac{\partial}{\partial X})\frac{\partial}{\partial t} + (1 + \frac{n}{4})\frac{\partial^2}{\partial t^2}$$

is the Laplace operator on our Riemannian space $\mathfrak{T}$.

Let us set

$$B = \Lambda K + \frac{n+1}{2} K, \qquad A^{(1)} = B,$$

and assume that for $j = 1,2,\ldots,h$ we have determined the differential operators $A^{(j)}$ in such a way that

$$A^{(j)} = B^j + \text{terms of lower degree}$$

and

$$\overset{*}{A}{}^{(j)} = (ZC' + D')^{-1}\{(CZ + D)A^{(j)\prime}\}'.$$

We write $A = A^{(h)}$ and compute

$$\overset{**}{\mathrm{BA}} = (ZC' + D')^{-1}(\Lambda + \tfrac{n+1}{2}E).$$

$$\{K(ZC' + D') - \tfrac{n+1}{2}(ZC' + D') + \tfrac{n+1}{2}(\overline{Z}C' + D')\}\cdot(ZC' + D')^{-1}\{(CZ + D)A'\}'$$

$$= (ZC' + D')^{-1}(\Lambda + \tfrac{n+1}{2}E)K((CZ + D)A')' -$$

$$- \tfrac{n+1}{2}(ZC' + D')^{-1}\{(CZ + D)((\Lambda + \tfrac{n+1}{2}E)A)'\}' + \tfrac{n+1}{2}\overset{**}{\Lambda A} +$$

$$+ (\tfrac{n+1}{2})^2(ZC' + D')^{-1}(Z - \overline{Z})C'(ZC' + D')^{-1}\{(CZ + D)A'\}' +$$

$$+ (\tfrac{n+1}{2})^2(ZC' + D')^{-1}(\overline{Z}C' + D')(ZC' + D')^{-1}\{(CZ + D)A'\}'.$$

It follows that

$$\overset{**}{\mathrm{BA}} - \tfrac{n+1}{2}(\overset{*}{\Lambda} + \tfrac{n+1}{2}E)\overset{*}{A} =$$

$$= (ZC' + D')^{-1}\{(CZ + D)(\mathrm{BA} - \tfrac{n+1}{2}(\Lambda + \tfrac{n+1}{2}E)A)'\}' + S_1 ,$$

where we set

$$S_1 = \tfrac{1}{2}(ZC' + D')^{-1}(\Lambda + \tfrac{n+1}{2}E)(Z - \overline{Z})(E\sigma(A) + A')C'$$

$$= \tfrac{1}{2}(ZC' + D')^{-1}(\Lambda + \tfrac{n+1}{2}E)(ZC' + D' + \overline{Z}C' - D')\sigma(A) +$$

$$+ \tfrac{1}{2}(ZC' + D')^{-1}(\Lambda + \tfrac{n+1}{2}E)(Z - \overline{Z})A'C'$$

$$= \tfrac{1}{2}(ZC' + D')^{-1}\{(CZ + D)\Lambda'\sigma(A)\}' - \tfrac{1}{2}\overset{*}{\Lambda}\sigma(\overset{*}{A}) + S_2$$

with

$$S_2 = \tfrac{1}{2}(ZC' + D')^{-1}(\Lambda + \tfrac{n+1}{2}E)(Z - \overline{Z})A'C'.$$

In the following relations we denote by underlining that a matrix has to be considered as a constant with respect to the differential operators which appear as left factors. Then we can continue writing

$$S_2 = \tfrac{1}{2}(ZC' + D')^{-1}\Lambda(Z - \underline{Z})A'(\underline{Z} - \underline{Z})^{-1}(\underline{Z}C' + D') -$$

$$- \tfrac{1}{2}(ZC' + D')^{-1}\Lambda(Z - \underline{Z})A'(\underline{Z} - \underline{Z})^{-1}(\underline{Z}C' + D')$$

$$= - \tfrac{1}{2}(ZC' + D')^{-1}\Lambda(\underline{Z}C' + D')(\underline{Z}C' + D')^{-1}(Z - \underline{Z})A'(\underline{Z} - \underline{Z})^{-1}(\underline{Z}C' + D') +$$

$$+ \tfrac{1}{2}(ZC' + D')^{-1}\{(CZ + D)(\Lambda(Z - \underline{Z})A'(\underline{Z} - \underline{Z})^{-1})'\}'$$

$$= - \tfrac{1}{2}\overset{*}{\Lambda}(Z^* - \underline{Z}^*)\overset{*}{A}'(\underline{Z}^* - \underline{Z}^*)^{-1} +$$

$$+ \tfrac{1}{2}(ZC' + D')^{-1}\{(CZ + D)(\Lambda(Z - \underline{Z})A'(\underline{Z} - \underline{Z})^{-1})'\}'.$$

Because of $\Lambda(Z - \underline{Z}) = (Z - \overline{Z})\Lambda'$ we get

$$S_2 = - \tfrac{1}{2}(Z^* - \overline{Z}^*)((Z^* - \overline{Z}^*)^{-1}(\overset{*}{\Lambda}'\overset{*}{A}')')' +$$

$$+ \tfrac{1}{2}(ZC' + D')^{-1}\{(CZ + D)(Z - \overline{Z})^{-1}((Z - \overline{Z})\Lambda'A')'\}'.$$

This shows that

$$P = BA - \tfrac{n+1}{2}(\Lambda + \tfrac{n+1}{2}E)A + \tfrac{1}{2}\Lambda\sigma(A) + \tfrac{1}{2}(Z - \overline{Z})\{(Z - \overline{Z})^{-1}(\Lambda'A')'\}'$$

satisfies the transformation formula

$$\overset{*}{P} = (ZC' + D')^{-1}\{(CZ + D)P'\}'$$

so that

$$\sigma(\overset{*}{P}) = \sigma(P).$$

We can choose

$$A^{(h+1)} = BA^{(h)} - \tfrac{n+1}{2}\Lambda A^{(h)} + \tfrac{1}{2}\Lambda\sigma(A^{(h)}) + \tfrac{1}{2}(Z - \overline{Z})\{(Z - \overline{Z})^{-1}(\Lambda'A^{(h)\prime})'\}'$$

$$= P + (\tfrac{n+1}{2})^2 A^{(h)}$$

and obtain

$$A^{(h+1)} = B^{h+1} + \text{terms of lower degree}$$

$$= (\Lambda K)^{h+1} + \text{terms of lower degree}$$

and

$$\sigma(\overset{*}{A}{}^{(h+1)}) = \sigma(A^{(h+1)}) = \sigma(\Lambda K)^{h+1} + \text{terms of lower degree.}$$

The operators $A^{(j)}$ for $1 \leqq j \leqq n$ are defined by recursion, and we see that if we define the required invariant operators by $H_1 = \sigma(A^{(1)})$ and

$$H_j = \sigma(A^{(j)}) = \sigma(BA^{(j-1)}) - \frac{n}{2}\sigma(\Lambda A^{(j-1)}) + \frac{1}{2}\sigma(\Lambda)\sigma(A^{(j-1)})$$

for $1 < j \leqq n$, then they satisfy indeed the condition

$$H_j = \sigma(\Lambda K)^j + \text{terms of lower degree.}$$

We get thus one part of the

*Theorem. The ring* $L$ *of invariant differential operators on the Riemannian space* $\mathcal{G}_n \times \mathbb{R}/2\pi\mathbb{Z}$ *with respect to the group* $\Omega \times \mathbb{R}/2\pi\mathbb{Z}$ *of isometries is isomorphic with the ring* $\mathbb{C}[x_0, x_1, \ldots, x_n]$ *of polynomials in* $n+1$ *indeterminates.*

It remains to prove that the operators $\frac{\partial}{\partial t}$, $H_1$, ..., $H_n$, which generate $L$, are algebraically independent. This will be a consequence of the following considerations.

As in §6, we shall denote by $T$ triangular matrices of the form

$$T = (t_{\mu\nu}), \quad t_{\mu\nu} = 0 \text{ for } \mu < \nu, \quad t_{\nu\nu} > 0 \text{ for } 1 \leqq \nu \leqq n.$$

The subgroup $G_o$ of $G = \Omega \times \mathbb{R}/2\pi\mathbb{Z}$, defined by

$$G_0 = \left\{ \begin{pmatrix} T & ST'^{-1} \\ 0 & T'^{-1} \end{pmatrix}_a \mid S' = S, \quad T \text{ real}, \quad a \in \mathbb{R} \right\}$$

operates transitively on $\mathcal{T} = \mathfrak{H} \times \mathbb{R}/2\pi\mathbb{Z}$. If, as before, we set $Z = X + iY$, $Y = TT'$, then obviously

$$(Z,t) \leftrightarrow \begin{pmatrix} T & XT'^{-1} \\ 0 & T'^{-1} \end{pmatrix}_t$$

gives a one-to-one correspondence between $\mathcal{T}$ and $G_0$. Thus for $r \in \mathbb{Z}$ and $s = (s_1, s_2, \ldots, s_n) \in \mathbb{C}^n$ we can define on $\mathcal{T}$ the function $f_{r,s}$ by

$$f_{r,s}(Z,t) = \Phi_{r,s}(T,t) = \prod_{\nu=1}^{n} t_{\nu\nu}^{2s_\nu + \nu - \frac{1}{2}(n+1)} e^{irt}.$$

As was the case for $\Phi_s$ in §6, $\Phi_{r,s}$ is a multiplicative function, i.e.,

$$\Phi_{r,s}(T_1T_2, t_1 + t_2) = \Phi_{r,s}(T_1,t_1)\Phi_{r,s}(T_2,t_2).$$

From here we obtain, exactly as in §6, that $f_{r,s}$ is an eigenfunction of $L$.

We prove that the eigenvalues $\lambda_j'(r,s)$, $1 \leqq j \leqq n$, defined by

$$H_j f_{r,s}(Z,t) = \lambda_j'(r,s) f_{r,s}(Z,t),$$

are analytically independent, from where, of course, the algebraic independence of $\frac{\partial}{\partial t}, H_1, \ldots, H_n$ follows. Since $f_{r,s}$ is independent of $X$, we are only interested in the term of highest degree in $H_j$ with respect to $\frac{\partial}{\partial Y}$ and we see easily that

$$H_j = (-1)^j \sigma(Y \tfrac{\partial}{\partial Y})^{2j} + \text{terms of lower degree.}$$

This shows that

$$\lambda_j'(r,s) = (-1)^j \sum_{\nu=1}^{n} s_\nu^{2j} + \text{terms of lower degree},$$

and the Jacobian determinant

$$\frac{\partial(\lambda_1',\ldots,\lambda_n')}{\partial(s_1,\ldots,s_n)} = |(-1)^k 2ks_i^{2k-1} + \ldots|$$

$$= (-1)^{\frac{1}{2}n(n+1)}\, 2^n n!\, s_1 s_2 \ldots s_n |s_i^{2(k-1)}| + \ldots$$

$$= (-1)^{\frac{1}{2}n(n+1)}\, 2^n n!\, s_1 s_2 \ldots s_n \prod_{i<k} (s_k^2 - s_i^2) + \ldots,$$

where the dots indicate terms of lower degree, does not vanish identically. Therefore $\lambda_1',\ldots,\lambda_n'$ are indeed analytically independent and this concludes the proof of the theorem.

Let us observe that $L \in L$ is uniquely determined by its effect on the functions $f_{r,s}$ (cf. §6, p. 76).

Any function $f(Z,\overline{Z},t) \in C^\infty(\mathfrak{H} \times \mathbb{R}/2\pi\mathbb{Z})$ can be considered periodic with period $2\pi$ in the variable $t$ and has therefore a Fourier expansion

$$f(Z,\overline{Z},t) = \sum_{r=-\infty}^{\infty} f_r(Z,\overline{Z})e^{-irt}.$$

Assume now that $f$ is an eigenfunction of $L$, in particular of $\frac{\partial}{\partial t}$, then we have

$$f(Z,\overline{Z},t) = g(Z,\overline{Z})e^{-irt} \quad \text{for some} \quad r \in \mathbb{Z}.$$

With an arbitrary $q \in \mathbb{C}$ set

$$g(Z,\overline{Z}) = |Y|^{\frac{1}{2}q}\, h(Z,\overline{Z}).$$

Since

$$(Z-\overline{Z})\frac{\partial}{\partial Z}|Y|^{\frac{1}{2}q} = |Y|^{\frac{1}{2}q}(\tfrac{q}{2}E + (Z-\overline{Z})\frac{\partial}{\partial Z}),$$

we get

$$Kf = |Y|^{\frac{1}{2}q}\, e^{-irt}\, K_\alpha h$$

with

$$K_\alpha = \alpha E + (Z-\overline{Z})\frac{\partial}{\partial Z}, \qquad \alpha = \frac{r+q}{2}$$

and similarly

$$\Lambda f = |Y|^{\frac{1}{2}q}\, e^{-irt}\, \Lambda_\beta h$$

with

$$\Lambda_\beta = -\beta E + (Z-\overline{Z})\frac{\partial}{\partial \overline{Z}}, \qquad \beta = \frac{-r+q}{2}.$$

Denote by $L_o$ the subring of $L$ generated by $H_1, H_2, \ldots, H_n$. The operators $L \in L_o$ are polynomials with constant coefficients in the elements of K and $\Lambda$ only, i.e., $L = L(\Lambda, K)$. Obviously

$$L(\Lambda,K)f = |Y|^{\frac{1}{2}q}\, e^{-irt}\, L(\Lambda_\beta, K_\alpha)h$$

and so $L(\Lambda,K)f = \lambda f$, $\lambda \in \mathbb{C}$ implies

$$L(\Lambda_\beta, K_\alpha)h = \lambda h.$$

If we replace $\Lambda$ and K by $\Lambda_\beta$ and $K_\alpha$, then the construction which gave $A^{(j)}$ will lead to new operators which we shall call $A^{(j)}_{\alpha,\beta}$. Set

$$\Omega_{\alpha,\beta} = A^{(1)}_{\alpha,\beta} + \alpha(\beta - \frac{n+1}{2})E$$

$$= \Lambda_{\beta-\frac{n+1}{2}} K_\alpha + \alpha(\beta - \frac{n+1}{2})E$$

$$= (Z-\overline{Z})((Z-\overline{Z})\frac{\partial}{\partial \overline{Z}})'\frac{\partial}{\partial Z} - \beta(Z-\overline{Z})\frac{\partial}{\partial Z} + \alpha(Z-\overline{Z})\frac{\partial}{\partial \overline{Z}}.$$

The parameter $q$ can be chosen so that for a given eigenfunction $h$ we have

$$\sigma(\Omega_{\alpha\beta})h = \sigma(A^{(1)}_{\alpha\beta})h + n\alpha(\beta - \frac{n+1}{2})h$$

$$= (\lambda_1 + n\alpha(\beta - \frac{n+1}{2}))h = 0.$$

Let us determine how $A^{(j)}_{\alpha,\beta}$ transforms under

$$M = \begin{pmatrix} A & B \\ C & D \end{pmatrix} \in \Omega.$$

If $h = h(Z,\overline{Z})$ is an arbitrary function, then

$$A^{(j)}|Y|^{\frac{1}{2}q} e^{-irt} h(Z,\overline{Z}) = |Y|^{\frac{1}{2}q} e^{-irt} A^{(j)}_{\alpha,\beta} h(Z,\overline{Z}).$$

Moreover

$$|Y^*|^{\frac{1}{2}q} e^{-irt^* - ira}|CZ+D|^{\alpha}|C\overline{Z}+D|^{\beta} = |Y|^{\frac{1}{2}q} e^{-irt},$$

where the stars indicate the transforms with respect to $M_a \in \Omega \times \mathbb{R}/2\pi\mathbb{Z}$. These formulas together with

$$\overset{*}{A}^{(j)} = (ZC' + D')^{-1}\{(CZ+D)A^{(j)'}\}'$$

imply

$$|CZ+D|^{-\alpha}|C\overline{Z}+D|^{-\beta} \overset{*}{A}^{(j)}_{\alpha,\beta}|CZ+D|^{\alpha}|C\overline{Z}+D|^{\beta}$$

$$= (Z'C' + D')^{-1}\{(CZ+D)A^{(j)'}_{\alpha,\beta}\}',$$

in particular

$$|CZ+D|^{-\alpha}|C\overline{Z}+D|^{-\beta} \overset{*}{\Omega}_{\alpha,\beta}$$

$$= (Z'C' + D')^{-1}\{(CZ+D)\Omega'_{\alpha,\beta}\}'|CZ+D|^{-\alpha}|C\overline{Z}+D|^{-\beta}.$$

If we apply this identity to the constant function 1, we see that $|CZ+D|^{-\alpha}|C\bar{Z}+D|^{-\beta}$, and more generally the "non-analytic Eisenstein series"

$$\sum_{M\in\Omega} c_M |CZ+D|^{-\alpha}|C\bar{Z}+D|^{-\beta}$$

is annihilated by the differential operator $\Omega_{\alpha,\beta}$, i.e., the series is a solution of a system of $n^2$ partial differential equations of order 2. Such series arise in the theory of indefinite quadratic forms. The signature $\mu,\nu$ of the quadratic form is related to $\alpha,\beta$ by $\mu = 2\alpha$, $\nu = 2\beta$.

Since

$$\sigma(\mathrm{A}^{*(j)}_{\alpha,\beta}) = |CZ+D|^{\alpha}|C\bar{Z}+D|^{\beta}\sigma(\mathrm{A}^{(j)}_{\alpha,\beta})|CZ+D|^{-\alpha}|C\bar{Z}+D|^{-\beta},$$

we have

$$L^*_{\alpha,\beta} = |CZ+D|^{\alpha}|C\bar{Z}+D|^{\beta}L_{\alpha,\beta}|CZ+D|^{-\alpha}|C\bar{Z}+D|^{-\beta}$$

for

$$L_{\alpha,\beta} = L(\Lambda_\beta, \mathrm{K}_\alpha), \qquad L(\Lambda,\mathrm{K}) \in L_o.$$

If we introduce the notation

$$f|M_{\alpha,\beta}(Z,\bar{Z}) = f(M\langle Z\rangle, M\langle\bar{Z}\rangle)|CZ+D|^{-\alpha}|C\bar{Z}+D|^{-\beta},$$

the transformation formula for $L_{\alpha,\beta}$ yields

$$L_{\alpha,\beta}(f|M_{\alpha,\beta}) = (L_{\alpha,\beta}f)|M_{\alpha,\beta}.$$

## §9. *The reduction theory of positive quadratic forms*

In this section we want to present the reduction theory due to Minkowski and Siegel.

We shall call a matrix $U = U^{(n)}$ *unimodular* if its entries are elements of $\mathbb{Z}$ and if the determinant $|U| = \pm 1$. The discrete group $\Gamma = \Gamma_n$ of all unimodular matrices is a subgroup of $\Omega = GL(n, \mathbb{R})$ and operates discontinuously on the space $\mathcal{Y} = \{Y \mid Y = Y^{(n)} = Y' > 0\}$ of positive matrices according to $Y \mapsto Y[V] = V'YV$.

The goal of reduction theory is to determine a fundamental domain for $\Gamma$ in $\mathcal{Y}$, i.e., a subset of $\mathcal{Y}$ which contains a single "reduced" matrix from each equivalence class $\{Y[U] \mid U \in \Gamma\}$, characterized by some minimal conditions.

We write $U \in \Gamma$ in the form $U = (\check{\mathfrak{u}}_1, \check{\mathfrak{u}}_2, \ldots, \check{\mathfrak{u}}_n)$, where $\check{\mathfrak{u}}_1, \check{\mathfrak{u}}_2, \ldots, \check{\mathfrak{u}}_n$ denote column vectors. For a given $Y \in \mathcal{Y}$ we choose the first column $\check{\mathfrak{u}}_1$ so that $Y[\check{\mathfrak{u}}_1]$ is minimal; this can be done since $Y$ is a positive matrix. Next we determine $\check{\mathfrak{u}}_2$ so that $Y[\check{\mathfrak{u}}_2]$ is minimal when $\check{\mathfrak{u}}_2$ runs through the second columns of all unimodular matrices $U$ whose first column is the already chosen $\check{\mathfrak{u}}_1$. Replacing $\check{\mathfrak{u}}_2$ by $-\check{\mathfrak{u}}_2$ if necessary, we may assume that $\check{\mathfrak{u}}_1' Y \check{\mathfrak{u}}_2 \geqq 0$. In the next step we choose $\check{\mathfrak{u}}_3$ among the third columns of the matrices $U = (\check{\mathfrak{u}}_1, \check{\mathfrak{u}}_2, *, \ldots, *)$ so that $Y[\check{\mathfrak{u}}_3]$ shall be minimal and $\check{\mathfrak{u}}_2' Y \check{\mathfrak{u}}_3 \geqq 0$. Proceeding in this fashion we pick a certain unimodular matrix $U = (\check{\mathfrak{u}}_1, \check{\mathfrak{u}}_2, \ldots, \check{\mathfrak{u}}_n) \in \Gamma$ and call $R = (r_{\mu\nu}) = Y[U]$ the reduced matrix.

Let us determine explicitly the conditions for a matrix to be reduced. For $1 \leqq k \leqq n$ denote by $U_k$ any unimodular matrix which has the same first $k-1$ columns as $U$. These $U_k$ are given by

$$U_k = U\begin{pmatrix} E & A \\ 0 & B \end{pmatrix},$$

where $E = E^{(k-1)}$, $A$ is an integral matrix and $B$ is unimodular. Denote by $\mathfrak{g}_k$ the $k$-th column of the matrix $U^{-1}U_k$. The first $k-1$ elements $g_1,\dots,g_{k-1}$ of $\mathfrak{g}_k$ are the entries in the first column of $A$, and the last $n-k+1$ elements $g_k,g_{k+1},\dots,g_n$ of $\mathfrak{g}_k$ form the first column of $B$. Since $B$ is unimodular, the integers $g_k,g_{k+1},\dots,g_n$ have no common divisor. Conversely, given any set of integers $g_k,g_{k+1},\dots,g_n$ without a common divisor, there exists a unimodular matrix $B$ whose first column is formed by these integers. Since $U\mathfrak{g}_k$ is the $k$-th column of the matrix $U_k$, we have according to our minimal conditions

$$Y[U\mathfrak{g}_k] = R[\mathfrak{g}_k] \geqq R[\mathfrak{e}_k] = r_{kk} = r_k,$$

where $\mathfrak{e}_k$ is the $k$-th unit column vector. Obviously $r_{k,k+1} = \mathfrak{u}_k' Y \mathfrak{u}_{k+1} \geqq 0$ and we see that the reduced matrices $R = (r_{\mu\nu})$ are characterized by the following properties

$$\text{(1)} \quad \begin{cases} r_{k,k+1} \geqq 0 \quad \text{for } k = 1,2,\dots,n-1, \\ R[\mathfrak{g}_k] \geqq r_k \quad \text{for } k = 1,2,\dots,n \text{ and for any integral} \\ \text{vector } \mathfrak{g}_k' = (g_1,g_2,\dots,g_n) \text{ such that } g_k,g_{k+1},\dots,g_n \\ \text{are without common divisor.} \end{cases}$$

We shall omit those inequalities which give no condition, i.e., we assume that $\mathfrak{g}_k \neq \pm\mathfrak{e}_k$. Taking for $\mathfrak{g}_k$ the $\ell$-th unit column vector $\mathfrak{e}_\ell$ with $\ell > k$ we obtain

$$\text{(2)} \qquad r_k \leqq r_\ell \qquad \text{for } k < \ell.$$

Next we take $\mathfrak{g}_k = \mathfrak{e}_k \pm \mathfrak{e}_\ell$ with $\ell < k$. This yields

$$r_k \pm 2r_{k\ell} + r_\ell \geqq r_k$$

and thus

$$(3) \qquad -r_\ell \leqq 2r_{k\ell} \leqq r_\ell \qquad \text{for } k \neq \ell.$$

Next we want to prove by induction on $n$ that any reduced matrix $R$ satisfies the inequality

$$(4) \qquad r_1 r_2 \ldots r_n < c_1 |R|,$$

where $c_1$, and later $c_2, c_3, \ldots,$ denote positive constants which depend only on $n$. The assertion is clearly valid for $n = 1$ taking any $c_1 > 1$; let us assume that it is true for $n-1$ instead of $n$.

For $R = R^{(n)}$ let us write

$$R_\ell = R \begin{bmatrix} E^{(\ell)} \\ 0 \end{bmatrix} .$$

If $R$ is positive and reduced then so is $R_\ell$. This can be easily verified if we observe that $R_\ell$ is the square matrix of order $\ell$ which is formed by the elements which are in the upper left-hand corner of $R$.

According to the induction hypothesis we have

$$r_1 r_2 \ldots r_{n-1} < c_2 |R_{n-1}|.$$

Denote by $\rho_{k\ell}$ the algebraic complement of $r_{k\ell}$ in $R_{n-1}$, i.e., $R_{n-1}^{-1} \cdot |R_{n-1}| = (\rho_{k\ell})$. By virtue of (3) we have

$$\pm \rho_{k\ell} r_\ell < c_3 r_1 r_2 \ldots r_{n-1},$$

hence

$$\pm \rho_{k\ell} |R_{n-1}|^{-1} \leqq c_3 r_1 r_2 \ldots r_{n-1} \frac{1}{|R_{n-1}|} \cdot \frac{1}{r_\ell}$$

and thus

(5) $$\pm\ \rho_{k\ell}|R_{n-1}|^{-1} \leqq c_2 c_3 r_\ell^{-1}.$$

We set

$$R = \begin{pmatrix} R_{n-1} & \mathfrak{r} \\ \mathfrak{r}' & r_n \end{pmatrix},$$

so that

$$R = \begin{pmatrix} R_{n-1} & 0 \\ 0 & r \end{pmatrix} \begin{bmatrix} E & R_{n-1}^{-1}\mathfrak{r} \\ 0 & 1 \end{bmatrix},$$

where $r = r_n - R_{n-1}^{-1}[\mathfrak{r}]$. It follows from (5) that

$$R_{n-1}^{-1}[\mathfrak{r}] \leqq \sum_{\mu,\nu=1}^{n-1} \frac{c_2 c_3}{r_\mu} r_\mu r_\nu,$$

hence by (2)

$$R_{n-1}^{-1}[\mathfrak{r}] < c_4 r_{n-1}$$

and therefore

$$r_n = r + R_{n-1}^{-1}[\mathfrak{r}] < r + c_4 r_{n-1}.$$

Since $|R| = |R_{n-1}|\cdot r$, we have by the induction hypothesis

$$r_1 r_2 \ldots r_n < c_2 |R_{n-1}| r_n = c_2 |R| \frac{r_n}{r}$$

and consequently

$$r_1 r_2 \ldots r_n < c_2 (1 + c_4 \frac{r_{n-1}}{r}) |R|.$$

(4) will be a consequence of this inequality as soon as we have proved

$$r_{n-1} < c_5 r.$$

Let us set

$$c_6 = 4(n-1)^2, \qquad c_7 = (2n-2)^{n-1} = c_6^{\frac{1}{2}(n-1)},$$

and assume that

$$r_{\ell+1} < c_6 r_\ell \tag{6}$$

is valid for $\ell = n-2, n-3, \ldots, k+1, k$, but not for $\ell = k-1$. Here $k$ is an integer such that $1 \leqq k \leqq n-1$, and in the two limit cases $k = 1$ and $k = n-1$ some parts of the assertion must be omitted. With a variable $\mathfrak{x}' = (\mathfrak{z}', x_n) = (x_1, \ldots, x_{n-1}, x_n)$ we write

$$R[\mathfrak{x}] = \begin{pmatrix} R_{n-1} & 0 \\ 0 & r \end{pmatrix} \begin{bmatrix} \mathfrak{z} + R_{n-1}^{-1}\mathfrak{w}\cdot x_n \\ x_n \end{bmatrix} = R_{n-1}[\mathfrak{z} + R_{n-1}^{-1}\mathfrak{w}\cdot x_n] + r x_n^2 .$$

Let $x_\nu + a_\nu \cdot x_n$, $1 \leqq \nu \leqq n-1$, be the components of the column vector $\mathfrak{z} + R_{n-1}^{-1}\mathfrak{w}\cdot x_n$. For each integer $x_n'$ in the interval $0 \leqq x_n' \leqq c_7^{n-k}$ we determine integers $x_\nu'$ $(\nu = k, k+1, \ldots, n-1)$ so that $0 \leqq x_\nu' + a_\nu x_n' < 1$. We subdivide the unit interval $[0,1]$ into $c_7$ equal subintervals; this gives a subdivision of the $(n-k)$-dimensional unit cube into $c_7^{n-k}$ equal subcubes. Since there are $c_7^{n-k} + 1$ systems of numbers $(x_\nu' + a_\nu x_n')_{k \leqq \nu \leqq n-1}$, by the drawer principle two of them must lie in the same subcube, i.e.,

$$|x_\nu' - x_\nu'' + a_\nu(x_n' - x_n'')| < \frac{1}{c_7} .$$

Thus we proved that there exist integers $x_k, x_{k+1}, \ldots, x_{n-1}, x_n$ so that

$$|x_\nu + a_\nu x_n| < c_7^{-1}, \qquad 0 < x_n < c_7^{n-k} \qquad (\nu = k, \ldots, n-1).$$

We may assume that $x_k, x_{k+1}, \ldots, x_n$ are without common divisor. Let us furthermore determine the integers $x_1, \ldots, x_{k-1}$ so that

$$|x_\nu + a_\nu x_n| < 1 \qquad (\nu = 1,\dots,k-1).$$

Since $R$ is reduced, it follows from (1) that

$$R[\mathfrak{y}] \geq r_k.$$

On the other hand,

$$R[\mathfrak{y}] = R_{n-1}[\mathfrak{z} + R_{n-1}^{-1}\mathfrak{w}\cdot x_n] + rx_n^2$$

$$\leq (k-1)^2 r_{k-1} + (k-1)(n-k)r_{k-1}\cdot\frac{1}{c_7} + (n-k)^2 c_6^{n-k-1} r_k\cdot\frac{1}{c_7^2} + c_7^{2(n-k)} r,$$

since the elements of $R_{n-1}$ in the first $k-1$ rows and first $k-1$ columns are $\leq r_{k-1}$ by (2) and (3), the elements in the last $n-k$ rows and first $k-1$ columns, and again in the first $k-1$ rows and last $n-k$ columns are $\leq \frac{1}{2} r_{k-1}$ by (3), and finally the elements in the last $n-k$ rows and last $n-k$ columns are $\leq c_6^{n-k-1} r_k$ by (2), (3) and the assumption (6). Continuing we get, since $r_k \geq c_6 r_{k-1}$,

$$R[\mathfrak{y}] \leq \frac{(n-1)^2}{c_6} r_k + \frac{(n-1)^2}{c_6 c_7} r_k + \frac{(n-1)^2}{c_7^2} c_6^{n-2} r_k + c_7^{2(n-k)} r$$

$$\leq \frac{1}{4} r_k + \frac{1}{4} r_k + \frac{1}{4} r_k + c_7^{2(n-k)} r = \frac{3}{4} r_k + c_7^{2(n-k)} r.$$

From $r_k \leq R[\mathfrak{y}]$ we get therefore $r_k \leq 4c_7^{2(n-k)} r$, and taking into account that by (6) we have $r_{n-1} \leq c_6^{n-k-1} r_k$, we obtain $r_{n-1} \leq c_5 r$. As was pointed out already, (4) follows from the last inequality, thus its proof is complete.

If $Y = (y_{\mu\nu})$ is a positive matrix, then we have

$$|Y| \leq \prod_{h=1}^{n} y_{hh}. \tag{7}$$

To prove this inequality, we choose $Q = (\mathfrak{q}_1, \mathfrak{q}_2, \dots, \mathfrak{q}_n)$ such that

$Y = Q'Q$. By the well-known inequality

$$\|Q\| \leqq \prod_{\nu=1}^{n} (\mathfrak{q}_\nu' \mathfrak{q}_\nu)^{\frac{1}{2}}$$

we obtain indeed

$$|Y| = |Q|^2 \leqq \prod_{\nu=1}^{n} \mathfrak{q}_\nu' \mathfrak{q}_\nu = \prod_{\nu=1}^{n} y_{\nu\nu}.$$

Assume again that $R = (r_{\mu\nu})$ is a reduced matrix and set $R_0 = (\delta_{\mu\nu} r_{\nu\nu})$. Then there exists a constant $c_9 > 0$ depending only on $n$ such that

(8) $$\frac{1}{c_9} R_0 < R < c_9 R_0.$$

*Proof*. Denote by $\rho_1, \rho_2, \ldots, \rho_n$ the characteristic roots of $R[R_0^{-\frac{1}{2}}]$. Then we have

$$\rho_1 + \rho_2 + \ldots + \rho_n = \sigma(R[R_0^{-\frac{1}{2}}]) = \sigma(RR_0^{-1}) = n,$$

and

$$\rho_1 \rho_2 \ldots \rho_n = |R| \cdot |R_0|^{-1} > \frac{1}{c_1}$$

by (4). Thus

$$\rho_\nu < n, \qquad \rho_\nu > \frac{1}{n^{n-1} c_1} \qquad (\nu = 1, 2, \ldots, n).$$

If $V$ is an orthogonal matrix such that

$$R[R_0^{-\frac{1}{2}}][V] = \begin{pmatrix} \rho_1 & & \\ & \ddots & \\ & & \rho_n \end{pmatrix},$$

then

$$\frac{1}{n^{n-1}c_1} E < R[R_o^{-\frac{1}{2}}][V] < nE.$$

Transforming with $V^{-1}$ we get

$$\frac{1}{n^{n-1}c_1} E < R[R_o^{-\frac{1}{2}}] < nE$$

and consequently

$$\frac{1}{n^{n-1}c_1} R_o < R < nR_o.$$

Thus we have proved (8) with $c_9 = \max(n, n^{n-1}c_1)$.

Any matrix $Y > 0$ has a unique representation

$$Y = D[B],$$

where $D = (\delta_{\mu\nu} d_\nu)$ is a diagonal matrix with $d_\nu > 0$ and $B = (b_{\mu\nu})$ is an upper triangular matrix with $b_{\nu\nu} = 1$ for all $\nu$ and $b_{\mu\nu} = 0$ for $\mu > \nu$. The entries $d_\nu$, $b_{\mu\nu}$ $(\mu < \nu)$ are called the *Jacobian coordinates* of $Y$.

Assume that for $Y$ we choose the reduced matrix $R = (r_{\mu\nu}) = D[B]$, where we write again $r_{\nu\nu} = r_\nu$. From

$$r_\nu = d_\nu + \sum_{\mu=1}^{\nu-1} d_\mu b_{\mu\nu}^2 \qquad (\nu = 1,2,\ldots,n)$$

and

$$|R| = d_1 d_2 \ldots d_n$$

it follows by virtue of (4) that

$$1 \leqq \frac{r_\nu}{d_\nu} \leqq \prod_{\mu=1}^{n} \frac{r_\mu}{d_\mu} < c_1 \qquad (\nu = 1,2,\ldots,n).$$

Moreover from

$$1 \leqq \frac{r_\nu}{d_\nu} < c_1 \leqq c_1 \frac{r_\mu}{d_\mu}$$

we get, using (2),

$$0 < \frac{d_\mu}{d_\nu} < c_1 \frac{r_\mu}{r_\nu} < c_1 \qquad (\mu < \nu).$$

We prove by induction on $\mu$ that all $b_{\mu\nu}$ lie between bounds which depend only on $n$. Assume that

$$\pm\, b_{p\nu} < c_{10} \quad \text{for} \quad \nu > p, \ 1 \leqq p < \mu.$$

Then

$$r_{\mu\nu} = d_\mu b_{\mu\nu} + \sum_{p=1}^{\mu-1} d_p b_{p\mu} b_{p\nu} \qquad (\mu < \nu)$$

yields

$$\pm\, b_{\mu\nu} \leqq \frac{1}{2} \frac{r_\mu}{d_\mu} + \sum_{p=1}^{\mu-1} \frac{d_p}{d_\mu} c_{10}^2 < \frac{1}{2} c_1 + (n-1) c_1 c_{10}^2 = c_{11},$$

which proves our assertion. We formulate the result in

Lemma 1. *There exists a positive constant* $c_{12}$ *depending only on* $n$ *such that the Jacobian coordinates* $d_\nu$, $b_{\mu\nu}$ $(\mu < \nu)$ *of any reduced positive matrix* $R$ *satisfy*

(9) $\quad d_\nu < c_{12} d_{\nu+1} \quad (1 \leqq \nu < n), \qquad \pm\, b_{\mu\nu} < c_{12} \quad (1 \leqq \mu < \nu \leqq n).$

Inequalities (9) do not imply conversely that $R$ is reduced. But if $R$ satisfies (9) then the entries of any $U \in \Gamma$ such that $R[U]$ is reduced lie between bounds which depend only on $n$. This is a consequence of

*Lemma* 2. *Denote the Jacobian coordinates of the matrix* $Y > 0$ *by* $d_\nu^*$, $b_{\mu\nu}^*$. *Let* $G = (g_{\mu\nu})$ *be an integral matrix such that* $|G| \neq 0$ *and that* $Y[G]$ *is reduced. If for some* $m > 0$ *we have the inequalities*

$$\pm\ |G| < m, \qquad \frac{d_\nu^*}{d_{\nu+1}^*} < m \quad (1 \leqq \nu < n), \qquad \pm\ b_{\mu\nu}^* < m \quad (1 \leqq \mu < \nu \leqq n)$$

*then* $\pm\ g_{\mu\nu} < m_1$ $(\mu,\nu = 1,2,\ldots,n)$, *where* $m_1$ *depends only on* $n$ *and* $m$.

*Proof*. The assertion is true for $n = 1$; we assume that it is true for all orders less than $n$. We denote by $m_1$, $m_2$, ... constants which depend only on $n$ and $m$. Write $Y = D^*[B^*]$ and

$$Y[G] = D^*[B^*G] = D[B],$$

where

$$D^* = (\delta_{\mu\nu}d_\nu^*), \quad B^* = (b_{\mu\nu}^*), \quad b_{\mu\nu}^* = 0 \quad \text{for} \quad \mu > \nu, \quad b_{\nu\nu}^* = 1,$$

$$D = (\delta_{\mu\nu}d_\nu), \quad B = (b_{\mu\nu}), \quad b_{\mu\nu} = 0 \quad \text{for} \quad \mu > \nu, \quad b_{\nu\nu} = 1.$$

The Jacobian coordinates $d_\nu$ and $b_{\mu\nu}$ satisfy (9) since $Y[G]$ is assumed to be reduced.

Introduce the matrices

$$B^*GB^{-1} = Q = (q_{\mu\nu}) \quad \text{and} \quad B^{*-1} = (\beta_{\mu\nu}).$$

Then we have

$$D^*[Q] = D, \qquad D[Q^{-1}] = D^*,$$

hence

$$d_\nu = \sum_{\mu=1}^{n} d_\mu^* q_{\mu\nu}^2 \qquad (1 \leqq \nu \leqq n)$$

and therefore

$$d_\mu^* q_{\mu\nu}^2 \leqq d_\nu \qquad (1 \leqq \mu,\nu \leqq n).$$

Since $B$ and $B^*$ are triangular matrices, we obtain for the entries of $G = B^{*-1}QB$ the expression

$$g_{\mu\nu} = \sum_{\kappa=\mu}^{n} \sum_{\lambda=1}^{\nu} \beta_{\mu\kappa} q_{\kappa\lambda} b_{\lambda\nu} ,$$

and therefore

$$d_\mu^* g_{\mu\nu}^2 = \sum_{\kappa=\mu}^{n} \sum_{\lambda=1}^{\nu} \sum_{\kappa'=\mu}^{n} \sum_{\lambda'=1}^{\nu} \beta_{\mu\kappa}(\sqrt{d_\mu^*}\, q_{\kappa\lambda}) b_{\lambda\nu} \beta_{\mu\kappa'}(\sqrt{d_\mu^*}\, q_{\kappa'\lambda'}) b_{\lambda'\nu} .$$

Since $\mu \leqq \kappa,\kappa'$ we have $\sqrt{d_\mu^*}\, q_{\kappa\lambda} \leqq m^{\frac{1}{2}(n-1)} \sqrt{d_\kappa^*}\, q_{\kappa\lambda} \leqq m^{\frac{1}{2}(n-1)} \sqrt{d_\lambda}$, $\sqrt{d_\mu^*}\, q_{\kappa'\lambda'} \leqq m^{\frac{1}{2}(n-1)} \sqrt{d_{\kappa'}^*}\, q_{\kappa'\lambda'} \leqq m^{\frac{1}{2}(n-1)} \sqrt{d_{\lambda'}}$, and since $\lambda,\lambda' \leqq \nu$ we have $d_\lambda \leqq c_{12}^{n-1} d_\nu$, $d_{\lambda'} \leqq c_{12}^{n-1} d_\nu$. It follows that

$$d_\mu^* q_{\mu\nu}^2 \leqq m_2 d_\nu \qquad (1 \leqq \mu,\nu \leqq n).$$

For the elements of $G^{-1} = (f_{\mu\nu})$ we get by an analogous consideration

$$d_\mu f_{\mu\nu}^2 = m_3 d_\nu^* \qquad (1 \leqq \mu,\nu \leqq n).$$

Since the determinant $|f_{\mu\nu}| \neq 0$, there exists a permutation $(\nu_1,\ldots,\nu_n)$ of $(1,\ldots,n)$ such that $\prod_{\mu=1}^{n} f_{\mu\nu_\mu} \neq 0$. Moreover $|G| f_{\mu\nu}$ is an integer and $\pm\, |G| < m$, hence

$$d_\mu < m_4 d_{\nu_\mu}^* \qquad (1 \leqq \mu \leqq n).$$

Since $\nu_\mu, \nu_{\mu+1}, \dots, \nu_n$ are $n+1-\mu$ numbers belonging to $\{1,2,\dots,n\}$, we have $\min(\nu_\mu, \nu_{\mu+1}, \dots, \nu_n) \leqq \mu$. Let $\rho$ be such that $\mu \leqq \rho \leqq n$ and $\nu_\rho \leqq \mu$. Then from $d_\nu < c_{12} d_{\nu+1}$ $(\mu \leqq \nu < \rho)$, $d_\rho < m_4 d^*_{\nu_\rho}$ and $d^*_\nu < m d^*_{\nu+1}$ $(\nu_\rho \leqq \nu < \mu)$ it follows that

$$d_\mu < m_5 d^*_\mu \qquad (\mu = 1,2,\dots,n),$$

and consequently

$$(10) \qquad d_\mu g^2_{\mu\nu} < m_6 d_\nu \qquad (1 \leqq \mu,\nu \leqq n),$$

with $m_6 = m_2 m_5$.

Denote by $p$ the largest number in the sequence $1,2,\dots,n$ such that

$$(11) \qquad d_\mu \geqq m_6 d_\nu \quad \text{for } 1 \leqq \nu < p, \ p \leqq \mu \leqq n.$$

If no such $p$ exists, we take $p = 1$. Since $G$ is an integral matrix, it follows from (10) and (11) that $g_{\mu\nu} = 0$ for $p \leqq \mu \leqq n$, $1 \leqq \nu < p$. Thus $G$ has the form

$$G = \begin{pmatrix} G_1 & G_{12} \\ 0 & G_2 \end{pmatrix},$$

with $G_1 = G_1^{(p-1)}$.

Next we prove that the elements of $G_2$ are bounded. By (11) for any $g$ in the sequence $p+1, p+2, \dots, n$ there exist $\mu = \mu(g) \geqq g$ and $\nu = \nu(g) < g$ such that

$$(12) \qquad d_\mu < m_6 d_\nu.$$

According to (9) we get

$$d_g < m_7 d_{g-1} \qquad \text{for } g = p+1, p+2, \dots, n,$$

hence

$$d_\nu < m_8 d_\mu \qquad (p \leqq \mu,\nu \leqq n),$$

and thus by (10)

$$g_{\mu\nu}^2 < m_6 m_8 \qquad (p \leqq \mu,\nu \leqq n).$$

In the case when $p = 1$ we have $G = G_2$ and then the proof is finished. Assume therefore that $p > 1$. We decompose into blocks similarly as we have done for $G$:

$$D = \begin{pmatrix} D_1 & 0 \\ 0 & D_2 \end{pmatrix}, \quad D^* = \begin{pmatrix} D_1^* & 0 \\ 0 & D_2^* \end{pmatrix}, \quad B = \begin{pmatrix} B_1 & B_{12} \\ 0 & B_2 \end{pmatrix}, \quad B^* = \begin{pmatrix} B_1^* & B_{12}^* \\ 0 & B_2^* \end{pmatrix}$$

and obtain

$$\begin{pmatrix} D_1^* & 0 \\ 0 & D_2^* \end{pmatrix} \begin{bmatrix} B_1^* G_1 & B_1^* G_{12} + B_{12}^* G_2 \\ 0 & B_2^* G_2 \end{bmatrix} = \begin{pmatrix} D_1 & 0 \\ 0 & D_2 \end{pmatrix} \begin{bmatrix} B_1 & B_{12} \\ 0 & B_2 \end{bmatrix}$$

or

$$D_1^*[B_1^* G_1] = D_1[B_1]$$

and

$$G_1' B_1^{*\prime} D_1^* (B_1^* G_{12} + B_{12}^* G_2) = B_1' D_1 B_{12}.$$

Now together with $D[B]$ also $D_1[B_1]$ is reduced. Since $|G| = |G_1| \cdot |G_2|$, we have $\pm |G_1| < m$, and so since $p-1 < n$, we have by our initial induction hypothesis that the entries of $G_1$ are bounded in absolute value by some constant $m_9$. Furthermore since $G_1' B_1^{*\prime} D_1^* = D_1[B_1] G_1^{-1} B_1^{*-1}$, we get

$$D_1[B_1] G_1^{-1} B_1^{*-1} (B_1^* G_{12} + B_{12}^* G_2) = B_1' D_1 B_{12},$$

and thus

$$G_{12} = G_1 B_1^{-1} B_{12} - B_1^{*-1} B_{12} G_2 .$$

It follows from what we have already proved and from our assumptions that also the entries of $G_{12}$ have an absolute value less than $m_{10}$. So our lemma is proved.

For any compact subset $\mathcal{L}$ of $\mathcal{Y}$ we can determine a positive number $m = m(\mathcal{L})$ such that the Jacobian coordinates of the $Y \in \mathcal{L}$ satisfy the conditions of Lemma 2. So we get

*Corollary* 1. *Denote by* $\mathcal{R}$ *the set of all reduced matrices in* $\mathcal{Y} = \{Y \mid Y = Y^{(n)} = Y' > 0\}$, *and let* $\mathcal{L}$ *be a compact subset of* $\mathcal{R}$. *Then there exist only a finite number of unimodular matrices* $U$ *such that* $\mathcal{L} \cap \mathcal{R}[U] \neq \emptyset$.

Another obvious consequence of Lemma 2 is the

*Corollary* 2. *If* $Y$ *and* $Y[U]$ *are both reduced positive matrices and* $U$ *is unimodular, then the entries of* $U$ *have absolute values less than some constant* $c_{13}$ *independent of* $Y$.

The set $\mathcal{R}$ of all reduced positive matrices $R = (r_{\mu\nu})$ is characterized by condition (1) and it is a convex cone with vertex at 0 if we consider the $r_{k\ell}$ with $k \leqq \ell$ as cartesian coordinates.

Denote by $\mathfrak{Y}$ the set $\{Y \mid Y = Y^{(n)} = Y'\}$ of all symmetric matrices of order $n$. We have $\mathcal{R} \subset \mathcal{Y} \subset \mathfrak{Y}$ and we propose to determine the boundary $\partial(\mathcal{Y})$ of $\mathcal{Y}$ in $\mathfrak{Y}$. Let $G_k$ be a sequence of elements of $\mathfrak{Y} - \mathcal{Y}$ and assume that $G_k$ converges to a matrix $G \in \mathfrak{Y}$ as $k \to \infty$. Since $G_k$ is not positive, for every $k$ there exists a column vector $\mathfrak{x}_k$ such that $\mathfrak{x}_k' \mathfrak{x}_k = 1$ and $G_k[\mathfrak{x}_k] \leqq 0$. Taking, if necessary, a subsequence, we may assume that $\mathfrak{x}_k$ converges to a vector $\mathfrak{x}$. Then $\mathfrak{x}' \cdot \mathfrak{x} = 1$ and $G[\mathfrak{x}] \leqq 0$. Thus $G$ belongs to $\mathfrak{Y} - \mathcal{Y}$, hence $\mathfrak{Y} - \mathcal{Y}$

is closed and so $\mathcal{Y}$ is an open subset of $\mathcal{Y}$ .

If $G$ belongs to $\partial(\mathcal{Y})$, then there exists a sequence $Y_k$ of elements of $\mathcal{Y}$ which converges to $G$. For any column vector $\mathfrak{x}$ we have $Y_k[\mathfrak{x}] > 0$ and so $G[\mathfrak{x}] \geqq 0$. On the other hand $G$ is not positive, hence there exists a vector $\mathfrak{x} \neq 0$ such that $G[\mathfrak{x}] = 0$. Matrices of this kind are called *semi-positive*. Conversely, every semi-positive matrix $G$ belongs to $\partial(\mathcal{Y})$ since $G + \varepsilon E \in \mathcal{Y}$ for $\varepsilon > 0$ and tends to $G$ as $\varepsilon \to 0$.

If $R$ is an interior point of $\mathcal{R}$, then the inequalities (1) hold in a neighborhood of $R$ and therefore in $R$ the strict inequality $>$ holds throughout in (1). A boundary point $R_0$ of $\mathcal{R}$ is either positive or semi-positive. Assume first that $R_0$ is positive. There exists a sequence of positive matrices $Y_k \in \mathcal{Y}$ such that $Y_k \notin \mathcal{R}$ and $Y_k \to R_0$ as $k \to \infty$. Let $R_0 = D[B]$, $D = (\delta_{\mu\nu} d_\nu)$, $B = (b_{\mu\nu})$ with $b_{\mu\nu} = 0$ for $\mu > \nu$ and $b_{\nu\nu} = 1$. The correspondence $R_0 \leftrightarrow \{d_\nu, b_{\mu\nu}\}$ defines a homeomorphism from a neighborhood of $R_0$ onto a neighborhood of the point $\{d_\nu, b_{\mu\nu}\}$. We set analogously $Y_k = D_k[B_k]$. Since $D_k \to D$, $B_k \to B$, and since $D[B]$ is reduced, the Jacobian coordinates of $Y_k$ satisfy the inequalities of Lemma 1 for sufficiently large $k$. For each $k$ determine $U_k \in \Gamma$ so that $Y_k[U_k] \in \mathcal{R}$. Since $Y_k \notin \mathcal{R}$ we have $U_k \neq \pm E$. It follows from Lemma 2 that for large $k$ the $U_k$ belong to a finite set of matrices. Passing, if necessary, to a subsequence, we may assume that for all $k$ we have $U_k = U \neq \pm E$ and $Y_k[U] \to R_0[U]$. Since $\mathcal{R}$ is closed and the $Y_k[U]$ are reduced, it follows that $R_0[U] \in \mathcal{R}$. We show that $R_1 = R_0[U]$ is also a boundary point of $\mathcal{R}$, or more generally: If $R_0 \in \mathcal{R}$, $R_1 = R_0[U] \in \mathcal{R}$, where $U \in \Gamma$, $U \neq \pm E$, then both $R_0$ and $R_1$ belong to the boundary $\partial(\mathcal{R})$ of $\mathcal{R}$. In the first place, we observe that by Corollary 2 the matrix $U$ belongs to a finite set of matrices. Let us first assume that $U = (\mathfrak{u}_1, \mathfrak{u}_2, \ldots, \mathfrak{u}_n)$ is not a diagonal matrix. Then there exists a first column $\mathfrak{y}_k$ which is $\neq \pm \mathfrak{n}_k$. Then the column $\mathfrak{f}_k$ of $U^{-1} = (\mathfrak{f}_1, \mathfrak{f}_2, \ldots, \mathfrak{f}_n)$ has the same property. Since $U$ has the form

$$U = \left(\begin{array}{ccc|c} \pm 1 & & & g_1 \\ & \ddots & & \vdots \\ & & \pm 1 & g_{k-1} \\ \hline & & & g_k \\ & 0 & & \vdots \\ & & & g_n \end{array}\right), \qquad \mathfrak{y}_k = \begin{pmatrix} g_1 \\ g_2 \\ \vdots \\ g_n \end{pmatrix},$$

and since $|U| = \pm 1$, we see developing $|U|$ according to the $k^{\text{th}}$ column that $g_k, g_{k+1}, \ldots, g_n$ are without common divisor. Writing $R_0 = (r_{\mu\nu})$, $R_1 = (s_{\mu\nu})$, we obtain from (1) that $R_0[\mathfrak{y}_k] \geqq r_k$, i.e., $s_k \geqq r_k$. It follows similarly from $R_0 = R_1[U^{-1}]$ that $R_1[\mathfrak{f}_k] \geqq s_k$, i.e., $r_k \geqq s_k$. Hence we have

$$R_0[\mathfrak{y}_k] = r_k = s_k = R_1[\mathfrak{f}_k],$$

i.e., $R_0, R_1 \in \partial(\mathcal{R})$ and $\mathfrak{y}_k, \mathfrak{f}_k$ belong to a finite set of possible columns. Next we consider the case when $U$ is a diagonal matrix $U = (\pm\, \delta_{\mu\nu}) \neq \pm E$. There must be a first change of signs, assume that it occurs as we go from the $k^{\text{th}}$ to the $(k+1)^{\text{th}}$ diagonal element. Then $s_{k,k+1} = \mathfrak{y}_k' R_0 \mathfrak{y}_{k+1} = - r_{k,k+1}$. But by (1) we have $r_{k,k+1} \geqq 0$, $s_{k,k+1} \geqq 0$ and consequently $r_{k,k+1} = s_{k,k+1} = 0$. It follows again that $R_0, R_1 \in \partial(\mathcal{R})$.

In particular we have proved that the positive points of $\partial(\mathcal{R})$ lie on a finite set of hyperplanes. These hyperplanes bound a closed convex cone $\mathfrak{C}$ which contains $\mathcal{R}$. We prove that $\mathcal{R}$ *contains interior points*. Indeed, assume that this is not the case and let $\mathcal{L}$ be a compact domain in $\mathcal{T}$ which contains interior points. Then for $Y = D^*[B^*] \in \mathcal{L}$ Lemma 2 is valid with a certain $m = m(\mathcal{L})$, where we choose $G = U \in \Gamma$. It follows that there exists a finite subset of $\Gamma$

with the property that for every $Y \in \mathcal{L}$ we can pick an element $U$ of this finite subset and have $Y[U] \in \mathcal{R}$. Now we have $\mathcal{R} = \partial(\mathcal{R})$ by assumption and therefore a finite set of linear maps $Y \mapsto Y[U]$ does map $\mathcal{L}$ into a finite set of hyperplanes, which is impossible.

Next let $T = (t_{\mu\nu})$ be a point of $\mathcal{E}$ which does not belong to $\mathcal{R}$, and let $R = (r_{\mu\nu})$ be an interior point of $\mathcal{R}$. The line segment

$$(1-\lambda)T + \lambda R \qquad (0 \leqq \lambda \leqq 1)$$

joining $T$ and $R$ belongs to $\mathcal{E}$ and its points corresponding to $0 < \lambda \leqq 1$ are interior points of $\mathcal{E}$. On the segment we can find a boundary point $R_0 = (1-\lambda_0)T + \lambda_0 R \in \partial(\mathcal{R})$. This matrix $R_0 = (\mathring{r}_{\mu\nu})$ cannot be positive because if it were, we would have $R_0 \in \partial(\mathcal{E})$, hence $\lambda_0 = 0$ and $R_0 = T$, in contradiction with $T \notin \mathcal{R}$. Thus $R_0$ is semi-positive and $|R_0| = 0$. By condition (4) we have $\mathring{r}_1\mathring{r}_2 \ldots \mathring{r}_n < c_1|R_0| = 0$ and since $0 \leqq \mathring{r}_1 \leqq \mathring{r}_2 \leqq \ldots \leqq \mathring{r}_n$ we have $\mathring{r}_1 = 0$, hence

$$(1-\lambda_0)t_{11} + \lambda_0 r_{11} = \mathring{r}_1 = 0.$$

Furthermore by (3) we have

$$(1-\lambda_0)t_{1\nu} + \lambda_0 r_{1\nu} = \mathring{r}_{1\nu} = 0 \qquad (2 \leqq \nu \leqq n)$$

and therefore

$$\begin{vmatrix} t_{11} & r_{11} \\ t_{1\nu} & r_{1\nu} \end{vmatrix} = 0$$

or

$$t_{1\nu} = t_{11}\frac{r_{1\nu}}{r_{11}} \qquad (2 \leqq \nu \leqq n).$$

Now we keep $T$ fixed and let $R$ vary in a small open set. It follows that $t_{11} = 0$, hence $\lambda_0 = 0$ and $R_0 = T$, i.e., $T$ is a semi-positive boundary point of $\mathcal{R}$. Thus we have proved that $\mathcal{E} - \mathcal{R}$ consists of all semi-positive boundary points of $\mathcal{R}$.

Finally we can state the

*Theorem. The domain $\mathcal{R}$ of all reduced positive matrices is a convex cone with vertex 0. Its boundary lies on a finite collection of hyperplanes. $\mathcal{R}$ is a fundamental domain for the action of $\Gamma$ on $\mathcal{Y} = \{Y \mid Y = Y' > 0\}$, i.e., $\mathcal{Y} = \bigcup_{U \in \Gamma} \mathcal{R}[U]$. If $U \in \Gamma$, $U \neq \pm E$ and $\mathcal{R} \cap \mathcal{R}[U]$ is not empty, then $\mathcal{R} \cap \mathcal{R}[U] \subset \partial(\mathcal{R})$. Only for a finite set of matrices $U \in \Gamma$ is it possible that $\mathcal{R} \cap \mathcal{R}[U] \neq \emptyset$.*

## §10. *Größen-characters of quadratic forms*

Hecke's Größen-characters for number fields are characterized mainly by two facts: they are eigenfunctions of a commutative ring of differential operators which, in a certain sense, are invariant, and they are invariant under a transformation group defined by the group of units of the given field.

In analogy we define Größen-characters for positive quadratic forms, which can be represented by matrices $Y > 0$, as follows. Let $\mathcal{Y} = \mathcal{Y}_n = \{Y \mid Y = Y^{(n)} = Y' > 0\}$ be the space of all positive matrices, $\mathcal{R} = \mathcal{R}_n$ Minkowski's domain of reduced matrices in $\mathcal{Y}$ and $\Gamma = \Gamma_n$ the group of unimodular matrices of order $n$. The function $u(Y)$ is called a Größen-character of quadratic forms if

1. $u(Y) \in C^\infty(\mathcal{Y})$,

2. $u(Y)$ is an eigenfunction of the ring $L$ of invariant differential operators on $\mathcal{Y}$, so that in particular $\sigma(Y \frac{\partial}{\partial Y})u(Y) = \lambda u(Y)$ with a constant $\lambda$,

3. $u(Y)$ is homogeneous of degree 0, i.e., $\lambda = 0$,

4. $u(Y[U]) = u(Y)$ for every $U \in \Gamma$,

5. for $h \geqq 0$ there exist positive constants $C_h$ and $\kappa_h$ such that

$$\left| \frac{\partial^h u(Y)}{\partial y_{\alpha\beta} \cdots \partial y_{\mu\nu}} \right| \leqq C_h (\sigma(Y))^{\kappa_h}$$

for $Y > 0$, $|Y| = 1$.

In order to construct a Größen-character, by means of a given eigenfunction $f(Y)$ of $L$ which is homogeneous of degree 0, one can proceed as follows. Determine the largest subgroup $\Sigma$ of $GL(n,\mathbb{R})$ such that $f(Y[U]) = f(Y)$ for $U \in \Sigma$, and introduce $\mathrm{A} = \Sigma \cap \Gamma$. If we set

$$u(Y) = \sum_{U:\Gamma/\mathrm{A}} f(Y[U]),$$

where $U$ runs through a complete set of representatives of the cosets of $\Gamma$ modulo A, i.e., $\Gamma = \bigcup_{U:\Gamma/\mathrm{A}} U\mathrm{A}$, then formally $u(Y)$ is invariant under $\Gamma$, and it remains essentially to prove the convergence of the series and to check the fifth of the above conditions.

Accordingly, we start with the function

$$f(Y) = |Y|^{\frac{1}{n}\sum_{\nu=1}^{r} k_\nu z_\nu} \prod_{\nu=1}^{r} \left| Y \begin{bmatrix} E^{(k_\nu)} \\ 0 \end{bmatrix} \right|^{-z_\nu}$$

where $k_1, k_2, \ldots, k_r$ are given integers such that

$$0 = k_0 < k_1 < \ldots < k_r < k_{r+1} = n.$$

We obtain $f(Y)$ from the function $f_s(Y)$ introduced in §6 by specializing the variables $s_1, s_2, \ldots, s_n$. We change the notation of §6 insofar as we use the variables $z_1, z_2, \ldots, z_r$ and also $s_1, s_2, \ldots, s_{r+1}$ here with a new meaning. We require the relations

$$z_\nu = s_{\nu+1} - s_\nu + \frac{h_{\nu+1} + h_\nu}{4}, \qquad h_\nu = k_\nu - k_{\nu-1}, \quad 1 \leqq \nu \leqq r,$$

and get $h_1 + h_2 + \ldots + h_{r+1} = n$ if $h_{r+1} = k_{r+1} - k_r$.

It is obvious that $f(Y)$ is homogeneous of degree 0 and invariant under

$$\Sigma = \{(C_{\mu\nu}) \in GL(n,\mathbb{R}) \mid C_{\mu\nu} = C_{\mu\nu}^{(h_\mu, h_\nu)}, \quad |C_{\nu\nu}|^2 = 1 \quad \text{for} \quad 1 \leqq \mu,\nu \leqq r+1,$$

$$C_{\mu\nu} = 0 \quad \text{for} \quad 1 \leqq \nu < \mu \leqq r+1\}.$$

We leave aside the question whether $\Sigma$ is maximal and define the group $A = \Sigma \cap \Gamma$ which is identical with

$$\Gamma_{h_1,\ldots,h_{r+1}} = \{(B_{\mu\nu}) \in \Gamma_n \mid B_{\mu\nu} = 0 \quad \text{for} \quad 1 \leqq \nu < \mu \leqq r+1,$$

$$B_{\nu\nu} \in \Gamma_{h_\nu} \quad \text{for} \quad 1 \leqq \nu \leqq r+1\}.$$

So we get

$$u(Y) = u_{k_1,\ldots,k_r}(Y) = |Y|^{\frac{1}{n}\sum_{\nu=1}^{r} k_\nu z_\nu} \sum_{U:\Gamma/\Gamma_{h_1,\ldots,h_{r+1}}} \prod_{\nu=1}^{r} |Y[U]_{k_\nu}|^{-z_\nu},$$

where in general $Y_k = Y\begin{bmatrix} E^{(k)} \\ 0 \end{bmatrix}$. In the special case $r = n-1$, $h_1 = h_2 = \ldots = h_n = 1$, $k_\nu = \nu$ $(0 \leqq \nu \leqq n)$, this function was introduced by Atle Selberg as the Eisenstein series of the group $\Gamma$.

*Assume that $u(E)$ converges for a given system of numbers $z_\nu > 0$ $1 \leqq \nu \leqq r$. Then there exists a positive constant $C$, depending only on $n, z_1, \ldots, z_r$ such that*

$$u(Y) < C\left(\frac{\sigma(Y)}{\sqrt[n]{|Y|}}\right)^{\kappa} u(E) \qquad \textit{for} \quad Y > 0$$

*with $\kappa = \sum_{\nu=1}^{r}(n-k_\nu)z_\nu$. The series for $\left(\frac{\sigma(Y)}{\sqrt[n]{|Y|}}\right)^{-\kappa} u(Y)$ converges uniformly in $\mathcal{R}$.*

*Proof*. 1. Assume that $Y$ belongs to $\mathcal{R}$ and denote by $c_1, c_2, \ldots$ positive constants which depend only on $n, z_1, \ldots, z_r$. If we set $Y = (y_{\mu\nu})$ and $Y_0 = (\delta_{\mu\nu} y_{\nu\nu})$, then by formula (8) of §9 we have $Y > c_1 Y_0$ and therefore $Y[U]_{k_\nu} > c_1 Y_0[U]_{k_\nu}$. Set $U = (\mathfrak{C}_1, \mathfrak{C}_2, \ldots, \mathfrak{C}_n)$

and, in order to estimate $|Y_0[U]_{k_\nu}|$, assume that $Y_0[U]_{k_\nu} \in \mathcal{R}_{k_\nu}$. It follows that

$$|Y[U]_{k_\nu}| > c_2|Y_0[U]_{k_\nu}| = c_2|Y_0[P_{k_\nu}]|$$

$$= c_2 \sum_{\alpha_1<\ldots<\alpha_{k_\nu}} \begin{pmatrix} \alpha_1 & \alpha_2 & \ldots & \alpha_{k_\nu} \\ 1 & 2 & \ldots & k_\nu \end{pmatrix}^2_{P_{k_\nu}} y_{\alpha_1} y_{\alpha_2} \ldots y_{\alpha_{k_\nu}}$$

$$\geqq c_2 y_1 y_2 \ldots y_{k_\nu} |P'_{k_\nu} P_{k_\nu}| \geqq c_2 |Y_{k_\nu}| |E[U]_{k_\nu}|,$$

where $y_1, y_2, \ldots, y_n$ are the diagonal elements of $Y$ and $P_k = (\mathfrak{e}_1, \mathfrak{e}_2, \ldots, \mathfrak{e}_k)$. Thus we obtain

$$u(Y) < C|Y|^{\frac{1}{n}\sum_{\nu=1}^{r} k_\nu z_\nu} \prod_{\nu=1}^{r} |Y_{k_\nu}|^{-z_\nu} u(E) = Cf_t(Y)u(E).$$

The parameters $t_1, t_2, \ldots, t_n$ of the function $f_t(Y)$ are uniquely determined by $(k_\nu, z_\nu)$, $1 \leqq \nu \leqq r$. They satisfy the relations

$$t_\nu - t_{\nu+1} - \frac{1}{2} = \begin{cases} -z_\mu & \text{if } \nu = k_\mu < n \\ \frac{1}{n}\sum_{\nu=1}^{r} k_\nu z_\nu & \text{if } \nu = n \\ 0 & \text{if } \nu \neq k_1, k_2, \ldots, k_{r+1} = n \end{cases}$$

$$t_{n+1} = -\frac{n+1}{4}.$$

At the beginning of §7 we proved that

$$f_t(Y) \leqq \left(\frac{\sigma(Y)}{\sqrt[n]{|Y|}}\right)^{\kappa}$$

with $\kappa = \max\limits_{1\leqq\nu\leqq n} \frac{n}{2}(\frac{n+1}{2} - \nu - 2t_\nu) = \sum\limits_{\nu=1}^{r} (n - k_\nu)z_\nu$. The asserted inequality is now proved for $Y \in \mathcal{R}$, as well as the uniform convergence in $\mathcal{R}$.

2. For an arbitrary $Y^* > 0$ we determine $V = (v_{\mu\nu}) \in \Gamma$ so that $Y = Y^*[V^{-1}] \in \mathcal{R}$ and set $Y_0 = (\delta_{\mu\nu}y_\nu)$. Then

$$\sigma(Y^*) = \sigma(Y[V]) > c_3\sigma(Y_0[V]) = c_3 \sum_{\mu=1}^{n} y_\mu \sum_{\nu=1}^{n} v_{\mu\nu}^2 \geqq c_3\sigma(Y).$$

Since $u(Y^*) = u(Y)$ and $|Y^*| = |Y|$, the asserted inequality is true for $Y^*$ because we proved it for $Y$.

*Theorem. Assume that* $\mathcal{Re}\, z_\nu > \frac{1}{2}(h_{\nu+1} + h_\nu) = \frac{1}{2}(k_{\nu+1} - k_{\nu-1})$ *for* $1 \leqq \nu \leqq r$ *and denote by*

$$u_{k_1\ldots k_r}(Y) = \left(\frac{\sigma(Y)}{\sqrt[n]{|Y|}}\right)^{\kappa} \sum_{U:\Gamma/\Gamma_{h_1,\ldots,h_{r+1}}} \chi_U(Y),$$

$$\kappa = \sum_{\nu=1}^{r} (n - k_\nu)\, \mathcal{Re}\, z_\nu ,$$

*a short form of the Eisenstein series*

$$u_{k_1\ldots k_r}(Y) = |Y|^{\frac{1}{n}\sum\limits_{\nu=1}^{r} k_\nu z_\nu} \sum_{U:\Gamma/\Gamma_{h_1,\ldots,h_{r+1}}} \prod_{\nu=1}^{r} |Y[U]_{k_\nu}|^{-z_\nu}.$$

*Then* $\sum_U |\chi_U(Y)|$ *is bounded in the whole region*

$$\mathcal{Y} = \{Y \mid Y = Y' > 0\}$$

*and converges uniformly in any region*

$$\{Y > \varepsilon(\delta_{\mu\nu} y_\nu) \mid \varepsilon y_{\nu-1} < y_\nu \quad \text{for} \quad 1 < \nu \leqq n\} \qquad (\varepsilon > 0).$$

There is no loss of generality if we prove the theorem for real variables $z_1, z_2, \ldots, z_r$. In view of the preceding assertion, we only need to show that $u(E)$ converges, because the uniform convergence of $\sum_U \chi_U(Y)$ in the regions described above follows then in the same way as for $\mathcal{R}$. The following lemmas will be useful.

*Lemma 1. Denote by* $\rho(Y, Y^*)$ *the distance of two points* $Y, Y^* \in \mathcal{Y}$ *measured in the Riemannian metric (§6). Then*

$$\rho(Y_h, Y_h^*) \leqq \rho(Y, Y^*) \qquad \text{for} \quad 1 \leqq h \leqq n.$$

*Proof.* In §3 we showed that for the metric fundamental forms of the domains $\mathcal{Y}_n = \{Y > 0\}$ and $\mathcal{Y}_h = \{Y_h = Y \begin{bmatrix} E^{(h)} \\ 0 \end{bmatrix} \mid Y \in \mathcal{Y}_n\}$ the estimate

$$\sigma(Y_h^{-1} dY_h)^2 \leqq \sigma(Y^{-1} dY)^2$$

holds. From here the assertion is obvious.

*Lemma 2. Let* $\phi(y)$ *denote a complex-valued function which is continuous in* $0 < y < \infty$. *Then*

$$\int\limits_{\substack{\mathcal{R}_n \\ y_1 < |Y| < y_2}} \phi(|Y|) |Y|^{-\frac{1}{2}(n+1)} [dY] = \frac{n+1}{2} v_n \int_{y_1}^{y_2} \phi(y) y^{-1} dy \qquad (y_1 > 0),$$

*where*

$$\int\limits_{\substack{\mathcal{R}_n \\ |Y|<1}} [dY] = v_n < \infty.$$

*Proof*. The inequality $v_n < \infty$ is an immediate consequence of the inequalities of reduction theory; we shall take it for granted. Then

$$\int\limits_{\substack{\mathcal{R}_n \\ |Y|<y}} [dY] = v_n y^{\frac{1}{2}(n+1)}$$

follows by a change of variables. In order to find the derivative of

$$f(y) = \int\limits_{\substack{\mathcal{R} \\ y_1<|Y|<y}} \phi(|Y|)|Y|^{-\frac{1}{2}(n+1)} [dY]$$

we form

$$\frac{f(y+h) - f(y)}{h} = \phi(y + \theta h)(y + \theta h)^{-\frac{1}{2}(n+1)} \frac{1}{h} \int\limits_{\substack{\mathcal{R} \\ y<|Y|<y+h}} [dY]$$

$$= \phi(y + \theta h)(y + \theta h)^{-\frac{1}{2}(n+1)} v_n \frac{(y+h)^{\frac{1}{2}(n+1)} - y^{\frac{1}{2}(n+1)}}{h} \quad (h > 0)$$

and obtain

$$f'(y) = \frac{n+1}{2} v_n \phi(y) y^{-1}.$$

The integral relation of Lemma 2 now follows.

A first application yields

$$\int\limits_{\substack{\mathcal{R} \\ y_1<|Y|<y_2}} |Y|^{s} dv = \frac{n+1}{2s} v_n (y_2^s - y_1^s) \qquad (s \neq 0),$$

(1)

$$\int\limits_{\substack{\mathcal{R} \\ |Y|>y}} |Y|^{-s} dv = \frac{n+1}{2s} v_n y^{-s} \qquad (s > 0),$$

where $dv = |Y|^{-\frac{1}{2}(n+1)}[dY]$ denotes the invariant volume element of $\mathcal{Y}$.

Now we can prove the convergence of $u(E)$. First we introduce the ball

$$\mathcal{L}(Y^*) = \{Y \in \mathcal{Y} \mid \rho(Y,Y^*) \leqq \rho_o\},$$

where $\rho_o$ is a positive number. Clearly $Y \in \mathcal{L}(Y^*)$ implies $Y^*[\sqrt{Y}^{-1}] \in \mathcal{L}(E)$, so that $\frac{1}{c_1} < |Y^*[\sqrt{Y}^{-1}]| < c_1$ or

$$\frac{1}{c_1} < \frac{|Y^*|}{|Y|} < c_1 \qquad \text{for } Y \in \mathcal{L}(Y^*).$$

Here $c_1$, and later $c_2, c_3, \ldots,$ denote positive constants which depend on $n$, $\rho_o$ and possibly $z_1, z_2, \ldots, z_r$. It follows from Lemma 1 that $Y \in \mathcal{L}(Y^*)$ implies $Y[U] \in \mathcal{L}(Y^*[U])$ and $Y[U]_k \in \mathcal{L}(Y^*[U]_k)$. Hence

$$\frac{1}{c_1} < \frac{|Y^*[U]_k|}{|Y[U]_k|} < c_1 \qquad \text{for } Y \in \mathcal{L}(Y^*), \quad 1 \leqq k \leqq n,$$

thus

$$u(E) < c_2 u(Y) \qquad \text{for } Y \in \mathcal{L}(E)$$

and

$$u(E) \int_{\mathcal{L}(E)} dv < c_2 \int_{\mathcal{L}(E)} u(Y)dv$$

$$= c_2 \sum_{U:\Gamma/\Gamma_{h_1,\ldots,h_{r+1}}} \int_{\mathcal{L}(E)} \prod_{\nu=1}^{r+1} |Y[U]_{k_\nu}|^{-z_\nu} dv$$

$$= c_2 \sum_{U:\Gamma/\Gamma_{h_1,\ldots,h_{r+1}}} \int_{\mathcal{L}(U'U)} \prod_{\nu=1}^{r+1} |Y_{k_\nu}|^{-z_\nu} dv,$$

where $z_{r+1}$ is defined by

$$\sum_{\nu=1}^{r+1} k_\nu z_\nu = 0.$$

We observe that the integrand is invariant under $\Gamma_{h_1,h_2,\ldots,h_{r+1}}$, so that the domain of integration $\mathcal{L}(U'U)$ or parts of it can be replaced by point sets which are equivalent with respect to $\Gamma_{h_1,h_2,\ldots,h_{r+1}}$.

We have seen that

$$\frac{1}{c_1} < \frac{|Y_k|}{|E[U]_k|} < c_1 \qquad \text{for} \quad Y \in \mathcal{L}(U'U).$$

This implies $\frac{1}{c_1} < |Y| < c_1$ and

$$|Y_k| > \frac{1}{c_1}|E[U]_k| \geq \frac{1}{c_1} \qquad \text{for all } k,$$

hence we may assert that

$$(2) \qquad \mathcal{L}(U'U) \subset \{Y \mid Y > 0,\ \frac{1}{c_1} < |Y| < c_1,\ |Y_{k_\nu}| > \frac{1}{c_1} \quad (1 \leq \nu \leq r)\}$$

for $U \in \Gamma$. We introduce new coordinates for $Y$ by setting

$$Y = (\delta_{\mu\nu} Q_\nu)[(R_{\mu\nu})] \qquad (\mu,\nu = 1,2,\ldots,r+1),$$

$$Q_\nu = Q_\nu^{(h_\nu)} > 0, \quad R_{\mu\nu} = R_{\mu\nu}^{(h_\mu,h_\nu)}, \quad R_{\nu\nu} = E^{(h_\nu)}, \quad R_{\mu\nu} = 0 \quad \text{for} \quad \mu > \nu.$$

In the special case when $r = n-1$, i.e., $h_\nu = 1$ for $1 \leqq \nu \leqq n$, the matrices $Q_\nu$, $R_{\mu\nu}$ $(\mu < \nu)$ are the Jacobian coordinates of $Y$. We write

$$dv_Y = dv = |Y|^{-\frac{1}{2}(n+1)}[dY]$$

and define

$$\mathfrak{N}_{\alpha\beta} = \{R_{\alpha\beta} = (r_{\mu\nu}^{\alpha\beta}) \mid 0 \leqq r_{\mu\nu}^{\alpha\beta} \leqq 1 \quad \text{for} \quad \mu = 1,2,\ldots,h_\alpha; \quad \nu = 1,2,\ldots,h_\beta\}.$$

In the special case $r = 1$ we have

$$Y = \begin{pmatrix} Q_1 & 0 \\ 0 & Q_2 \end{pmatrix}\begin{bmatrix} E & R_{12} \\ 0 & E \end{bmatrix}$$

and

$$[dY] = |Q_1|^{h_2}[dQ_1][dQ_2][dR_{12}],$$

as we have seen in § 3. This yields

$$dv_Y = |Q_1|^{\frac{1}{2}h_2}|Q_2|^{-\frac{1}{2}h_1} dv_{Q_1} dv_{Q_2}[dR_{12}].$$

The effect of

$$U = \begin{pmatrix} U_1 & U_{12} \\ 0 & U_2 \end{pmatrix} \in \Gamma_{h_1,h_2}$$

on $Q_1$, $Q_2$, $R_{12}$, corresponding to the action $Y \mapsto Y[U]$, is given by

$$Q_1 \mapsto Q_1[U_1], \quad Q_2 \mapsto Q_2[U_2], \quad R_{12} \mapsto U_1^{-1}(U_{12} + R_{12}U_2),$$

and it can be easily verified that

$$\mathcal{F}_1 = \{Y \mid Q_1 \in \mathcal{R}_{h_1}, \ Q_2 \in \mathcal{R}_{h_2}, \ R_{12} \in \mathcal{U}_{12}\}$$

is a fundamental domain of $\Gamma_{h_1,h_2}$ in $\mathcal{Y}$.

For an arbitrary $r \geqq 1$, using the relations

$$Y = Y_{k_{r+1}} = \begin{pmatrix} Y_{k_r} & 0 \\ 0 & Q_{r+1} \end{pmatrix} \begin{bmatrix} E & R \\ 0 & E \end{bmatrix}, \qquad R = \begin{pmatrix} R_{1,r+1} \\ \vdots \\ R_{r,r+1} \end{pmatrix},$$

and the above statements, we can prove by induction on $r$ that

$$dv_Y = \prod_{\nu=1}^{r+1} \{|Q_\nu|^{\frac{1}{2}(n - k_\nu - k_{\nu-1})} dv_{Q_\nu}\} \prod_{1 \leqq \mu < \nu \leqq r+1} [dR_{\mu\nu}]$$

and that

$$\mathcal{F}_r = \{Y \mid Q_\nu \in \mathcal{R}_{h_\nu} \ (1 \leqq \nu \leqq r+1), \quad R_{\mu\nu} \in \mathcal{U}_{\mu\nu} \ (1 \leqq \mu < \nu \leqq r+1)\}$$

is a fundamental domain of $\Gamma_{h_1,h_2,\ldots,h_{r+1}}$ in $\mathcal{Y}$.

Because of $|Y_{k_\nu}| = \prod_{\mu=1}^{\nu} |Q_\mu| \quad (1 \leqq \nu \leqq r+1)$, the inclusion (2) can be rewritten as

$$(3) \qquad \mathfrak{L}(U'U) \subset \{Y \mid Q_\nu > 0, \ \prod_{\mu=1}^{\nu} |Q_\mu| > \frac{1}{c_1} \ (1 \leqq \nu \leqq r+1), \quad \prod_{\mu=1}^{r+1} |Q_\mu| < c_1\}$$

for $U \in \Gamma$. We denote by $U_\mu$ $(\mu = 1,2,\ldots)$ a fixed complete set of representatives of the cosets of $\Gamma$ modulo $\Gamma_{h_1,h_2,\ldots,h_{r+1}}$. Let us intersect a given ball $\mathcal{L}(U_\mu' U_\mu)$ with the images of $\mathcal{F}_r$ under $\Gamma_{h_1,h_2,\ldots,h_{r+1}}$ and choose a minimal system of matrices $V_{\mu\nu} \in \Gamma_{h_1,h_2,\ldots,h_{r+1}}$ $(1 \leqq \nu < \sigma_\mu)$ such that

$$\mathcal{O}_{\mu\nu} = \mathcal{L}(U_\mu' U_\mu) \cap \mathcal{F}_r[V_{\mu\nu}] \neq \emptyset \quad \text{and} \quad \mathcal{L}(U_\mu' U_\mu) = \bigcup_{\nu<\sigma_\mu} \mathcal{O}_{\mu\nu} .$$

The region on the right-hand side of inclusion (3) is invariant under $\Gamma_{h_1,h_2,\ldots,h_{r+1}}$, so that $\mathcal{O}_{\mu\nu}[V_{\mu\nu}^{-1}] \subset \vartheta$, where

$$\vartheta = \{Y \mid Q_\nu \in \mathcal{R}_{h_\nu}, \ \prod_{\mu=1}^{\nu} |Q_\mu| > \frac{1}{c_1} \ (1 \leqq \nu \leqq r+1),$$

$$\prod_{\mu=1}^{r+1} |Q_\mu| < c_1, \ R_{\mu\nu} \in \mathcal{Q}_{\mu\nu} \ (1 \leqq \mu < \nu \leqq r+1)\}.$$

Our last estimate of $u(E)$ leads now to

$$u(E) < c_3 \sum_{\mu=1}^{\infty} \sum_{\kappa<\sigma_\mu} \int_{\mathcal{O}_{\mu\kappa}[V_{\mu\kappa}^{-1}]} \prod_{\nu=1}^{r+1} |Y_{k_\nu}|^{-s_\nu} dv$$

$$< c_3 N \int_{\vartheta} \prod_{\nu=1}^{r+1} |Y_{k_\nu}|^{-s_\nu} dv$$

$$= c_3 N \int \prod_{\mu=1}^{r+1} \{|Q_\mu|^{s_\mu - s_{r+1} - s_{r+1} + \frac{1}{2}(h_{r+1} - k_\mu - k_{\mu-1})} dv_{Q_\mu}\} \prod_{1 \leqq \mu<\nu \leqq r+1} [dR_{\mu\nu}],$$

where $N$ is a natural number such that a given point $Y$ is contained in at most $N$ different sets $\mathfrak{A}_{\mu\nu}[V_{\mu\nu}^{-1}]$. We still have to prove that such an $N$ exists. Assuming this, let us finish the proof of the finiteness of $u(E)$. One sees immediately that

$$u(E) < c_3 N \int_{\vartheta_1} \prod_{\mu=1}^{r+1} \{|Q_\mu|^{s_\mu - s_{r+1} - z_{r+1} + \frac{1}{2}(h_{r+1} - k_\mu - k_{\mu-1})} dv_{Q_\mu}\},$$

where

$$\vartheta_1 = \{(Q_1,\ldots,Q_{r+1}) \mid Q_\nu \in \mathcal{R}_{h_\nu},\ \prod_{\mu=1}^{\nu} |Q_\mu| > \frac{1}{c_1} \quad (1 \leqq \nu \leqq r+1),$$

$$\prod_{\mu=1}^{r+1} |Q_\mu| < c_1\}.$$

The integration can be carried out explicitly with the help of formulas (1), if we first integrate with respect to $Q_{r+1}$ and then proceed with $Q_r, Q_{r-1},\ldots,Q_1$, in this order. The existence of the integrals amounts to the assumption

$$z_\nu > \frac{h_{\nu+1} + h_\nu}{2} \qquad \text{for} \quad 1 \leqq \nu \leqq r,$$

and we get

$$u(E) < c_4 N < \infty.$$

To conclude, we prove the existence of $N$. Assume that a given $Y$ belongs to both $\mathfrak{A}_{\mu\nu}[V_{\mu\nu}^{-1}]$ and $\mathfrak{A}_{\kappa\lambda}[V_{\kappa\lambda}^{-1}]$. Then

$$\mathcal{L}(E[U_\mu V_{\mu\nu}^{-1}]) \cap \mathcal{L}(E[U_\kappa V_{\kappa\lambda}^{-1}]) \supset \mathfrak{A}_{\mu\nu}[V_{\mu\nu}^{-1}] \cap \mathfrak{A}_{\kappa\lambda}[V_{\kappa\lambda}^{-1}] \neq \emptyset,$$

hence

$$\rho(E[U_\mu V_{\mu\nu}^{-1}], E[U_\kappa V_{\kappa\lambda}^{-1}]) \leqq 2\rho_0$$

or

$$\rho(E,E[W]) \leqq 2\rho_0 \quad \text{with} \quad WU_\kappa V_{\kappa\lambda}^{-1} = U_\mu V_{\mu\nu}^{-1}.$$

We consider $\kappa$ and $\lambda$ as fixed. Then $U_\mu$, $V_{\mu\nu}$ are uniquely determined by $W$ because the $U_\mu$'s form a set of representatives of the cosets of $\Gamma$ modulo $\Gamma_{h_1,h_2,\ldots,h_{r+1}}$. Thus we can take for $N$ the number of $W \in \Gamma$ which satisfy the last inequality. This number is finite because $\Gamma$ is discontinuous.

On the basis of our results concerning convergence and those obtained in §7, we prove that $u_{k_1\ldots k_r}(Y)$ is an eigenfunction of the integral operator $J_s(X,u)$ discussed in §7, provided that $\mathcal{R}e\, s$ is sufficiently large. Indeed,

$$\int_{Y>0} e^{-\sigma(YX^{-1})}|Y|^s u_{k_1\ldots k_r}(Y)\,dv =$$

$$= \sum_{U:\Gamma/\Gamma_{h_1,h_2,\ldots,h_{r+1}}} \int_{Y>0} e^{-\sigma(YX^{-1})}|Y|^s \left(\frac{\sigma(Y)}{\sqrt[n]{|Y|}}\right)^{\kappa} \chi_U(Y)\,dv$$

$$= \sum_{U:\Gamma/\Gamma_{h_1,h_2,\ldots,h_{r+1}}} \int_{Y>0} e^{-\sigma(YX^{-1})}|Y|^s \prod_{\nu=1}^{r+1} |Y[U]_{k_\nu}|^{-s_\nu}\,dv$$

$$= \pi^{\frac{1}{4}n(n-1)} \prod_{\nu=1}^{n} \Gamma(s-\alpha_\nu) \sum_{U:\Gamma/\Gamma_{h_1,h_2,\ldots,h_{r+1}}} |X|^s \prod_{\nu=1}^{r+1} |X[U]_{k_\nu}|^{-s_\nu}$$

$$= \pi^{\frac{1}{4}n(n-1)} \prod_{\nu=1}^{n} \Gamma(s-\alpha_\nu)|X|^s u_{k_1\ldots k_r}(X).$$

As a solution of this integral equation, $u_{k_1 \dots k_r}(Y)$ belongs to $C^{\infty}(\mathcal{Y})$ and satisfies the fifth of the defining properties of Größen-characters. The partial derivatives of $u_{k_1 \dots k_r}(Y)$ can be found by differentiating the series termwise. This implies that $u_{k_1 \dots k_r}$ is an eigenfunction of the ring $L$ of invariant differential operators. So, after all, the function $u_{k_1 \dots k_r}(Y)$ turns out to be a Größen-character, provided that the conditions for convergence are fulfilled.

## §11. *The modular group of degree n*

We recall the notation introduced in §4, according to which $I$ is the square matrix of order $2n$

$$I = \begin{pmatrix} 0 & E \\ -E & 0 \end{pmatrix}$$

and $\Omega = \Omega(I)$ is the symplectic group of degree $n$. The matrices

$$M = \begin{pmatrix} A & B \\ C & D \end{pmatrix} \in \Omega$$

with entries in $\mathbb{Z}$ form a subgroup $\Gamma = \Gamma_n$ of $\Omega$ which is called the *modular group* of degree $n$. The group $\Gamma$ is discrete and therefore it operates discontinuously on the Siegel upper half-plane

$$\mathfrak{H} = \{Z \mid Z = Z^{(n)} = Z' = X + iY, \quad Y > 0\}.$$

First of all, we want to characterize the pairs of matrices $C = C^{(n)}$, $D = D^{(n)}$ which occur as the second matrix row of a matrix $M \in \Gamma$. A necessary condition is that $C, D$ be a *symmetric pair*, i.e.,

$$CD' = DC'.$$

Moreover, the pair $C, D$ must be *coprime* in the following sense: if $GC$ and $GD$ are both matrices with integral entries, then so is $G$. Indeed, if $GC$ and $GD$ are integral matrices, then $(GC, GD)$ is an integral matrix. Now $M^{-1} = -IM'I$ is also an integral matrix and $(GC, GD)\cdot M^{-1} = (0, G) = G\cdot(0, E)$.

We prove that conversely *any coprime symmetric pair* $(C, D)$ *is the second matrix row of a matrix* $M \in \Gamma$. The pair $(C,D)$ is coprime if and only if for any pair $U_1 = U_1^{(n)}$, $U_2 = U_2^{(2n)}$ of unimodular matrices the pair $(C_1,D_1)$ defined by

$$(C_1,D_1) = U_1(C,D)U_2$$

is coprime. By elementary divisor theory we can choose the matrices $U_1$ and $U_2$ so that $D_1 = 0$ and $C_1$ is a diagonal matrix with non-negative elements along the diagonal. If we assume that $C,D$ is coprime, then necessarily $C_1 = E$ since otherwise there exists a non-integral $G$ such that $GC_1$ is integral. Thus we have

$$(U_1^{-1},0) = (C,D)U_2$$

and therefore $\text{rank}(C,D) = n$. Defining the integral matrices $X,Y$ by

$$U_2\begin{pmatrix} U_1 \\ 0 \end{pmatrix} = \begin{pmatrix} X \\ Y \end{pmatrix}$$

we get

$$CX + DY = E.$$

By assumption $CD' = DC'$, therefore the matrices

$$A = Y' + X'YC, \qquad B = -X' + X'YD$$

satisfy

$$\begin{aligned} AB' - BA' &= (Y' + X'YC)(-X + D'Y'X) - (-X' + X'YD)(Y + C'Y'X) \\ &= -Y'X + X'Y - X'Y(CX + DY) + (Y'D' + X'C')Y'X = 0, \\ AD' - BC' &= (Y' + X'YC)D' - (-X' + X'YD)C' = E, \end{aligned}$$

i.e.,

$$M = \begin{pmatrix} A & B \\ C & D \end{pmatrix} \in \Gamma,$$ Q.E.D.

We introduce the subgroup

$$\mathrm{T} = \left\{ \begin{pmatrix} E & S \\ 0 & E \end{pmatrix} \mid S = S' \text{ integral} \right\}$$

of $\Gamma$. For two matrices

$$M_\nu = \begin{pmatrix} A_\nu & B_\nu \\ C_\nu & D_\nu \end{pmatrix} \in \Gamma \qquad (\nu = 1,2)$$

we have $\mathrm{T}M_1 = \mathrm{T}M_2$ if and only if $C_1 = C_2$, $D_1 = D_2$, i.e., the $M \in \Gamma$ with a given second matrix row $C,D$ form a right coset of $\Gamma$ modulo T.

We call two coprime symmetric pairs $(C_1,D_1)$ and $(C_2,D_2)$ *associated* if $C_1 D_2' = D_1 C_2'$. If these pairs are the second matrix rows of the matrices $M_1, M_2 \in \Gamma$, respectively, then it follows from

$$M_1 M_2^{-1} = \begin{pmatrix} A_1 & B_1 \\ C_1 & D_1 \end{pmatrix} \begin{pmatrix} D_2' & -B_2' \\ -C_2' & A_2' \end{pmatrix} = \begin{pmatrix} * & * \\ C_1 D_2' - D_1 C_2' & * \end{pmatrix}$$

that $M_1$ and $M_2$ have associated second matrix rows if and only if

$$\mathrm{A}M_1 = \mathrm{A}M_2,$$

where

$$\mathrm{A} = \left\{ \begin{pmatrix} U' & SU^{-1} \\ 0 & U^{-1} \end{pmatrix} \mid U \text{ unimodular}, \quad S = S' \text{ integral} \right\}.$$

This group A is a subgroup of $\Gamma$ which contains T as a normal subgroup.

In particular $(C_1, D_1)$ and $(C_2, D_2)$ are associated if and only if there exists a unimodular matrix $U$ such that

$$(C_2, D_2) = U(C_1, D_1).$$

We denote by $\{C, D\}$ the equivalence class of all coprime symmetric pairs which are associated with $(C, D)$. We want to determine a special representative in each class $\{C, D\}$. Assume that rank $C = r$, $0 \leqq r \leqq n$. If $r = 0$, i.e., $C = 0$, then $D$ is necessarily unimodular and we choose $(0, E)$ as a representative. Thus it is sufficient to discuss the cases $0 < r \leqq n$. Determine two unimodular matrices $U_1$ and $U_2$ so that

$$U_1 C = \begin{pmatrix} C_1 & 0 \\ 0 & 0 \end{pmatrix} U_2', \quad \text{where } C_1 = C_1^{(r)}, \quad |C_1| \neq 0.$$

If we set with an analogous block decomposition

$$U_1 D = \begin{pmatrix} D_1 & D_2 \\ D_3 & D_4 \end{pmatrix} U_2^{-1}, \qquad D_1 = D_1^{(r)},$$

then the symmetry $CD' = DC'$ of $(C, D)$ yields that of $(U_1 C, U_1 D)$, i.e.,

$$\begin{pmatrix} C_1 & 0 \\ 0 & 0 \end{pmatrix} \begin{pmatrix} D_1' & D_3' \\ D_2' & D_4' \end{pmatrix} = \begin{pmatrix} D_1 & D_2 \\ D_3 & D_4 \end{pmatrix} \begin{pmatrix} C_1' & 0 \\ 0 & 0 \end{pmatrix}.$$

It follows that

$$C_1 D_1' = D_1 C_1' \quad \text{and} \quad D_3 = 0,$$

in particular the pair $(C_1, D_1)$ is symmetric.

Next we show that $D_4$ is unimodular. For this, it is sufficient to show that if $G = G^{(s, n-r)}$ is such that $GD_4$ is integral, then $G$ is integral. We complete $G$ by 0-columns to a matrix $(0, G)$ with $n$ columns.

Then $(0,G)U_1C = 0$ and $(0,G)U_1D = (0,GD_4)U_2^{-1}$ are integral. But $C$ and $D$ are coprime, therefore $(0,G)U_1$ is integral, and since $U_1$ is unimodular, also $G$ is integral.

If we replace $U_1$ by

$$\begin{pmatrix} E & D_2 \\ 0 & D_4 \end{pmatrix} U_1$$

we obtain

$$U_1C = \begin{pmatrix} C_1 & 0 \\ 0 & 0 \end{pmatrix} U_2', \qquad U_1D = \begin{pmatrix} D_1 & 0 \\ 0 & E \end{pmatrix} U_2^{-1}$$

because

$$\begin{pmatrix} E & D_2 \\ 0 & D_4 \end{pmatrix}\begin{pmatrix} D_1 & 0 \\ 0 & E \end{pmatrix} = \begin{pmatrix} D_1 & D_2 \\ 0 & D_4 \end{pmatrix}.$$

Also the pair $(C_1,D_1)$ is coprime. Indeed, if $GC_1$ and $GD_1$ are integral, then, completing $G = G^{(s,r)}$ to a matrix $(G,0)$ with $n$ columns, we obtain that $(G,0)U_1C = (GC_1,0)U_2'$ and $(G,0)U_1D = (GD_1,0)U_2^{-1}$ are integral. Since $C$ and $D$ are coprime, it follows that $(G,0)U_1$ is integral and thus also $G$.

Let $Q = Q^{(n,r)}$ be the matrix formed by the first $r$ columns of $U_2$. If we replace $U_2$ by the matrix

$$U_2\begin{pmatrix} U_3 & 0 \\ 0 & E \end{pmatrix},$$

where $U_3 = U_3^{(r)}$ is unimodular, then the form of the above equations remains unchanged and $C_1$, $D_1$, $Q$ are replaced by $C_1U_3'$, $D_1U_3^{-1}$, $QU_3$ respectively. We introduce the equivalence class

$$\{Q\} = \{QU_3 \mid U_3 = U_3^{(r)} \text{ unimodular}\}$$

of primitive matrices, i.e., matrices $Q = Q^{(n,r)}$ which can be completed with $n-r$ columns to a unimodular matrix $(Q,*)$.

Finally, we multiply $U_1$ from the left by

$$\begin{pmatrix} U_4 & 0 \\ 0 & E \end{pmatrix},$$

where $U_4 = U_4^{(r)}$ is again unimodular. This replaces the pair $(C_1,D_1)$ by $(U_4C_1,U_4D_1)$, i.e., by an associated pair. Thus $C_1,D_1$ can be chosen as any given representative of the class $\{C_1,D_1\}$.

Thus we have associated with each class $\{C,D\}$ verifying rank $C = r$ the equivalence classes $\{C_1^{(r)},D_1^{(r)}\}$, with $|C_1| \neq 0$, and $\{Q^{(n,r)}\}$, such that for appropriate representatives $C_1,D_1$ and $Q$ we have

$$C = \begin{pmatrix} C_1 & 0 \\ 0 & 0 \end{pmatrix} U_2', \quad D = \begin{pmatrix} D_1 & 0 \\ 0 & E \end{pmatrix} U_2^{-1}, \quad U_2 = (Q,R).$$

The following lemma states that this correspondence is one-to-one:

*Lemma. Let* $Q = Q^{(n,r)}$ *run through a complete set of representatives of the classes* $\{Q\}$ *of primitive matrices, and let* $C_1 = C_1^{(r)}$, $D_1 = D_1^{(r)}$ *run over a complete set of representatives of the classes* $\{C_1,D_1\}$, *where* $C_1,D_1$ *are coprime symmetric pairs with* $|C_1| \neq 0$. *Complete each* $Q$ *in exactly one way to a unimodular matrix* $U_2 = U_2^{(n)} = (Q,*)$. *Then the pairs* $C,D$ *given by*

$$C = \begin{pmatrix} C_1 & 0 \\ 0 & 0 \end{pmatrix} U_2', \quad D = \begin{pmatrix} D_1 & 0 \\ 0 & E \end{pmatrix} U_2^{-1}$$

*are a complete set of representatives of the classes* $\{C,D\}$, *where* $C = C^{(n)}$, $D = D^{(n)}$ *are coprime symmetric pairs with* rank $C = r$.

*Proof.* We must only show that the different pairs $C,D$ so obtained are not associated. Assume that

$$C^* = \begin{pmatrix} C_1^* & 0 \\ 0 & 0 \end{pmatrix} U_2^{*\prime}, \qquad D^* = \begin{pmatrix} D_1^* & 0 \\ 0 & E \end{pmatrix} U_2^{*-1}$$

is associated with $C,D$:

$$C^*D' = D^*C'.$$

This means that

$$\begin{pmatrix} C_1^* & 0 \\ 0 & 0 \end{pmatrix} U_2^{*\prime} U_2^{\prime -1} \begin{pmatrix} D_1' & 0 \\ 0 & E \end{pmatrix} = \begin{pmatrix} D_1^* & 0 \\ 0 & E \end{pmatrix} U_2^{*-1} U_2 \begin{pmatrix} C_1' & 0 \\ 0 & 0 \end{pmatrix}.$$

With the decomposed matrices

$$U_2^{*\prime} U_2^{\prime -1} = \begin{pmatrix} V_1 & V_2 \\ V_3 & V_4 \end{pmatrix}, \qquad U_2^{*-1} U_2 = \begin{pmatrix} W_1 & W_2 \\ W_3 & W_4 \end{pmatrix}$$

we get

$$\begin{pmatrix} C_1^* V_1 D_1' & C_1^* V_2 \\ 0 & 0 \end{pmatrix} = \begin{pmatrix} D_1^* W_1 C_1' & 0 \\ W_3 C_1' & 0 \end{pmatrix}$$

and therefore $V_2 = 0$, $W_3 = 0$. From

$$U_2 = U_2^* \begin{pmatrix} W_1 & W_2 \\ 0 & W_4 \end{pmatrix}$$

it follows that $W_1$ is unimodular. Write

$$U_2 = (Q,R), \qquad U_2^* = (Q^*,R^*),$$

then $Q = Q^* W_1$, i.e., $\{Q\} = \{Q^*\}$, hence $Q = Q^*$ and $W_1 = E$.

Since

$$\begin{pmatrix} V_1 & 0 \\ V_3 & V_4 \end{pmatrix} \begin{pmatrix} W_1' & 0 \\ W_2' & W_4' \end{pmatrix} = U_2^{*\prime} U_2^{\prime -1} U_2' U_2^{*\prime -1} = E,$$

we have $V_1 W_1' = E$ and thus $V_1 = E$. It follows therefore from the above equation that $C_1^* D_1' = D_1^* C_1'$, which means that $\{C_1^*, D_1^*\} = \{C_1, D_1\}$, i.e., $C_1^* = C_1$ and $D_1^* = D_1$, q.e.d.

We also see from the above proof that the class $\{C, D\}$ does not depend on the choice of $R$, since if $C_1 = C_1^*$, $D_1 = D_1^*$, $Q = Q_1^*$, then

$$C^* D' - D^* C' = \begin{pmatrix} C_1^* D_1' - D_1^* C_1' & 0 \\ 0 & 0 \end{pmatrix} = 0.$$

*Remarks.* 1) It is well known [36] that the modular group $\Gamma_n$ can be generated by the matrices

$$\begin{pmatrix} E & S \\ 0 & E \end{pmatrix}, \quad \begin{pmatrix} U' & 0 \\ 0 & U^{-1} \end{pmatrix}, \quad \begin{pmatrix} 0 & E \\ -E & 0 \end{pmatrix},$$

where $S = S'$ is an arbitrary symmetric integral matrix and $U = U^{(n)}$ is an arbitrary unimodular matrix. We want to point out here that the matrices

$$\begin{pmatrix} U' & 0 \\ 0 & U^{-1} \end{pmatrix}$$

are not necessary to generate $\Gamma_n$. In the first place, the group $\{U^{(2)}\}$ can be generated by

$$\begin{pmatrix} 1 & 0 \\ 0 & -1 \end{pmatrix}, \quad \begin{pmatrix} 0 & 1 \\ 1 & 0 \end{pmatrix}, \quad \begin{pmatrix} 1 & 1 \\ 0 & 1 \end{pmatrix}.$$

Now

$$\begin{pmatrix} 5 & 3 \\ 3 & 2 \end{pmatrix}\begin{pmatrix} 0 & 1 \\ 1 & 0 \end{pmatrix}\begin{pmatrix} -3 & 2 \\ 2 & -1 \end{pmatrix} = \begin{pmatrix} 1 & 1 \\ 0 & 1 \end{pmatrix},$$

and thus $\{U^{(2)}\}$ has a finite set of symmetric generators. From here it follows easily, considering matrices of the form

$$\begin{array}{c} \\ \\ \\ i) \\ \\ j) \\ \\ \end{array}\begin{pmatrix} 1 & & & \vdots & & \vdots & & \\ & \ddots & & \vdots & & \vdots & & \\ & & 1 & \vdots & & \vdots & & \\ \cdots & \cdots & \cdots & * & \cdots & * & \cdots & \cdots \\ & & & \vdots & 1 \ddots & \vdots & & \\ \cdots & \cdots & \cdots & * & \cdots & * & \cdots & \cdots \\ & & & \vdots & & \vdots & 1 & \\ & & & \vdots & & \vdots & & \ddots \end{pmatrix},$$

(with the starred entries in columns $i$ and $j$)

that for $n \geqq 2$ the group $\{U^{(n)}\}$ also has a finite set of symmetric generators. Thus it is sufficient to generate

$$\begin{pmatrix} U' & 0 \\ 0 & U^{-1} \end{pmatrix}$$

with a symmetric unimodular $U$ by matrices of the type

$$\begin{pmatrix} E & S \\ 0 & E \end{pmatrix} \quad \text{and} \quad \begin{pmatrix} 0 & E \\ -E & 0 \end{pmatrix}.$$

But if $U = U'$ then we can use the identity

$$\begin{pmatrix} E & U \\ 0 & E \end{pmatrix}\begin{pmatrix} 0 & E \\ -E & 0 \end{pmatrix}\begin{pmatrix} E & U^{-1} \\ 0 & E \end{pmatrix}\begin{pmatrix} 0 & E \\ -E & 0 \end{pmatrix}\begin{pmatrix} E & U \\ 0 & E \end{pmatrix}\begin{pmatrix} 0 & E \\ -E & 0 \end{pmatrix} = \begin{pmatrix} U & 0 \\ 0 & U^{-1} \end{pmatrix}.$$

2) We state without proof that the index of the commutator subgroup $\Gamma_n^*$ of $\Gamma_n$ has the following values:

$$(\Gamma_n : \Gamma_n^*) = \begin{cases} 12 & \text{for } n = 1, \\ 2 & \text{for } n = 2, \\ 1 & \text{for } n > 2. \end{cases}$$

In particular for $n > 2$ there is no non-trivial abelian character on $\Gamma_n$.

## §12. *The fundamental domain of the modular group*

In order to determine a fundamental domain of the modular group $\Gamma$ in Siegel's upper half-plane $\mathfrak{H}$, we consider the equivalence classes $\{C,D\}$ of coprime symmetric pairs of matrices $C,D$. We prove that for a given $Z \in \mathfrak{H}$ and a given number $m > 0$ only a finite number of classes $\{C,D\}$ satisfy

$$\| CZ + D \| < m.$$

We may assume that rank $C = r$. Then we have a representation

$$C = U_1 \begin{pmatrix} C_1 & 0 \\ 0 & 0 \end{pmatrix} U_2', \qquad D = U_1 \begin{pmatrix} D_1 & 0 \\ 0 & E \end{pmatrix} U_2^{-1}, \qquad U_2 = (Q,R),$$

where $C_1 = C_1^{(r)}$, $D_1 = D_1^{(r)}$ is a coprime symmetric pair with $|C_1| \neq 0$, $Q = Q^{(n,r)}$ is a primitive matrix and $U_1$, $U_2$ are unimodular matrices. Moreover, we can assume that $C_1,D_1$ and $Q$ are given fixed representatives of their classes $\{C_1,D_1\}$ and $\{Q\}$ (cf. Lemma in §11). We get

$$CZ + D = U_1 \left[ \begin{pmatrix} C_1 & 0 \\ 0 & 0 \end{pmatrix} \begin{pmatrix} Q' \\ R' \end{pmatrix} Z(Q,R) + \begin{pmatrix} D_1 & 0 \\ 0 & E \end{pmatrix} \right] U_2^{-1}$$

$$= U_1 \begin{pmatrix} C_1 Z[Q] + D_1 & * \\ 0 & E \end{pmatrix} U_2^{-1},$$

hence

$$|CZ + D| = \varepsilon |C_1 Z[Q] + D_1|, \qquad \varepsilon = |U_1 U_2^{-1}| = \pm 1,$$

and

$$\| CZ + D \| = \| C_1 \| \cdot \| Z[Q] + P \| ,$$

where $P = P^{(r)} = C_1^{-1} D_1$ is a rational symmetric matrix.

We claim that the equation $P = C_1^{-1} D_1$ establishes a one-to-one correspondence between the set of all classes $\{C_1, D_1\}$ with $|C_1| \neq 0$ and the set of all rational symmetric matrices $P$. Clearly $P$ is uniquely determined by the class $\{C_1, D_1\}$. Assume next that $\{C_1, D_1\}$ and $\{C_1^*, D_1^*\}$ are both mapped into the same $P = P'$. Then $C_1^{-1} D_1 = C_1^{*-1} D_1^*$, i.e., $C_1^{*-1} D_1^* = (C_1^{-1} D_1)' = D_1' C_1'^{-1}$, hence $C_1^* D_1' = D_1^* C_1'$ and thus $\{C_1, D_1\} = \{C_1^*, D_1^*\}$. So we have shown that the map $\{C_1, D_1\} \mapsto P$ is one-to-one. Finally assume that $P = P'$ is given. There exist unimodular matrices $U_3$ and $U_4$ such that

$$U_3 P U_4 = (\delta_{\mu\nu} \frac{a_\nu}{b_\nu}), \qquad b_\nu > 0, \qquad (a_\nu, b_\nu) = 1.$$

Then $P = C_1^{-1} D_1$ holds with

$$C_1 = (\delta_{\mu\nu} b_\nu) U_3, \qquad D_1 = (\delta_{\mu\nu} a_\nu) U_4^{-1}.$$

The condition $P = P'$ means precisely that $C_1, D_1$ is a symmetric pair, let us show that it is also coprime. If $(GC_1, GD_1)$ is integral then so are also $G(\delta_{\mu\nu} b_\nu)$ and $G(\delta_{\mu\nu} a_\nu)$. Denote by $\mathfrak{y}_\nu$ the $\nu^{\text{th}}$ column of $G$, then $\mathfrak{y}_\nu b_\nu$ and $\mathfrak{y}_\nu a_\nu$ are integral. Since $(a_\nu, b_\nu) = 1$, there exist $x_\nu, y_\nu \in \mathbb{Z}$ such that $b_\nu x_\nu + a_\nu y_\nu = 1$. It follows that $\mathfrak{y}_\nu = \mathfrak{y}_\nu b_\nu x_\nu + \mathfrak{y}_\nu a_\nu y_\nu$ is integral, hence $G$ is integral and $C_1, D_1$ are indeed coprime. Thus our claim is proved.

Since $Z$ is a given matrix in $\mathfrak{H}$, we may assume that the representative $Q$ is such that $Y[Q]$ is reduced. Indeed, this amounts to replacing $Q$, if necessary, by $QU_3$, where $U_3$ is unimodular. We set

$$T = Y[Q], \qquad S = X[Q] + P,$$

and determine the real matrix $F = F^{(r)}$ so that

$$T[F] = E, \qquad S[F] = H = (\delta_{\mu\nu} h_\nu).$$

Observing that $|T| = |F|^{-2}$ and $S + iT = (H + iE)[F^{-1}]$, we obtain

$$|Z[Q] + P| = |S + iT| = |T| \cdot \prod_{\nu=1}^{r} (h_\nu + i)$$

and

$$\| CZ + D \|^2 = |C_1|^2 \cdot |T|^2 \prod_{\nu=1}^{r} (h_\nu^2 + 1) < m^2 .$$

This shows that $|T|$ is bounded. Introduce the columns of $Q$ by $Q = (\mathfrak{q}_1, \ldots, \mathfrak{q}_r)$ so that $T = (\mathfrak{q}_\mu' Y \mathfrak{q}_\nu)$. Since $T$ is reduced, we have by (4) of §9 the inequality

$$\prod_{\nu=1}^{r} Y[\mathfrak{q}_\nu] < c|T|, \qquad \text{with} \quad c = c(r) > 0.$$

If $\lambda$ denotes the smallest eigenvalue of $Y$, then we have furthermore

$$Y[\mathfrak{q}_\nu] \geqq \lambda\, \mathfrak{q}_\nu' \mathfrak{q}_\nu \geqq \lambda > 0$$

and thus

$$Y[\mathfrak{q}_\nu] < c|T| \cdot \lambda^{1-r},$$

which shows that $Y[\mathfrak{q}_\nu]$ is bounded and therefore $\mathfrak{q}_\nu$ belongs to a finite set of integral vectors for $1 \leqq \nu \leqq r$. Thus we have only a finite number of representatives $Q$, hence $T = Y[Q]$ belongs to a finite set and we obtain from

$$|C_1|^2 \prod_{\nu=1}^{r} (h_\nu^2 + 1) < m^2 |T|^{-2}$$

that $\|C_1\|$ and $H$ are bounded. Furthermore it follows from $T = F'^{-1}F^{-1}$ that $F^{-1}$ is bounded, hence $S = H[F^{-1}]$ and $P = S - X[Q]$ are also bounded. Now $P$ is a bounded symmetric rational matrix having the bounded denominator $\|C_1\|$. Therefore $P$ belongs to a finite set and, by the one-to-one correspondence established above, the same is true about the classes $\{C_1, D_1\}$. This proves that there exist only finitely many classes $\{C, D\}$ such that $\|CZ + D\| < m$, q.e.d.

We define the height of a point $Z = X + iY \in \mathfrak{H}$ as the number $|Y| > 0$, and assert that in the set of points which are equivalent to $Z$ under $\Gamma$ there exists a point with maximal height. If this were not true, we could find a sequence $M_k \in \Gamma$ $(k = 1,2,\ldots)$ such that $Z_k = X_k + iY_k = M_k\langle Z\rangle$ and

$$|Y| < |Y_1| < |Y_2| < \ldots .$$

But since $|Y_k| = |Y| \cdot \|C_k Z + D_k\|^{-2}$ where

$$M_k = \begin{pmatrix} A_k & B_k \\ C_k & D_k \end{pmatrix},$$

this implies

$$1 > \|C_1 Z + D_1\| > \|C_2 Z + D_2\| > \ldots .$$

Now no two classes $\{C_k, D_k\}$ are equivalent since $(C_\ell, D_\ell) = U(C_k, D_k)$ with a unimodular $U$ implies $\|C_k Z + D_k\| = \|C_\ell Z + D_\ell\|$. But this is a contradiction with our preceding result.

Denote again by $\mathfrak{R}$ the set of all reduced matrices $Y > 0$ and by $\mathfrak{X}$ the unit cube in the space of all symmetric matrices $X = (x_{\mu\nu})$ given by $-\frac{1}{2} \leqq x_{\mu\nu} \leqq \frac{1}{2}$ $(1 \leqq \mu,\nu \leqq n)$.

*Theorem. The subset* $\mathcal{F}$ *of* $\mathfrak{H}$ *defined by*

1) $\|CZ+D\| \geqq 1$ *for all coprime symmetric pairs* $C,D$ *with* $C \neq 0$,

2) $Y \in \mathcal{R}$ $\qquad\qquad (Z = X+iY \in \mathfrak{H})$

3) $X \in \mathfrak{X}$

*is a fundamental domain of* $\Gamma$ *in* $\mathfrak{H}$. *The set* $\mathcal{F}$ *is closed in the space of all complex symmetric matrices.* $\mathcal{F}$ *is connected and the boundary of* $\mathcal{F}$ *consists of a finite number of algebraic hypersurfaces.*

*Proof.* 1. In a given set $\{M\langle Z\rangle \mid M \in \Gamma\}$ of points equivalent under $\Gamma$ we choose a point of maximal height. We may assume that $Z$ is such a point. Then obviously $\|CZ+D\| \geqq 1$ for all coprime symmetric pairs $C,D$. Without changing the height, we may replace $Z$ by $Z[U]+S$, where $U$ is unimodular and $S=S'$ integral. We determine $U$ so that $Y[U] \in \mathcal{R}$ and $S$ so that $X[U]+S \in \mathfrak{X}$, hence we have $Z[U]+S \in \mathcal{F}$. This shows that $\{M\langle Z\rangle \mid M \in \Gamma\}$ meets $\mathcal{F}$.

2. We prove that $\mathcal{F}$ is closed in the space of all complex symmetric matrices. Let $\{C,D\}$ be an equivalence class of coprime symmetric pairs with $\operatorname{rank} C = r$, and let $\{C_1,D_1\}$, $\{Q\}$ be the classes associated with it according to the Lemma of §11. We have then

$$\|CZ+D\| = \|C_1 Z[Q] + D_1\| .$$

In particular, if we assume that $Z \in \mathcal{F}$ and we choose $C_1 = E^{(r)}$, $D_1 = 0$, $Q = \begin{pmatrix} E^{(r)} \\ 0 \end{pmatrix}$, we obtain

$$\|Z_r\| \geqq 1 \quad \text{for} \quad Z \in \mathcal{F},$$

where $Z_r = Z\begin{bmatrix} E^{(r)} \\ 0 \end{bmatrix}$. The special case $r=1$ yields $|z_{11}| \geqq 1$, and since $|x_{11}| \leqq \frac{1}{2}$, we have $|y_{11}| \geqq \frac{1}{2}\sqrt{3}$. The inequalities of reduction theory

$$0 < y_{11} \leqq y_{22} \leqq \cdots \leqq y_{nn},$$

$$y_{11} y_{22} \cdots y_{nn} < c_1 |Y|$$

imply therefore

$$|Y| > c_1^{-1} (\tfrac{1}{2}\sqrt{3})^n \qquad \text{for} \quad Z \in \mathcal{F}.$$

Let $Z$ be a limit point of a sequence of points of $\mathcal{F}$. Then the last inequality shows that $Z$ belongs to $\mathfrak{H}$. But $Z$ clearly satisfies the inequalities which define $\mathcal{F}$, i.e., $Z \in \mathcal{F}$.

3. Next we show that $\mathcal{F}$ is connected. Let us use again the notation introduced at the beginning of this section:

$$P = C_1^{-1} D_1, \qquad S = X[Q] + P, \qquad T = Y[Q],$$

$$T[F] = E, \qquad S[F] = H = (\delta_{\mu\nu} h_\nu).$$

Besides the matrix $Z = X + iY$ we consider also the matrix $Z_\lambda = X + i\lambda Y$, where $\lambda \geqq 1$. If $Z \in \mathfrak{H}$ then $Z_\lambda \in \mathfrak{H}$, and if $Z$ satisfies conditions 2) and 3) in the definition of $\mathcal{F}$ then so does $Z_\lambda$. We have

$$Z[Q] + P = S + iT = (H + iE)[F^{-1}],$$

$$Z_\lambda[Q] + P = S + i\lambda T = (H + i\lambda E)[F^{-1}],$$

hence

$$\| CZ + D \|^2 = |C_1|^2 \cdot |T|^2 \prod_{\nu=1}^{r} (h_\nu^2 + 1),$$

$$\| CZ_\lambda + D \|^2 = |C_1|^2 \cdot |T|^2 \cdot \prod_{\nu=1}^{r} (h_\nu^2 + \lambda^2) \geqq \| CZ + D \|^2 .$$

We see therefore that if $Z \in \mathcal{F}$ then $Z_\lambda \in \mathcal{F}$ for $\lambda \geqq 1$.

Let us prove that there exists a constant $\lambda_o \geqq 1$ such that if $Z = X + iY \in \mathcal{F}$ and if $Q = Q^{(n,r)}$ is an arbitrary primitive matrix, then $|\lambda_o Y[Q]| \geqq 1$. Without loss of generality we may assume that $Y[Q]$ is reduced. Set $Q = (\mathfrak{q}_1, \mathfrak{q}_2, \ldots, \mathfrak{q}_r)$. It follows then from inequality (4) of § 9 that

$$|Y[Q]| \geqq c_1^{-1} \prod_{\nu=1}^{r} Y[\mathfrak{q}_\nu].$$

On the other hand $Y$ is also reduced, hence

$$Y[\mathfrak{q}_\nu] \geqq y_{11} \geqq \frac{1}{2}\sqrt{3} \; .$$

Hence

$$|Y[Q]| \geqq c_1^{-1}(\tfrac{1}{2}\sqrt{3}\,)^r$$

and the existence of a suitable $\lambda_o \geqq 1$ follows.

If now $X_o \in \mathfrak{X}$ and $Z = X + iY \in \mathcal{F}$, then $Z_o = X_o + i\lambda_o Y$ belongs to $\mathcal{F}$ since

$$\| CZ_o + D \| \geqq |\lambda_o Y[Q]| \geqq 1$$

for all coprime symmetric pairs $C,D$.

Let $Z_1 = X_1 + iY_1$ and $Z_2 = X_2 + iY_2$ be two points in $\mathcal{F}$. The two points can be joined by the segments

$$Z = X_1 + i\lambda Y_1 \qquad (1 \leqq \lambda \leqq \lambda_o),$$

$$Z = (1-\lambda)(X_1 + i\lambda_o Y_1) + \lambda(X_2 + i\lambda_o Y_2) \qquad (0 \leqq \lambda \leqq 1),$$

$$Z = X_2 + i(\lambda_o - \lambda)Y_2 \qquad (0 \leqq \lambda \leqq \lambda_o - 1).$$

The first and the third segment belong to $\mathcal{F}$ by what we already proved. But the second segment also belongs to $\mathcal{F}$. Indeed $(1-\lambda)X_1 + \lambda X_2 \in \mathfrak{X}$ and $(1-\lambda)Y_1 + \lambda Y_2 \in \mathcal{R}$ since $\mathfrak{X}$ and $\mathcal{R}$ are convex.

Furthermore $|\lambda_o Y_1[Q]| \geqq 1$, $|\lambda_o Y_2[Q]| \geqq 1$ for any primitive matrix $Q = Q^{(n,r)}$. We can transform $\lambda_o Y_1[Q]$ and $\lambda_o Y_2[Q]$ simultaneously into diagonal matrices $(\delta_{\mu\nu} a_\nu)$ and $(\delta_{\mu\nu} b_\nu)$ by a matrix with determinant 1; then the stated inequalities mean that $\prod_{\nu=1}^{r} a_\nu \geqq 1$, $\prod_{\nu=1}^{r} b_\nu \geqq 1$. It follows that

$$\prod_{\nu=1}^{r} ((1-\lambda)a_\nu + \lambda b_\nu) \geqq \prod_{\nu=1}^{r} a_\nu^{1-\lambda} b_\nu^{\lambda} \geqq 1,$$

i.e., $|(1-\lambda)\lambda_o Y_1[Q] + \lambda\lambda_o Y_2[Q]| \geqq 1$. As we have seen above, these conditions imply that the second segment belongs to $\mathcal{F}$ and so we have proved that $\mathcal{F}$ is connected.

4. Now we show that the boundary $\partial(\mathcal{F})$ of $\mathcal{F}$ in $\mathcal{Y}$ is contained in the union of a finite number of algebraic surfaces. The defining equations of these hypersurfaces are obtained by taking some of the inequalities which define $\mathcal{F}$ with only the equal sign.

First let us find a lower estimate for the smallest eigenvalue $\lambda$ of a reduced matrix $Y$. Denote by $Y_h$ the matrix one gets from $Y$ by cancelling the $h^{\text{th}}$ row and the $h^{\text{th}}$ column. We set $Y = (y_{\mu\nu})$, $y_{\nu\nu} = y_\nu$ and denote by $\lambda_1, \lambda_2, \ldots, \lambda_n$ the eigenvalues of $Y$. With a $c_1$ depending only on $n$ we have

$$|Y_\nu| \leqq y_1 \cdots y_{\nu-1} y_{\nu+1} \cdots y_n$$

because $Y_\nu$ is positive, and

$$y_1 y_2 \cdots y_n < c_1 |Y|$$

because $Y$ is reduced. If we denote by $\lambda$ the smallest of the eigenvalues $\lambda_1, \lambda_2, \ldots, \lambda_n$, we have

$$\frac{1}{\lambda} \leqq \frac{1}{\lambda_1} + \frac{1}{\lambda_2} + \ldots + \frac{1}{\lambda_n} = \sigma(Y^{-1})$$

$$= \frac{|Y_1| + |Y_2| + \ldots + |Y_n|}{|Y|}$$

$$< c_1\left(\frac{1}{y_1} + \frac{1}{y_2} + \ldots + \frac{1}{y_n}\right) \leqq \frac{nc_1}{y_1},$$

i.e., $\lambda \geqq \frac{1}{nc_1} y_1$.

If now $Z = X + iY \in \mathcal{F}$, then $y_1 \geqq \frac{1}{2}\sqrt{3}$ and so

$$\lambda \geqq \frac{\sqrt{3}}{2nc_1}.$$

Since

$$\lambda = \min_{\mathfrak{r}'\mathfrak{r}=1} Y[\mathfrak{r}] \qquad (\mathfrak{r} \text{ real column-vector}),$$

we obtain

$$Y[\mathfrak{r}] \geqq \lambda \mathfrak{r}'\mathfrak{r} \geqq \frac{\sqrt{3}}{2nc_1} \mathfrak{r}'\mathfrak{r}$$

for $Z = X + iY \in \mathcal{F}$ and real $\mathfrak{r}$.

Now let $Z$ be a point in $\partial(\mathcal{F})$, and let $\{Z_k\}$ be a sequence of points in $\mathfrak{H}$ such that

$$Z_k \notin \mathcal{F}, \qquad Z_k \to Z \qquad \text{as} \quad k \to \infty.$$

Choose

$$M_k = \begin{pmatrix} A_k & B_k \\ C_k & D_k \end{pmatrix} \in \Gamma$$

so that

$$W_k = M_k\langle Z_k\rangle \in \mathcal{F} \quad \text{for} \quad k = 1,2,3,\ldots .$$

Obviously $M_k \neq \pm E^{(2n)}$.

First we consider the case that for infinitely many $k$ we have $C_k \neq 0$. Passing to a subsequence we may assume without loss of generality that all $C_k$ have the same rank $r > 0$. To each class $\{C_k, D_k\}$ there corresponds a class $\{C_o^{(r)}, D_o^{(r)}\}$ of coprime symmetric pairs with $|C_o^{(r)}| \neq 0$ and a class $\{Q^{(n,r)}\}$ of primitive matrices so that if for $Z_k = X_k + iY_k$ we write

$$T = Y_k[Q], \qquad S = X_k[Q] + C_o^{-1}D_o,$$

$$T[F] = E, \qquad S[F] = H = (\delta_{\mu\nu}h_\nu),$$

with a real matrix $F = F^{(r)}$, then

$$\| C_kZ_k + D_k\|^2 = \| C_oZ_k[Q] + D_o\|^2 = |C_o|^2 \cdot |T|^2 \prod_{\nu=1}^{r} (h_\nu^2 + 1).$$

The matrices $C_o$, $D_o$, $Q$, $S$, $T$, $H$, $F$ depend of course on $k$, even if we do not indicate it in our notation.

Choose $Q = (\mathfrak{q}_1, \mathfrak{q}_2, \ldots, \mathfrak{q}_r)$ so that $T = (\mathfrak{q}_\mu' Y \mathfrak{q}_\nu)$ is reduced, and hence

$$\prod_{\nu=1}^{r} Y_k[\mathfrak{q}_\nu] < c_1|T|.$$

Since

$$\lambda^{(k)} = \min_{\mathfrak{x}'\mathfrak{x}=1} Y_k[\mathfrak{x}]$$

is a continuous function of $Y_k$ and since $Z_k \to Z = X + iY \in \mathcal{F}$, we have

$$\lim_{k\to\infty} \lambda^{(k)} = \lambda \geq \frac{\sqrt{3}}{2nc_1}$$

and thus

$$\lambda^{(k)} \geqq \frac{\sqrt{3}}{4nc_1}$$

for sufficiently large $k$. We may assume that this inequality holds for all $k$. Then

$$Y_k[\mathfrak{y}_\nu] \geqq \frac{\sqrt{3}}{4nc_1}\mathfrak{y}_\nu'\mathfrak{y}_\nu \geqq \frac{\sqrt{3}}{4nc_1} = c_2$$

and

$$0 < c_2^r c_1^{-1} < |T|.$$

Since $W_k \in \mathcal{F}$ and $(-C_k', A_k')$ is a coprime symmetric pair, we have $\|-C_k'W_k + A_k'\| \geqq 1$ and so it follows from

$$(-C_k'W_k + A_k')(C_kZ_k + D_k) = E$$

that

$$\|C_kZ_k + D_k\| \leqq 1.$$

This shows that $|C_0|$, $|T|$, $h_1^2, h_2^2, \ldots, h_r^2$ have upper bounds which depend only on $n$. From

$$c_2\mathfrak{y}_\nu'\mathfrak{y}_\nu \leqq Y_k[\mathfrak{y}_\nu] \leqq \frac{c_1|T|}{\prod\limits_{\mu\neq\nu} Y_k[\mathfrak{y}_\mu]} \leqq c_1c_2^{1-r}|T|$$

it follows that the $\mathfrak{y}_\nu$ are bounded, and so the $Q$ belong to a finite set independent of $Z$ and $k$. Now $T$ is a positive matrix, therefore its elements outside the diagonal are bounded by the diagonal elements $Y[\mathfrak{y}_\nu]$, and since the latter are bounded, we see that all the elements of $T$ have a bound which depends only on $n$.

It follows from $T = E[F^{-1}]$ and $S = H[F^{-1}]$ that $F^{-1}$ and hence $S$ are bounded. Since $X_k \to X \in \mathfrak{X}$, we may assume that $X_k$ is bounded. But then the rational matrix $C_0^{-1}D_0 = S - X_k[Q]$ is bounded and has a bounded denominator $|C_0|$, hence we have only finitely many matrices $C_0, D_0$. Thus we see that the classes $\{C_0, D_0\}$, $\{Q\}$ and therefore also $\{C_k, D_k\}$ are selected from a finite set of classes independent of $Z$ and $k$. In particular, we may assume that $C_k = C \neq 0$ and $D_k = D$ for every $k$, and then it follows from $\|CZ_k + D\| \leqq 1$, $\|CZ + D\| \geqq 1$, $Z_k \to Z$ that $\|CZ + D\| = 1$. Thus we proved that in the case considered, $Z$ lies on one of the algebraic hypersurfaces

$$\|CZ + D\| = 1,$$

determined by a finite set of pairs $C, D$.

It remains to consider the case when $C_k = 0$ for all but finitely many subscripts $k$, and we may clearly assume that the equality holds for all $k$. Then

$$W_k = Z_k[U_k] + S_k,$$

where $U_k$ is a unimodular matrix and $S_k = S_k'$ is an integral matrix. Since the positive matrices $Y_k[U_k]$ and $Y$ are reduced and $Y_k \to Y$, it follows from Lemma 2 of §9 that the matrices $U_k$ belong to a finite set. Passing to a subsequence, we may assume that all the $U_k$ are equal to the same unimodular matrix $U$ independent of $k$.

If $U \neq \pm E$, then since $Y_k[U]$ is reduced, both $Y$ and $Y[U]$ will be reduced and so $Y$ belongs to the boundary of $\mathfrak{R}$. Thus $Z$ lies in a hyperplane belonging to a finite set of hyperplanes.

If $U = \pm E$, then $W_k = Z_k + S_k$ and it follows from $X_k \to X \in \mathfrak{X}$, $X_k + S_k \in \mathfrak{X}$ that $S_k$ belongs to a finite set. We may again assume that $S_k = S$ independently of $k$. Then $S \neq 0$ because $M_k \neq \pm E^{(2n)}$. It follows therefore from $X \in \mathfrak{X}$, $X + S \in \mathfrak{X}$ that $X$ lies on one of the $n(n+1)$ sides of the cube $\mathfrak{X}$. Hence $Z$ lies again in a hyperplane belonging to a finite set of hyperplanes.

Thus we have proved that any point $Z \in \partial(\mathcal{F})$ lies on one of finitely many algebraic hypersurfaces.

5. Finally we show that if two distinct points $Z \in \mathcal{F}$ and $Z_1 \in \mathcal{F}$ are equivalent under $\Gamma$, then $Z \in \partial(\mathcal{F})$ and $Z_1 \in \partial(\mathcal{F})$. Let

$$Z_1 = (AZ + B)(CZ + D)^{-1}, \qquad \begin{pmatrix} A & B \\ C & D \end{pmatrix} \in \Gamma.$$

Then the relations

$$(-C'Z_1 + A')(CZ + D) = E, \qquad \|-C'Z_1 + A'\| \geqq 1, \qquad \|CZ + D\| \geqq 1$$

imply that

$$\|-C'Z_1 + A'\| = \|CZ + D\| = 1.$$

If $C \neq 0$ then $Z$ and $Z_1$ clearly belong to $\partial(\mathcal{F})$.

If $C = 0$, then

$$Y_1 = Y[U], \qquad X_1 = X[U] + S,$$

where $U$ is unimodular and $S = S'$ integral. If $U \neq \pm E$, then $Y$ and $Y_1$ are boundary points of $\mathfrak{R}$, and therefore $Z$ and $Z_1$ are boundary points of $\mathcal{F}$. If $U = \pm E$, then $S \neq 0$ and so $X$ and $X_1$ are boundary points of $\mathfrak{X}$, hence $Z$ and $Z_1$ are again boundary points of $\mathcal{F}$.

Thus the Theorem is completely proved.

Occasionally we shall write $\Gamma = \Gamma_n$, $\mathcal{F} = \mathcal{F}_n$ when different degrees $n$ are considered simultaneously.

*Lemma. The quantity*

$$S_n = \sup_{X+iY \in \mathcal{F}_n} \sigma(Y^{-1})$$

*is finite and*

$$S_1 \leqq S_2 \leqq S_3 \leqq \dots \ .$$

*Proof.* The first assertion follows from the inequality

$$\sigma(Y^{-1}) \leqq \frac{nc_1}{y_{11}} \leqq \frac{2nc_1}{\sqrt{3}}$$

which was seen in the course of the preceding proof.

Choose $Z_1 = X_1 + iY_1 \in \mathcal{F}_{n-1}$ and set

$$Z = X + iY = \begin{pmatrix} Z_1 & 0 \\ 0 & i\lambda \end{pmatrix}.$$

We prove that $Z \in \mathcal{F}_n$ for sufficiently large $\lambda$.

First we show that given a coprime symmetric pair $C = C^{(n)}$, $D = D^{(n)}$, we have $\|CZ + D\| \geqq 1$. Let rank $C = r$ and let $\{C_1^{(r)}, D_1^{(r)}\}$, $\{Q^{(n,r)}\}$ be the classes associated with $\{C,D\}$ in §11. Then

$$\|CZ + D\| = \|C_1 Z[Q] + D_1\| \geqq \|C_1\| \cdot |Y[Q]|.$$

We choose the representative $Q$ of $\{Q\}$ so that $Y[Q]$ shall be reduced, and write

$$Q = (\mathfrak{q}_1, \mathfrak{q}_2, \dots, \mathfrak{q}_r) = \begin{pmatrix} Q_1 \\ \mathfrak{r} \end{pmatrix}, \qquad Q_1 = (\mathfrak{r}_1, \mathfrak{r}_2, \dots, \mathfrak{r}_r),$$

$$\mathfrak{r} = (s_1, s_2, \dots, s_r).$$

Then

$$|Y[Q]| > \frac{1}{c_1} \prod_{\nu=1}^{r} Y[\mathfrak{q}_\nu] = \frac{1}{c_1} \prod_{\nu=1}^{r} (Y_1[\mathfrak{r}_\nu] + \lambda s_\nu^2) \qquad (c_1 > 1).$$

If $\mathfrak{r} \neq 0$ and if we take $\lambda \geqq \frac{1}{2}\sqrt{3}$, then we have

$$Y_1[\mathfrak{r}_\nu] + \lambda s_\nu^2 \geqq \min(y_{11}, \lambda) \geqq \frac{1}{2}\sqrt{3}$$

for all $\nu$ and

$$Y_1[\mathfrak{r}_\nu] + \lambda s_\nu^2 \geqq \lambda$$

for at least one $\nu$, so that

$$|Y[Q]| \geqq \frac{1}{c_1}\left(\frac{1}{2}\sqrt{3}\right)^{r-1} \lambda .$$

Choosing $\lambda \geqq c_1 \left(\frac{2}{\sqrt{3}}\right)^{r-1}$ we get $\| CZ + D \| \geqq 1$.

If $\mathfrak{r} = 0$, then the matrix $Q_1$ must be primitive so that $r < n$, and we have $Z[Q] = Z_1[Q_1]$. Let $\{C_0^{(n-1)}, D_0^{(n-1)}\}$ be the class of coprime symmetric pairs of degree $n-1$ which corresponds to $\{C_1^{(r)}, D_1^{(r)}\}$, $\{Q_1^{(n-1,r)}\}$. The condition $Z_1 \in \mathcal{F}_{n-1}$ implies

$$\| CZ + D \| = \| C_1 Z[Q] + D_1 \|$$

$$= \| C_1 Z_1[Q_1] + D_1 \| = \| C_0 Z_1 + D_0 \| \geqq 1.$$

Since $X_1 \in \mathfrak{X}_{n-1}$, we have obviously $X \in \mathfrak{X}_n$. Thus it remains to show that $Y \in \mathcal{R}_n$ for sufficiently large $\lambda$. We consider columns

$$\mathfrak{g}_k = \begin{pmatrix} g_1 \\ \vdots \\ g_n \end{pmatrix} = \begin{pmatrix} \mathfrak{f}_k \\ g_n \end{pmatrix},$$

where $g_k, g_{k+1}, \ldots, g_n$ are without common divisor. If $g_n = 0$, then the reduction conditions for $Y_1 \in \mathcal{R}_{n-1}$ yield $Y[\mathfrak{y}_k] = Y_1[\mathfrak{f}_k] \geqq y_{kk}$. If $g_n \neq 0$, then $Y[\mathfrak{y}_k] \geqq \lambda g_n^2 \geqq \lambda$, so that $Y[\mathfrak{y}_k] \geqq y_{kk}$ will hold provided that $\lambda \geqq y_{kk}$. If $\lambda \geqq y_{n-1,n-1}$, then $\lambda \geqq y_{kk}$ is true for $1 \leqq k < n$, and for $k = n$ we do not get a new condition since $y_{nn} = \lambda$. The conditions $y_{k,k+1} \geqq 0$ are trivially satisfied, so we see that $Y \in \mathcal{R}_n$ if $\lambda \geqq y_{n-1,n-1}$, and that

$$Z \in \mathcal{F}_n \quad \text{if} \quad \lambda \geqq \max(c_1(\tfrac{2}{\sqrt{3}})^{n-1}, y_{n-1,n-1}).$$

We have

$$Y^{-1} = \begin{pmatrix} Y_1^{-1} & 0 \\ 0 & \lambda^{-1} \end{pmatrix}$$

and therefore $\sigma(Y^{-1}) = \sigma(Y_1^{-1}) + \frac{1}{\lambda}$. For an arbitrary $Z_1 = X_1 + iY_1 \in \mathcal{F}_{n-1}$ we have $S_n \geqq \sigma(Y^{-1}) = \sigma(Y_1^{-1}) + \frac{1}{\lambda}$ for all sufficiently large $\lambda$, and so, letting $\lambda \to \infty$, we get $S_n \geqq \sigma(Y_1^{-1})$. Thus $S_n \geqq S_{n-1}$ for $n > 1$, q.e.d.

## §13. *Modular forms of degree n*

For a function $f(Z)$ defined on Siegel's upper half-plane $\mathfrak{H} = \mathfrak{H}_n$ we shall use the notation

$$f|M(Z) = f(M\langle Z\rangle)|CZ + D|^{-k},$$

where

$$M = \begin{pmatrix} A & B \\ C & D \end{pmatrix}$$

is a symplectic matrix and $k \in \mathbb{Z}$ a given integer.

*Definition. A function $f(Z)$ defined on $\mathfrak{H} = \mathfrak{H}_n$ is called a modular form of degree $n$ and of weight $k$ if*

1. *$f(Z)$ is holomorphic on $\mathfrak{H}$,*
2. *$f|M = f$ for all $M$ belonging to the modular group $\Gamma = \Gamma_n$ of degree $n$,*
3. *$f(Z)$ is bounded in any domain $Y \geqq Y_0 > 0$ in the case $n = 1$.*

We shall see that for $n > 1$ the last condition can be proved as a consequence of the first two. It is reasonable to assume that $k$ is an integer, because for $n > 1$ there exist no non identically zero modular forms with weight $k \notin \mathbb{N}$.

If we choose

$$M = \begin{pmatrix} E & S \\ 0 & E \end{pmatrix} \in \Gamma,$$

where $S = S'$ is integral, then $f(Z+S) = f(Z)$, i.e., the function is periodic in the elements of $X$, and has therefore a Fourier expansion of the form

$$f(Z) = \sum_{T} a(T,Y)e^{2\pi i\sigma(TX)},$$

where $T$ runs through all *semi-integral* symmetric matrices; the word semi-integral is meant to indicate that the linear form

$$X \mapsto \sigma(TX) = \sum_{\nu=1}^{n} t_{\nu\nu}x_{\nu\nu} + 2\sum_{\mu<\nu} t_{\mu\nu}x_{\mu\nu}$$

has integral coefficients, i.e., that the $t_{\nu\nu}$ are integers and for $\mu < \nu$ the $2t_{\mu\nu}$ are integers, where $T = (t_{\mu\nu}) = T'$.

Since $f(Z)$ is holomorphic, it is annihilated by the operators

$$\frac{\partial}{\partial \bar{z}_{\mu\nu}} = \frac{1}{2}\left(\frac{\partial}{\partial x_{\mu\nu}} + i\frac{\partial}{\partial y_{\mu\nu}}\right)$$

and so we get the conditions

$$e_{\mu\nu}\frac{\partial a(T,Y)}{\partial y_{\mu\nu}} = -2\pi t_{\mu\nu}a(T,Y),$$

where $e_{\mu\nu} = \frac{1}{2}(1 + \delta_{\mu\nu})$. This yields

$$a(T,Y) = a(T)e^{-2\pi\sigma(TY)},$$

and thus we have the expansion

$$f(Z) = \sum_{T} a(T)e^{2\pi i\sigma(TZ)},$$

which is valid in the whole domain $\mathfrak{H}$.

If we choose

$$M = \begin{pmatrix} U' & 0 \\ 0 & U^{-1} \end{pmatrix} \in \Gamma,$$

where $U$ is unimodular, then $f(Z[U])|U|^k = f(Z)$. Since $\sigma(TZ[U]) = \sigma(T[U']Z)$ and

$$\sum a(T)e^{-2\pi\sigma(T[U']Z)} = \sum a(T[U'^{-1}])e^{-2\pi\sigma(TZ)},$$

it follows from the uniqueness of the Fourier expansion that the last condition is equivalent to

$$a(T[U]) = |U|^k a(T).$$

In particular, if $U = -E$ we get $a(T) = (-1)^{nk}a(T)$ so that if $f(Z)$ does not vanish identically the integer $nk$ must be even. We shall always assume that this is the case.

*Lemma. A modular form $f(Z)$ with Fourier coefficients $a(T)$ is bounded in every domain $Y \geqq Y_0 > 0$ if and only if $a(T) \neq 0$ only for $T \geqq 0$.*

*Proof.* 1. Assume that $|f(Z)| \leqq C(Y_0)$ for $Y \geqq Y_0 > 0$. Then from

$$a(T)e^{-2\pi\sigma(TY)} = \int_{\mathfrak{X}}\!\!\cdots\!\int f(Z)e^{-2\pi i\sigma(TX)}[dX],$$

where $\mathfrak{X}$ is the unit cube in the $X$-space, it follows that

$$|a(T)| \leqq C(Y_0)e^{2\pi\sigma(TY)} \qquad \text{for } Y \geqq Y_0 > 0$$

and thus

$$|a(T)| \leqq C(Y_0)e^{2\pi\lambda\sigma(TY_0)} \qquad \text{for } \lambda \geqq 1.$$

If $a(T) \neq 0$ then $\sigma(TY_0) \geqq 0$ since otherwise we get a contradiction letting $\lambda$ tend to $\infty$. Setting $Y_0 = RR'$, where $R = R^{(n)} = (\mathfrak{r}_1, \mathfrak{r}_2, \ldots, \mathfrak{r}_n)$ is a real matrix with $|R| \neq 0$, we get

$$\sigma(TY_0) = \sigma(T[R]) = \sum_{\nu=1}^{n} T[\mathfrak{r}_\nu] \geqq 0.$$

Since $\sigma(TY_0)$ is a continuous function of $Y_0$, we have $\sigma(T[R]) \geqq 0$ for all real square matrices $R$, in particular for $R = (\mathfrak{r}, 0, \ldots, 0)$, i.e., $T[\mathfrak{r}] \geqq 0$ for an arbitrary real column vector $\mathfrak{r}$. This proves that $a(T) \neq 0$ implies $T \geqq 0$.

2. Assume that the expansion of $f(Z)$ has the form

$$f(Z) = \sum_{T \geqq 0} a(T) e^{2\pi i \sigma(TZ)}.$$

Since the series converges in $\mathfrak{H}$ we have

$$|a(T)e^{2\pi i \sigma(TZ)}| = |a(T)| e^{-2\pi\sigma(TY)} \leqq C(Y),$$

hence $|a(T)| \leqq C(\frac{1}{2}Y) e^{\pi\sigma(TY)}$ and

$$|a(T)e^{2\pi i \sigma(TZ)}| \leqq C(\tfrac{1}{2}Y_0) e^{\pi\sigma(TY_0) - 2\pi\sigma(TY)}.$$

Let $Y \geqq Y_0 > 0$ and set $Y - Y_0 = RR'$ with a square matrix $R = R^{(n)}$. Then

$$\sigma(TY) - \sigma(TY_0) = \sigma(T[R]) \geqq 0 \qquad \text{for } T \geqq 0$$

hence

$$|a(T)e^{2\pi i \sigma(TZ)}| \leqq C(\tfrac{1}{2}Y_0) e^{-\pi\sigma(TY)} \qquad \text{for } Y \geqq Y_0 > 0, \quad T \geqq 0.$$

It follows that in the domain $Y \geqq Y_0 > 0$ the function $f(Z)$ has, up to the factor $C(\frac{1}{2}Y_0)$, the majorant $\sum_{T \geqq 0} e^{-\pi\sigma(TY_0)}$. If $Y_0 \geqq \varepsilon E$

with $\varepsilon > 0$, then $\sigma(TY_0) \geqq \varepsilon\sigma(T)$ and therefore we have

$$|f(Z)| \leqq C(\tfrac{1}{2}Y_0) \sum_{\substack{T \geqq 0 \\ T \text{ semi-integral}}} e^{-\pi\varepsilon\sigma(T)} \qquad \text{for } Y \geqq Y_0 > 0.$$

To show that the last series converges, we must estimate the number of semi-positive, semi-integral matrices $T = (t_{\mu\nu})$ which satisfy $\sigma(T) = \sum_{\nu=1}^{n} t_{\nu\nu} \leqq m$ for a given $m \in \mathbb{N}$. Since $t_{\nu\nu} \geqq 0$, there are at most $(m+1)^n$ possible choices for the diagonal elements, and since $\pm 2t_{\mu\nu} \leqq 2t_{\nu\nu} \leqq 2m$, there are at most $4m+1$ possibilities for each element outside the diagonal. Thus the number in question is less than $(m+1)^n \cdot (4m+1)^{\frac{1}{2}n(n-1)} \leqq (4m+1)^{\frac{1}{2}n(n+1)}$. Since the series

$$\sum_{m=0}^{\infty} (4m+1)^{\frac{1}{2}n(n+1)} e^{-\pi\varepsilon m}$$

converges and is independent of $Z$, we proved that $f(Z)$ is bounded in $Y \geqq Y_0 > 0$ and also that the Fourier series of $f(Z)$ converges absolutely and uniformly in $Y \geqq Y_0 > 0$.

*Theorem* 1. *A modular form $f(Z)$ is bounded in any domain $Y \geqq Y_0 > 0$, in particular in $\mathcal{F}_n$.*

*Proof*. For $n = 1$ this is part of the definition, so that we have to consider only $n > 1$. The Fourier expansion

$$f(Z) = \sum_T a(T) e^{2\pi i\sigma(TZ)}$$

can be considered as a power series in the variables $\zeta_{\mu\nu} = e^{2\pi i z_{\mu\nu}}$. Setting $T = (t_{\mu\nu})$ we get

$$f(Z) = \sum_T a(T) \prod_{\nu} \zeta_{\nu\nu}^{t_{\nu\nu}} \prod_{\mu<\nu} \zeta_{\mu\nu}^{2t_{\mu\nu}}.$$

This series converges in the open domain $Y = -\frac{1}{2\pi}(\log|\zeta_{\mu\nu}|) > 0$ and converges therefore absolutely in this domain. The same is true for the partial series

$$g(Z,T) = \sum_{U} a(T[U])e^{2\pi i\sigma(T[U]Z)},$$

where $U$ runs through a complete set of unimodular matrices which yield different matrices $T[U]$ and which satisfy $|U| = 1$. We have $a(T[U]) = a(T)$ for all such $U$, as we saw above. For a given integer, let $c(T,m)$ be the number of unimodular matrices $U$ with $|U| = 1$ which yield different $T[U]$ and satisfy $\sigma(T[U]) = m$. Choosing $Z = iE$ we get

$$g(iE,T) = a(T) \sum_{m=-\infty}^{\infty} c(T,m)e^{-2\pi m}$$

$$\geqq a(T) \sum_{m=1}^{\infty} c(T,-m).$$

If $a(T) \neq 0$, then this is only possible if $c(T,-m) = 0$ for $m \geqq m_0$, where $m_0$ is sufficiently large. We shall prove that if $T$ does not satisfy $T \geqq 0$, then $c(T,-m) \geqq 1$ for infinitely many $m \geqq 1$ and consequently $a(T) = 0$. This will prove the theorem by virtue of the Lemma.

Assume therefore that $T$ is not semi-positive. There exists an integral column-vector $\mathfrak{y}$ such that $T[\mathfrak{y}] < 0$. Let $j, d_1, d_2, \ldots, d_n \in \mathbb{Z}$ and set

$$U = E + j(d_1\mathfrak{y}, d_2\mathfrak{y}, \ldots, d_n\mathfrak{y}).$$

Since the rank of $A = U - E$ is one we have

$$|\lambda E + A| = \lambda^n + \lambda^{n-1}\sigma(A)$$

and setting $\lambda = 1$ yields

$$|U| = 1 + j\sigma(d_1\mathfrak{y}, d_2\mathfrak{y}, \ldots, d_n\mathfrak{y}).$$

Because $n > 1$, we can determine $d_\nu$ $(\nu = 1,2,\ldots,n)$ so that

$$\sigma(d_1\mathfrak{y}, d_2\mathfrak{y}, \ldots, d_n\mathfrak{y}) = \sum_{\nu=1}^{n} d_\nu g_\nu = 0, \qquad \sum_{\nu=1}^{n} d_\nu^2 > 0.$$

Then $|U| = 1$ and

$$\sigma(T[U]) = \sigma(T) + 2j\sigma(T(d_1\mathfrak{y}, d_2\mathfrak{y}, \ldots, d_n\mathfrak{y})) + j^2 T[\mathfrak{y}] \sum_{\nu=1}^{n} d_\nu^2 .$$

Since the coefficient of $j^2$ is negative, we see that for infinitely many $m > 0$ there exist $j \in \mathbb{Z}$ such that $\sigma(T[U]) = -m$, i.e., $c(T,-m) \geqq 1$ for infinitely many $m \geqq 1$.

It is meaningful to consider the complex constants as modular forms of weight 0.

*Theorem* 2. *To every modular form $f$ of degree $n \geqq 1$ there is attached a modular form $f|\phi$ of degree $n-1$ by the process*

$$f|\phi(Z_1) = \lim_{\lambda\to\infty} f\begin{pmatrix} Z_1 & 0 \\ 0 & i\lambda \end{pmatrix} \qquad Z_1 \in \mathfrak{H}_{n-1}.$$

*Proof.* In the proof of the above lemma we stated that the Fourier series

$$f(Z) = \sum_{T\geqq 0} a(T)e^{2\pi i\sigma(TZ)}$$

of a given form converges uniformly for $Y \geqq Y_0 > 0$. Setting

$$T = \begin{pmatrix} T_1 & * \\ * & t_{nn} \end{pmatrix} \qquad Z = \begin{pmatrix} Z_1 & 0 \\ 0 & i\lambda \end{pmatrix}$$

we have

$$\sigma(TZ) = \sigma(T_1 Z_1) + i\lambda t_{nn}$$

and

$$e^{2\pi i\sigma(TZ)} = e^{2\pi i\sigma(T_1 Z_1) - 2\pi\lambda t_{nn}}.$$

If $t_{nn} > 0$, then $\lim\limits_{\lambda\to\infty} e^{-2\pi\lambda t_{nn}} = 0$. If $t_{nn} = 0$, then because of $T \geqq 0$ we have

$$T = \begin{pmatrix} T_1 & 0 \\ 0 & 0 \end{pmatrix},$$

so that

$$f|\phi(Z_1) = \lim_{\lambda\to\infty} f(Z) = \sum_{T\geqq 0} a(T) \lim_{\lambda\to\infty} e^{2\pi i\sigma(TZ)}$$

$$= \sum_{T_1\geqq 0} a(T_1) e^{2\pi i\sigma(T_1 Z_1)},$$

where

$$a(T_1) = a\begin{pmatrix} T_1 & 0 \\ 0 & 0 \end{pmatrix}.$$

It remains to check the transformation properties of $g(Z_1) = f|\phi(Z_1)$ under

$$M_1 = \begin{pmatrix} A_1 & B_1 \\ C_1 & D_1 \end{pmatrix} \in \Gamma_{n-1}.$$

If we set

$$A = \begin{pmatrix} A_1 & 0 \\ 0 & 1 \end{pmatrix}, \quad B = \begin{pmatrix} B_1 & 0 \\ 0 & 0 \end{pmatrix}, \quad C = \begin{pmatrix} C_1 & 0 \\ 0 & 0 \end{pmatrix}, \quad D = \begin{pmatrix} D_1 & 0 \\ 0 & 1 \end{pmatrix},$$

then it is easy to see that

$$M = \begin{pmatrix} A & B \\ C & D \end{pmatrix} \in \Gamma_n$$

and

$$M\langle Z\rangle = \begin{pmatrix} M_1\langle Z_1\rangle & 0 \\ 0 & i\lambda \end{pmatrix}, \qquad |CZ + D| = |C_1Z_1 + D_1|$$

for

$$Z = \begin{pmatrix} Z_1 & 0 \\ 0 & i\lambda \end{pmatrix} \in \mathfrak{H}_n .$$

Thus

$$\begin{aligned} g|M_1(Z_1) &= g(M_1\langle Z_1\rangle)|C_1Z_1 + D_1|^{-k} \\ &= \lim_{\lambda\to\infty} f\begin{pmatrix} M_1\langle Z_1\rangle & 0 \\ 0 & i\lambda \end{pmatrix}|C_1Z_1 + D_1|^{-k} \\ &= \lim_{\lambda\to\infty} f(M\langle Z\rangle)|CZ + D|^{-k} \\ &= \lim_{\lambda\to\infty} f|M(Z) = \lim_{\lambda\to\infty} f(Z) = g(Z_1). \end{aligned}$$

*Theorem* 3. *Let*

$$S_n = \sup_{X+iY\in\mathcal{F}_n} \sigma(Y^{-1})$$

*and let*

$$f(Z) = \sum_{T \geqq 0} a(T) e^{2\pi i \sigma(TZ)}$$

*be a modular form of degree n and weight k. Assume that* $a(T) = 0$ *for the finitely many* $T \geqq 0$ *which satisfy* $\sigma(T) \leqq \frac{k}{4\pi} S_n$. *Then* $f(Z) \equiv 0$.

*Proof* by induction on $n$: 1. In the case $n = 1$, $Z = z$ it is known that $f(Z) = 0$ for $k \leqq 0$, and that in the case $k > 0$, $k \equiv 0$ (mod 2), the form $f(Z)$ vanishes identically if $a(t) = 0$ for

$$0 \leqq t \leqq \begin{cases} \left[\frac{k}{12}\right] & \text{in case } k \not\equiv 2 \pmod{12}, \\ \left[\frac{k}{12}\right] - 1 & \text{in case } k \equiv 2 \pmod{12}. \end{cases}$$

But $S_1 = \frac{2}{\sqrt{3}}$ and so $\left[\frac{k}{12}\right] \leqq \frac{k}{12} < \frac{k}{4\pi} S_1$ for $k > 1$. This proves the theorem for $n = 1$.

2. Assume that $n > 1$ and that Theorem 3 is true for $n$-1 instead of $n$. If $f(Z)$ satisfies the condition of the theorem, then

$$f|\phi(Z_1) = \sum_{T_1 \geqq 0} a(T_1) e^{2\pi i \sigma(T_1 Z_1)}$$

is a modular form of degree $n$-1 which satisfies $a(T_1) = 0$ for $\sigma(T_1) \leqq \frac{k}{4\pi} S_{n-1}$ since

$$a(T_1) = a\begin{pmatrix} T_1 & 0 \\ 0 & 0 \end{pmatrix}$$

and $S_{n-1} \leqq S_n$ by the Lemma of §12. By our induction hypothesis we have $f|\phi(Z_1) = 0$, i.e., $a(T_1) = 0$ for every $T_1 \geqq 0$.

If $T = T^{(n)} \geqq 0$ and $|T| = 0$, then there exists a unimodular matrix $U$ such that

$$T[U] = \begin{pmatrix} T_1 & 0 \\ 0 & 0 \end{pmatrix},$$

and therefore $a(T) = \pm a(T[U]) = 0$. Thus $a(T) \neq 0$ only if $T > 0$, and we can write

$$f(Z) = \sum_{T>0} a(T) e^{2\pi i \sigma(TZ)}.$$

Next we prove that there exist constants $C > 0$ and $\varepsilon > 0$ such that

$$|f(Z)| < Ce^{-\varepsilon \sqrt[n]{|Y|}} \qquad \text{for } Z = X + iY \in \mathcal{F}.$$

By (8) of §9 there exists a constant $c = c(n) > 0$ such that for $Z = X + iY \in \mathcal{F}$, i.e., $Y = (y_{\mu\nu}) \in \mathcal{R}$ we have $Y > cY_0$, where $Y_0 = (\delta_{\mu\nu} y_{\nu\nu})$. Setting $T = RR'$ with a real $R = R^{(n)}$ we have

$$\sigma(TY) = \sigma(Y[R]) > c\sigma(Y_0[R]) = c\sigma(TY_0)$$

$$= c \sum_{\nu=1}^{n} t_{\nu\nu} y_{\nu\nu} \geqq \frac{\sqrt{3}}{2} c\sigma(T)$$

and also

$$\sigma(TY) > c\sigma(TY_0) \geqq nc \sqrt[n]{\prod_{\nu=1}^{n} t_{\nu\nu} y_{\nu\nu}}$$

$$= nc\sqrt[n]{|Y|} \cdot \sqrt[n]{\prod_{\nu=1}^{n} t_{\nu\nu}}\ .$$

Since the series for $f(Z)$ converges at $Z = i\frac{\sqrt{3}}{4} cE$, there exists $C_0 > 0$ such that

$$|a(T)| \leq C_0 e^{\pi \frac{\sqrt{3}}{2} \sigma\sigma(T)}$$

for all $T > 0$ and thus

$$|f(Z)| \leq \sum_{T>0} C_0 e^{\pi \frac{\sqrt{3}}{2} \sigma\sigma(T) - \pi\{\frac{\sqrt{3}}{2} \sigma\sigma(T) + n\sigma\sqrt[n]{|Y|} \sqrt[n]{\prod_{\nu=1}^{n} t_{\nu\nu}}\}}$$

$$= \sum_{T>0} C_0 e^{-\pi n \sigma \sqrt[n]{|Y|} \sqrt[n]{\prod_{\nu=1}^{n} t_{\nu\nu}}}$$

$$\leq C_0 \sum_{t=1}^{n} t^n (4t + 1)^{\frac{1}{2}n(n-1)} e^{-\pi n \sigma \sqrt[n]{|Y|} \sqrt[n]{t}}$$

$$\leq C e^{-\varepsilon \sqrt[n]{|Y|}} ,$$

because, as we have seen in the proof of the Lemma, there are at most $t^n (4t + 1)^{\frac{1}{2}n(n-1)}$ semi-integral matrices $T = (t_{\mu\nu}) > 0$ with $\prod_{\nu=1}^{n} t_{\nu\nu} = t$.

It follows that

$$\lim_{|Y| \to \infty} |Y|^{\frac{1}{2}k} f(Z) = 0 \qquad \text{for } Z = X + iY \in \mathcal{F}.$$

The transformation $Z \mapsto Z^* = X^* + iY^* = M\langle Z\rangle$ maps $|Y|$ into $|Y^*| = |Y| \cdot \|CZ + D\|^{-2}$ and $f(Z)$ into $f(Z^*) = |CZ + D|^k f(Z)$ so that the function

$$\phi(Z) = |Y|^{\frac{1}{2}k} |f(Z)|$$

is invariant under $\Gamma$. Therefore $\phi(Z)$ achieves its maximum $M$ at some point $Z_0 = X_0 + iY_0$ of $\mathcal{F}$. Let $z = x + iy$ be a complex variable and set $Z = Z_0 + zE$, $t = e^{2\pi i z}$ and

$$g(t) = f(Z)e^{-i\lambda\sigma(Z)},$$

where $\lambda$ is determined by

$$\frac{n\lambda}{2\pi} = 1 + \left[\frac{k}{4\pi} S_n\right].$$

We have the expansion

$$g(t) = \sum a(T)e^{2\pi i\sigma(TZ_0) + 2\pi i\sigma(T)z - i\lambda\sigma(Z_0) - i\lambda nz}$$

$$= \sum a(T)e^{2\pi i\sigma(TZ_0) - i\lambda\sigma(Z_0)} t^{\sigma(T) - \frac{\lambda n}{2\pi}},$$

where, by hypothesis, the summation is extended over all semi-integral $T > 0$ such that $\sigma(T) > \frac{k}{4\pi} S_n$ and consequently the exponents of $t$ verify

$$\sigma(T) - \frac{\lambda n}{2\pi} \geqq 0.$$

There exists $\varepsilon > 0$ such that $Z \in \mathcal{G}$ for $y \geqq -\varepsilon$, and therefore the expansion converges, i.e., $g(t)$ is holomorphic in a circle $|t| = e^{-2\pi y} \leqq \rho$ with some $\rho > 1$. By the maximum principle, there exists a point $t$ with $|t| = \rho$ such that $|g(t)| \geqq |g(1)|$. Because of

$$|g(t)| = \phi(Z)|Y|^{-\frac{1}{2}k} e^{\lambda\sigma(Y)}$$

we have, after cancelling the factor $e^{\lambda\sigma(Y_0)}$,

$$M|Y_0|^{-\frac{1}{2}k} \leqq M|Y|^{-\frac{1}{2}k} e^{\lambda ny},$$

and thus

$$M \leqq Me^{\psi(y)},$$

where $y = \frac{1}{2\pi}\log\frac{1}{\rho} < 0$ and

$$\psi(y) = -\frac{k}{2}\log|Y| + \frac{k}{2}\log|Y_0| + \lambda n y$$

$$= n\lambda y - \frac{k}{2}\log|E + yY_0^{-1}|$$

and since $Y = Y_0 + yE$. Now $\psi(0) = 0$ and

$$\psi'(0) = n\lambda - \frac{k}{2}\sigma(Y_0^{-1}) \geqq n\lambda - \frac{k}{2}S_n$$

$$= 2\pi\left(1 + \left[\frac{k}{4\pi}S_n\right] - \frac{k}{4\pi}S_n\right) > 0,$$

hence we have $\psi(y) < 0$ if $\rho > 1$ is sufficiently close to 1. This shows that $M = 0$, i.e., $f(Z) = 0$, q.e.d.

The modular forms of degree $n$ and weight $k$ form a linear space $\{\Gamma_n,k\}$ over $\mathbb{C}$. As an application of Theorem 4 we assert

*Theorem* 4.

1. $\dim\{\Gamma_n,k\} \leqq Ck^{\frac{1}{2}n(n+1)}$ *with a constant* $C = C_n > 0$.
2. $\dim\{\Gamma_2,k\} \leqq 1$ *for* $k = 4,6,8,$
   $\dim\{\Gamma_2,k\} = 0$ *for* $k = 1,2,3,5,7.$

Part 1 follows from the preceding proof. Part 2 will be made more explicit by the next theorem. The linear spaces $\{\Gamma_2,k\}$ were completely determined first by J. Igusa [6] and later in a more elementary way by E. Freitag [2].

*Theorem* 5. *A modular form $f(Z)$ of degree $n \leqq 2$ and weight $k$ is uniquely determined by the Fourier coefficient $a(0)$ of $f(Z)$ if $0 \leqq k \leqq 8$. We have in particular $a(0) = 0$, and thus $f(Z) = 0$, if $k = 1,2,3,5,7$ and $n \leqq 2$.*

*Proof*. 1. $n = 1$. We know that $S_1 = \frac{2}{\sqrt{3}}$ so that $\frac{k}{4\pi} S_1 < 1$ for $k \leqq 8$. The condition $t \leqq \frac{k}{4\pi} S_1$ is therefore equivalent to $t = 0$ and thus $a(0) = 0$ implies $f(z) = 0$ by Theorem 3. The form $f(z)$ vanishes identically if $nk \not\equiv 0 \pmod 2$, i.e., if $k \equiv 1 \pmod 2$. But $f(z) = 0$ also for $k = 2$. Indeed $f(z)dz$ is then an invariant differential on the Riemann surface defined by $\mathcal{F}_1$. This surface has genus 0, hence there is no non-vanishing differential of the first kind (i.e., holomorphic) on it. From the series $f(z) = \sum_{t=0}^{\infty} a(t)e^{2\pi i t z}$ we get by setting $w = e^{2\pi i z}$, $dw = 2\pi i w\, dz$,

$$f(z)dz = \frac{1}{2\pi i} \sum_{t=0}^{\infty} a(t) w^{t-1} dw.$$

If $a(0) = 0$, then $f(z)dz$ is a differential of the first kind, hence $f(z) = 0$. If $a(0) \neq 0$, then $f(z)dz$ has a logarithmic singularity at $w = 0$ and no other singularity, so that the sum of the residues of $f(z)dz$ (counted on $\mathcal{F}_1$) does not vanish, which is impossible.

2. $n = 2$. If $Z = X + iY \in \mathcal{F}_2$, i.e.,

$$Y = \begin{pmatrix} y_0 & y_1 \\ y_1 & y_2 \end{pmatrix} \in \mathcal{R}_2,$$

then by the inequalities $\frac{\sqrt{3}}{2} \leqq y_0 \leqq y_2$, $\pm 2y_1 \leqq y_0$, $\pm 2y_1 \leqq y_2$ of reduction theory we get

$$\sigma(Y^{-1}) = \frac{y_0 + y_2}{y_0 y_2 - y_1^2} \leqq \frac{y_0 + y_2}{\frac{3}{4} y_0 y_2} \leqq \frac{4}{3}\left(\frac{2}{y_0}\right) \leqq \frac{16}{3\sqrt{3}},$$

i.e., $S_2 \leqq \frac{16}{3\sqrt{3}}$, hence $\frac{k}{4\pi} S_2 < 2$ for $k \leqq 8$. Thus $\sigma(T) \leqq \frac{k}{4\pi} S_2$, $k \leqq 8$, implies $t_{11} + t_{22} \leqq 1$, i.e., one of the elements $t_{11}$ or $t_{22}$ must be zero, and since $T \geqq 0$, necessarily $|T| = 0$. Assume $a(0) = 0$, $k \leqq 8$. Then $f|\phi$ is identically zero, i.e., $a\begin{pmatrix} t & 0 \\ 0 & 0 \end{pmatrix} = 0$ for $t \geqq 0$ and so $a(T) = 0$ for all $T$ such that $|T| = 0$. It follows therefore from Theorem 3 that $f(Z) = 0$.

Finally $a(0)$ is also the constant term in the Fourier expansion of $f|\phi$, therefore we know that necessarily $a(0) = 0$ if $k = 1,2,3,5,7$.

## §14. *Report on Eisenstein series of the modular group*

For the linear spaces $\{\Gamma_n, k\}$ of modular forms of degree $n$ and weight $k$ some structure theorems are known in the case when $k$ is even, which up to now can be proved only by use of Poincaré series or the general Eisenstein series introduced recently by H. Klingen [9]. This paper contains complete proofs and a sufficiently large number of further references, so that here we shall restrict ourselves to stating some of the results.

For a given integer $r$ with $0 \leqq r \leqq n$ we define the subgroup $\Delta_{n,r}$ of the modular group $\Gamma_n$ of degree $n$ as the group of all $M \in \Gamma_n$ whose elements in the first $n+r$ columns and last $n-r$ rows vanish, i.e.,

$$\Delta_{n,r} = \left\{ M = \begin{pmatrix} * & * \\ 0^{(n-r,n+r)} & * \end{pmatrix} \;\middle|\; M \in \Gamma_n \right\}.$$

If

$$M = \begin{pmatrix} A & B \\ C & D \end{pmatrix},$$

where we set

$$A^{(n)} = \begin{pmatrix} A_1^{(r)} & A_2 \\ A_3 & A_4 \end{pmatrix}, \qquad B^{(n)} = \begin{pmatrix} B_1^{(r)} & B_2 \\ B_3 & B_4 \end{pmatrix},$$

$$C^{(n)} = \begin{pmatrix} C_1^{(r)} & C_2 \\ C_3 & C_4 \end{pmatrix}, \qquad D^{(n)} = \begin{pmatrix} D_1^{(r)} & D_2 \\ D_3 & D_4 \end{pmatrix},$$

then $M \in \Delta_{n,r}$ implies that $A_2 = 0$, $C_2 = 0$, $C_3 = 0$, $C_4 = 0$, $D_3 = 0$ and

$$M_1 = \begin{pmatrix} A_1 & B_1 \\ C_1 & D_1 \end{pmatrix} \in \Gamma_r .$$

A modular form $f \in \{\Gamma_n, k\}$ is called a cusp form if $f|\phi = 0$. We introduce the new notation $\mathfrak{M}_n^k$ for $\{\Gamma_n, k\}$ and denote by $\mathfrak{I}_n^k$ the linear subspace of cusp forms in $\mathfrak{M}_n^k$, i.e., the kernel of the linear map $\phi: f \mapsto f|\phi$ from $\mathfrak{M}_n^k$ into $\mathfrak{M}_{n-1}^k$. For $n = 0$ we set $\mathfrak{M}_0^k = \mathfrak{I}_0^k = \mathbb{C}$. Finally for any $Z \in \mathfrak{H}_n$ we write

$$Z = \begin{pmatrix} Z^* & * \\ * & * \end{pmatrix} \quad \text{with} \quad Z^* = Z\begin{bmatrix} E^{(r)} \\ 0 \end{bmatrix} .$$

If $n \geqq 1$ and $0 \leqq r \leqq n$, then for any $f \in \mathfrak{I}_r^k$ we define the Eisenstein series

$$E_{n,r}^k(Z,f) = \sum_{M: \Delta_{n,r} \backslash \Gamma_n} f(M\langle Z\rangle^*)\,|CZ + D|^{-k} , \quad Z \in \mathfrak{H}_n ,$$

where the summation indicates that $M = \begin{pmatrix} A & B \\ C & D \end{pmatrix}$ runs through a complete set of representatives of the right cosets of $\Gamma_n$ modulo $\Delta_{n,r}$. For $r = 0$ we have in particular

$$E_{n,0}^k(Z,1) = \sum_{\{C,D\}} |CZ + D|^{-k}.$$

If we write $Z = X + iY$ and for $\varepsilon > 0$ define

$$\mathfrak{V}_n(\varepsilon) = \{Z \in \mathfrak{H}_n \mid \sigma(X^2) \leqq \frac{1}{\varepsilon} , \quad Y \geqq \varepsilon E\},$$

then we can prove in a similar way as we did in §10 for the Eisenstein series of the modular group:

*Theorem* 1. *The Eisenstein series* $E^k_{n,r}(Z,f)$ *converge absolutely and uniformly in any domain* $\mathfrak{V}_n(\varepsilon)$ *with* $\varepsilon > 0$ *if* $k > n+r+1$, $k \equiv 0 \pmod 2$. *They represent modular forms of degree* $n$ *and weight* $k$.

The proof is based on the fact that $|Y^*|^{\frac{1}{2}k} f(Z^*)$ for $Z^* = X^* + iY^* \in \mathfrak{H}_r$ is bounded, so that up to a constant factor

$$G^k_{n,r}(Z) = \sum_{M:\Delta_{n,r}\backslash\Gamma_n} |\mathfrak{Im}\, M\langle Z\rangle^*|^{-\frac{1}{2}k} \, \|CZ+D\|^{-k}$$

is a majorant of $E^k_{n,r}(Z,f)$ in $\mathfrak{H}_n$. Furthermore one has to observe that there exists a positive constant $\delta$ which depends only on $\varepsilon$ and $n$ such that

$$|\mathfrak{Im}\, M\langle Z\rangle^*| \cdot \|CZ+D\|^2 \geqq \delta\, |\mathfrak{Im}\, M\langle iE\rangle^*| \cdot \|iC+D\|^2$$

for all $Z \in \mathfrak{V}_n(\varepsilon)$ and for all symplectic matrices $M = \begin{pmatrix} A & B \\ C & D \end{pmatrix}$.

Thus it suffices to prove the convergence of $G^k_{n,r}(iE)$, and this can be done by means of symplectic geometry.

Now it is not difficult to determine the image of $E^k_{n,r}(Z,f)$ under the map $\phi$ because Theorem 1 allows to pass to the limit

$$\lim_{\lambda\to\infty} E^k_{n,r}(Z,f), \qquad Z = \begin{pmatrix} Z_1 & 0 \\ 0 & i\lambda \end{pmatrix}$$

in the series termwise. We mention that for $r = n$ we have $\Delta_{n,n} = \Gamma_n$ so that $E^k_{n,n}(Z,f) = f(Z)$. One obtains

*Theorem* 2. *For* $n > 0$, $0 \leqq r \leqq n$, $k > n+r+1$, $k \equiv 0 \pmod 2$ *and* $f \in \mathcal{T}_r^k$ *we have*

$$\phi: E_{n,r}^k(*,f) \to \begin{cases} 0 & \text{for } r = n, \\ E_{n-1,r}^k(*,f) & \text{for } r < n. \end{cases}$$

The main tool for proving the announced structure theorems is the generalization of Petersson's scalar product

$$(f,g) = \int_{\mathcal{F}_n} f(Z)\,\overline{g(Z)}\,|Y|^k dv,$$

where $f,g \in \mathfrak{M}_n^k$ and at least one of the forms is a cusp form. Here $\mathcal{F}_n$ denotes Siegel's fundamental domain of $\Gamma_n$ (§12) and $dv$ is the invariant volume element $|Y|^{-n-1}[dX][dY]$. We introduce the orthogonal complement of $\mathcal{T}_n^k$ in $\mathfrak{M}_n^k$:

$$\mathfrak{N}_n^k = \{f \in \mathfrak{M}_n^k \mid (f,g) = 0 \text{ for all } g \in \mathcal{T}_n^k\}.$$

Clearly the restriction of $\phi$ to $\mathfrak{N}_n^k$ is a one-to-one mapping. We define the subspaces $\mathfrak{M}_{nr}^k$ $(0 \leqq r \leqq n)$ of $\mathfrak{M}_n^k$ recurrently by $\mathfrak{M}_{00}^k = \mathfrak{M}_0^k = \mathbb{C}$ and for $n > 0$

$$\mathfrak{M}_{nr}^k = \begin{cases} \{f \mid f \in \mathfrak{N}_n^k, \ f \mid \phi \in \mathfrak{M}_{n-1,r}^k\} & \text{for } 0 \leqq r < n, \\ \mathcal{T}_n^k. & \text{for } r = n. \end{cases}$$

By induction on $n$ we can then prove:

*Theorem* 3. *Let* $n \geqq 0$, $k > 2n$, $k \equiv 0 \pmod 2$. *Then*

1. $\mathfrak{M}_n^k = \mathfrak{N}_n^k \oplus \mathcal{T}_n^k$, $\quad \mathfrak{N}_n^k = \bigoplus_{r=1}^{n-1} \mathfrak{M}_{nr}^k$, $\quad \mathcal{T}_n^k = \mathfrak{M}_{nn}^k$.

2. $\mathfrak{M}^k_{nr} = \{E^k_{nr}(Z,f) \mid f \in \mathfrak{T}^k_r\}$ *for* $0 \leqq r \leqq n$.

3. *The mapping* $\phi\colon \mathfrak{M}^k_{nr} \to \mathfrak{M}^k_{n-1,r}$ *is bijective for* $0 \leqq r < n$.

Under the assumptions of Theorem 3 we have

$$\dim \mathfrak{M}^k_n = \dim \mathfrak{M}^k_{n-1} + \dim \mathfrak{T}^k_n$$

$$\mathfrak{M}^k_n | \phi = \mathfrak{M}^k_{n-1} ,$$

and the inverse of the map $\phi^{n-r}\colon \mathfrak{M}^k_{nr} \to \mathfrak{M}^k_{rr} = \mathfrak{T}^k_r$ is the map $f \mapsto E^k_{nr}(*,f)$. We do not know whether these relations hold for small values of $k$.

Since for even $k \geqq 4$ the Eisenstein series $E^k_{2,0}(Z,1)$ has a constant term $a(0)$ equal to 1 in its Fourier expansion, it defines a non-vanishing modular form and so $\dim \mathfrak{M}^k_{2,0} \geqq 1$ for $k \geqq 4$. From §13 we know that $\dim \mathfrak{M}^k_2 \leqq 1$ for $k = 4,6,8$, hence $\dim \mathfrak{M}^k_2 = 1$ for $k = 4,6,8$. This result yields in particular $(E^4_{2,0}(Z,1))^2 = E^8_{2,0}(Z,1)$, which has an application to the number theory of quadratic forms.

## §15. *Dirichlet series corresponding to modular forms*

Let

$$f(Z) = \sum_{T \geq 0} a(T) e^{2\pi i \sigma(TZ)}$$

be a modular form of degree $n$ and weight $k$. We ask the question whether $f(Z)$ is uniquely determined by the coefficients $a(T)$ with $T > 0$. If this is not the case, we shall call $f(Z)$ a *singular form*. If $f \in \{\Gamma_n, k\}$ is a singular form, then there exists a form $g \in \{\Gamma_n, k\}$ such that $g \neq f$ but all the coefficients $a(T)$ with $T > 0$ are the same for $f$ and $g$. Thus $h = f-g$ does not vanish but all its coefficients $a(T)$ corresponding to $T > 0$ are zero, from where we see immediately that either all forms in $\{\Gamma_n, k\}$ are singular or $\{\Gamma_n, k\}$ contains no singular forms.

*Theorem. If* $\{\Gamma_n, k\}$ *consists of singular forms then* $nk \equiv 0 \pmod 2$ *and* $0 < k \leq \frac{n-1}{2}$.

*Proof.* Assume that $\{\Gamma_n, k\}$ consists of singular forms, and let $f \in \{\Gamma_n, k\}$ be a non-zero form whose Fourier coefficients $a(T)$ vanish for all $T > 0$. Since $f(-Z^{-1}) = |Z|^k f(Z)$, we get for $g(Y) = f(iY)$ the transformation formula

$$g(\hat{Y}) = (-1)^{\frac{1}{2}nk} |Y|^k g(Y) \qquad (\hat{Y} = Y^{-1}).$$

Furthermore

$$g(Y) = \sum_{T \geq 0} a(T) e^{-2\pi\sigma(TY)},$$

where $a(T) = 0$ for all $T > 0$ but $a(T_1) \neq 0$ for at least one

$T_1 \geqq 0$. It follows that

$$\left|\frac{\partial}{\partial Y}\right| g(Y) = \sum_{T \geqq 0} |-2\pi T| a(T) e^{-2\pi\sigma(TY)} = 0.$$

In §6 (p. 79) we proved the transformation formula

$$|\hat{Y}| \left|\frac{\partial}{\partial \hat{Y}}\right| = (-1)^n |Y|^{\frac{1}{2}(n+1)} \left|\frac{\partial}{\partial Y}\right| |Y|^{\frac{1}{2}(1-n)}$$

or

$$\hat{M}_n = (-1)^n |Y|^{\frac{1}{2}(n-1)} M_n |Y|^{\frac{1}{2}(1-n)},$$

where

$$M_n = |Y| \left|\frac{\partial}{\partial Y}\right|.$$

From $M_n g(Y) = 0$ we obtain $\hat{M}_n g(\hat{Y}) = 0$ and the transformation formulas yield

$$|Y|^{\frac{1}{2}(n-1) - k} M_n |Y|^{k - \frac{1}{2}(n-1)} g(Y) = 0.$$

We also recall from §6 that the operators $M_1, M_2, \ldots, M_n$ form a basis of the ring $L$ of invariant differential operators on the weakly Riemannian space of all positive matrices (p. 67). By the formula of p. 82 we have

$$M_n |Y|^{k - \frac{1}{2}(n-1)} = (-1)^n \varepsilon_n\left(\frac{n-1}{2} - k\right) |Y|^{k - \frac{1}{2}(n-1)}$$

where

$$\varepsilon_n(s) = s\left(s - \frac{1}{2}\right) \ldots \left(s - \frac{n-1}{2}\right),$$

(p. 80), so that the operator identity

$$|Y|^{\frac{1}{2}(n-1) - k} M_n |Y|^{k - \frac{1}{2}(n-1)} = (-1)^n \varepsilon_n\left(\frac{n-1}{2} - k\right) + p(M_1, M_2, \ldots, M_n)$$

holds, where $p$ is a polynomial with constant coefficients but without constant term. Applying the operator termwise to the expansion of $g(Y)$ we obtain

$$\sum_{T \geqq 0} a(T)\{(-1)^n \varepsilon_n(\tfrac{n-1}{2} - k) + q(Y,T)\} e^{-2\pi\sigma(TY)} = 0,$$

where $q(Y,T)$ is a polynomial in the entries of $Y$ without constant term: $q(0,T) = 0$. It follows necessarily that

$$a(T)\{(-1)^n \varepsilon_n(\tfrac{n-1}{2} - k) + q(Y,T)\} = 0.$$

In particular $T = T_1$, $Y = 0$ yields $\varepsilon(\frac{n-1}{2} - k) = 0$, i.e.,

$$k(k - \tfrac{1}{2}) \ldots (k - \tfrac{n-1}{2}) = 0.$$

This proves the theorem.

Let us show by an example that singular forms actually do exist. Let $S = S^{(m)} = S' = (s_{\mu\nu})$ be a positive integral matrix with even diagonal elements, i.e., $\frac{1}{2}S$ is semi-integral, and such that $|S| = 1$. Such matrices $S$ exist if and only if $m \equiv 0 \pmod 8$. The function

$$\theta(Z,S) = \sum_G e^{\pi i \sigma(S[G]Z)},$$

where $G = G^{(m,n)}$ runs through all integral matrices, is a modular form of weight $k = \frac{1}{2}m$. In the Fourier series

$$\theta(Z,S) = \sum_{T \geqq 0} a(T,S) e^{2\pi i \sigma(TZ)} \qquad (Z = Z^{(n)})$$

the coefficient $a(T,S)$ is the number of integral matrices $G$ which satisfy $S[G] = 2T$. Now choose $2k = m < n$, then since the rank of $S[G]$ is $\leqq m$, we have $|T| = 0$ whenever $a(T,S) > 0$, i.e., $a(T,S) = 0$ for all $T > 0$. Thus $\theta(Z,S)$ is a singular form.

We now estimate the Fourier coefficients of a modular form:

*Lemma. Let*

$$f(Z) = \sum_{T \geq 0} a(T) e^{2\pi i \sigma(TZ)}$$

*be a modular form of degree n and weight k. Then*

$$|a(T)| < C|T|^k \qquad \text{for } T > 0,$$

*where* $C = C(n,f) > 0$ *is a constant.*

*Proof.* Let $Z = X + iY \in \mathfrak{H}_n$ be given. There exists a matrix

$$M = \begin{pmatrix} A & B \\ C & D \end{pmatrix} \in \Gamma_n$$

such that $Z_1 = M\langle Z\rangle$ belongs to the fundamental domain $\mathcal{F}_n$ of $\Gamma_n$. If $S = S'$ is integral, the matrix

$$M_0 = \begin{pmatrix} S & -E \\ E & 0 \end{pmatrix}$$

belongs to $\Gamma_n$. We set

$$Z_0 = X_0 + iY_0 = M_0^{-1}\langle Z\rangle = (-Z+S)^{-1},$$

so that

$$Z_1 = MM_0\langle Z_0\rangle, \qquad MM_0 = \begin{pmatrix} * & * \\ CS + D & -C \end{pmatrix}.$$

The expression $|CS + D|$ is a polynomial in the entries of $S = (s_{\mu\nu})$, of degree at most 2 in each variable $s_{\mu\nu}$, and not vanishing identically. Setting $X = (x_{\mu\nu})$, we can determine $S$ so that

$$-2 < s_{\mu\nu} - x_{\mu\nu} < 2, \qquad |CS + D| \neq 0.$$

We have

$$f(Z_o) = |-Z+S|^k f(Z),$$

$$f(Z_1) = |(CS+D)Z_o - C|^k f(Z_o),$$

hence

$$|f(Z)| = \| -Z+S \|^{-k} \| (CS+D)Z_o - C \|^{-k} |f(Z_1)|$$

$$\leqq C_1 \| -Z+S \|^{-k} \| Z_o - (CS+D)^{-1}C \|^{-k}$$

$$\leqq C_1 \| -Z+S \|^{-k} \cdot |Y_o|^{-k}$$

$$= C_1 \| -Z+S \|^{k} \cdot |Y|^{-k}$$

because

$$|Y_o| = |Y| \cdot \| -Z+S \|^{-2}.$$

If $Y = (y_{\mu\nu})$, then $|-iY+S-X|$ is a polynomial in the $y_{\mu\nu}$ with bounded coefficients, and since $\pm y_{\mu\nu} \leqq (y_{\mu\mu} y_{\nu\nu})^{\frac{1}{2}} \leqq \sigma(Y)$ we have

$$\| -Z+S \| \leqq C_2(1+\sigma(Y))^n,$$

hence

$$|f(Z)| \leqq C_3(1+\sigma(Y))^{nk} |Y|^{-k}.$$

Let $T = (t_{\mu\nu}) > 0$ be semi-integral and reduced. Choose $Y = T^{-1}$ and observe that

$$\sigma(T^{-1}) = \sum_{\nu=1}^{n} \frac{|T_{\nu\nu}|}{|T|} \leqq \sum_{\nu=1}^{n} \frac{\frac{1}{t_\nu} \prod_{\mu=1}^{n} t_\mu}{c \prod_{\mu=1}^{n} t_\mu} = \frac{1}{c} \sum_{\nu=1}^{n} \frac{1}{t_\nu} \leqq \frac{n}{c}.$$

We get therefore

$$|f(X+iT^{-1})| \leqq C_4|T|^k$$

and

$$|a(T)| = \left|\int_{\mathfrak{X}}\!\!\cdots\!\!\int f(X+iT^{-1})e^{-2\pi i\sigma(X+iT^{-1})T}[dX]\right|$$

$$\leqq C_4|T|^k e^{2\pi n} = C|T|^k.$$

Since both sides of this inequality are invariant under $T \mapsto T[U]$, where $U$ is unimodular, we see that the inequality holds for arbitrary $T > 0$, q.e.d.

Let $f(Z)$ be a modular form of degree $n$ and weight $k$, where we now assume that $k \equiv 0 \pmod 2$. Denoting the Fourier coefficients of $f(Z)$ by $a(T)$, we associate with $f(Z)$ the *Dirichlet series*

$$D(s) = \sum_{\{T\}>0} \frac{a(T)}{\varepsilon(T)|T|^s},$$

where the summation indicates that $T$ runs through a complete set of representatives of the sets

$$\{T[U] \mid U \text{ unimodular}\},$$

where $T > 0$, and $\varepsilon(T)$ indicates the number of unimodular matrices $U$ which verify $T[U] = T$. The numbers $\varepsilon(T)$ are finite since for each $T = (t_{\mu\nu}) > 0$ the equations $T[\mathfrak{u}_\nu] = t_{\nu\nu}$ have only a finite number of integral solutions.

In order to determine the analytical properties of the function defined by $D(s)$, the right thing to do seems to follow E. Hecke and to proceed as follows. We set $g(Y) = f(iY)$, so that

$$g(Y) = \sum_{\nu=0}^{n} g_\nu(Y),$$

where

$$g_\nu(Y) = \sum_{\text{rank } T = \nu} a(T)e^{-2\pi\sigma(YT)}.$$

As we have seen at the beginning of the proof of the Theorem, we have

$$g(\hat{Y}) = (-1)^{\frac{1}{2}nk}|Y|^k g(Y) \qquad (\hat{Y} = Y^{-1}).$$

We introduce the function

$$\xi(s) = \int_{\mathcal{R}} g_n(Y)|Y|^s dv,$$

where

$$dv = |Y|^{-\frac{1}{2}(n+1)}[dY]$$

and $\mathcal{R}$ is Minkowski's domain of reduced positive matrices.

We want to obtain a functional equation for $\xi(s)$. We observe that the integrand is invariant under the group of unimodular matrices so that $\mathcal{R}$ could be replaced by any other fundamental domain of this group. We decompose $\mathcal{R}$ by the hypersurface $|Y| = 1$ and note that the image of the "lower part" $\{Y \in \mathcal{R} \mid |Y| < 1\}$ of $\mathcal{R}$ under the map $Y \mapsto \hat{Y}$ is equivalent with the "upper part" $\{Y \in \mathcal{R} \mid |Y| > 1\}$. Thus we obtain

$$\xi(s) = \int_{\substack{\mathcal{R} \\ |Y| \geqq 1}} \{g_n(Y)|Y|^s + g_n(\hat{Y})|Y|^{-s}\}dv.$$

The transformation formula for $g(Y)$ concerning $Y \mapsto \hat{Y}$ yields

$$g_n(\hat{Y}) = (-1)^{\frac{1}{2}nk}|Y|^k g_n(Y) + \sum_{\nu=0}^{n-1} \{(-1)^{\frac{1}{2}nk}|Y|^k g_\nu(Y) - g_\nu(\hat{Y})\}$$

so that

$$\xi(s) = \int\limits_{\substack{\mathcal{R} \\ |Y|\geqq 1}} g_n(Y)\{|Y|^s + (-1)^{\frac{1}{2}nk}|Y|^{k-s}\}dv$$

$$+ \int\limits_{\substack{\mathcal{R} \\ |Y|\geqq 1}} \sum_{\nu=0}^{n-1} \{(-1)^{\frac{1}{2}nk}|Y|^k g_\nu(Y) - g_\nu(\hat{Y})\}|Y|^{-s}dv.$$

Obviously the first integral is invariant under multiplication by $(-1)^{\frac{1}{2}nk}$ and simultaneous substitution $s \mapsto k-s$. The same is true about the second integral, i.e., $\xi(s)$ satisfies the functional equation

$$\xi(k-s) = (-1)^{\frac{1}{2}nk}\xi(s),$$

but for the second integral this can be checked by a direct computation, which also shows that it is a rational function of $s$, only in the case $n \leqq 2$. Such a treatment for a general $n$ seems to be extremely difficult.

We avoid these difficulties by using invariant differential operators which eliminate the disturbing terms $g_\nu(Y)$ with $0 \leqq \nu < n$. The method is based upon the observation (used already in the proof of the Theorem) that

$$\left|\frac{\partial}{\partial Y}\right| e^{-2\pi\sigma(TY)} = (-2\pi)^n |T| e^{-2\pi\sigma(TY)} \qquad (T \geqq 0)$$

vanishes if $T$ is not positive, since then $|T| = 0$.

For an arbitrary integer $k$ we introduce, using the notation of §6, the operator

$$P_k = |Y|^{-k}\hat{M}_n |Y|^k M_n .$$

Since the invariant differential operators commute, we get

$$\hat{P}_k = |Y|^k M_n |Y|^{-k}\hat{M}_n = \hat{M}_n |Y|^k M_n |Y|^{-k},$$

or

$$\hat{P}_k = |Y|^k P_k |Y|^{-k}.$$

Now it is obvious that

$$h(Y) = P_k g(Y)$$

satisfies the same transformation formula as $g(Y)$:

$$h(\hat{Y}) = (-1)^{\frac{1}{2}nk}|Y|^k h(Y),$$

and has the expansion

$$h(Y) = \sum_{T>0} a(T) P_k e^{-2\pi\sigma(TY)}.$$

Although $h(Y)$ is actually more complicated than $g_n(Y)$, we can proceed with Hecke's method and obtain the desired result for $D(s)$.

We apply to $h(Y)$ a more general transformation and consider

$$\eta(s,u) = \int_{\mathcal{R}} h(Y)|Y|^s \hat{u}(Y)dv,$$

where $u(Y)$ denotes an arbitrary Größen-character (§10) and $\hat{u}(Y) = u(Y^{-1})$. Observe that $\hat{u}$ is also a Größen-character: it satisfies obviously conditions 1 - 4 and condition 5 follows from the considerations at the end of §7. First of all, we determine a majorant of the series which appears as the integrand, and whose general term is

$$a(T)|Y|^s \hat{u}(Y) P_k e^{-2\pi\sigma(TY)} \qquad (T > 0).$$

By the Lemma we have

$$|a(T)| < C|T|^k$$

and by definition the Größen-character $\hat{u}(Y)$ satisfies an estimate of the type

$$|\hat{u}(Y)| < C_1\left(\frac{\sigma(Y)}{\sqrt[n]{|Y|}}\right)^{\kappa},$$

where $\kappa$, $C_1$, and later on $C_2$, $C_3$, ... denote positive constants which depend only on $n$, $k$, $u$, $f$. For a reduced $Y = (y_{\mu\nu})$ there exists a constant $c = c(n) > 0$ such that $Y_0 = (\delta_{\mu\nu}y_{\nu\nu}) < cY$ hence

$$\sigma(Y) \leqq \sigma(Y_0T) = \sigma(Y_0[\sqrt{T}]) < c\sigma(Y[\sqrt{T}]) = c\sigma(YT)$$

and

$$|\hat{u}(Y)| \leqq C_2(\sigma(YT))^{\kappa}|Y|^{-\frac{\kappa}{n}} \qquad (Y \in \mathcal{R}).$$

Since

$$\mathrm{P}_k = (-1)^n|Y|^{\frac{1}{2}(n+1)-k}\left|\frac{\partial}{\partial Y}\right||Y|^{\frac{1}{2}(3-n)+k}\left|\frac{\partial}{\partial Y}\right|$$

is an invariant differential operator of degree $2n$, we have

$$\mathrm{P}_k e^{-2\pi\sigma(Y)} = p(Y)e^{-2\pi\sigma(Y)},$$

with a polynomial $p(Y)$ of degree at most $2n$. Consequently

$$|\mathrm{P}_k e^{-2\pi\sigma(Y)}| < C_3(1+\sigma(Y))^{2n}e^{-2\pi\sigma(Y)},$$

and, with the help of the substitution $Y \mapsto Y[\sqrt{T}]$,

$$|\mathrm{P}_k e^{-2\pi\sigma(YT)}| < C_3(1+\sigma(YT))^{2n}e^{-2\pi\sigma(YT)}.$$

The inequality

$$|YT|^k(\sigma(YT))^\kappa|P_k e^{-2\pi\sigma(YT)}| < C_4(1+\sigma(YT))^{(k+2)n+\kappa} e^{-2\pi\sigma(YT)}$$

$$< C_5 e^{-\pi\sigma(YT)}$$

yields

$$|a(T)|Y|^s \hat{u}(Y) P_k e^{-2\pi\sigma(YT)}| < C_6 |Y|^{\mathcal{R}e\, s - k - \frac{\kappa}{n}} e^{-\pi\sigma(YT)}$$

for $Y \in \mathcal{R}$, where $C_6 = CC_2C_5$. The number of semi-integral matrices $T = (t_{\mu\nu}) > 0$ with given diagonal entries can be estimated by $C_7 \prod_{\nu=1}^{n} t_{\nu\nu}^{\frac{1}{2}(n-1)}$ since $\pm t_{\mu\nu} \leqq (t_{\mu\mu}t_{\nu\nu})^{\frac{1}{2}}$, so that the total number of possible choices for $T$ is $\leqq \prod_{\mu<\nu} \{5\sqrt{t_{\mu\mu}t_{\nu\nu}}\} = 5^{\frac{1}{2}n(n-1)} \prod_{\mu=1}^{n} t_{\mu\mu}^{\frac{1}{2}(n-1)}$.

Thus

$$\sum_{T>0} e^{-\pi\sigma(YT)} < \sum_{T>0} e^{-\frac{\pi}{c}\sigma(Y_0 T)}$$

$$< C_7 \prod_{\nu=1}^{n} \sum_{t=1}^{\infty} t^{\frac{n-1}{2}} e^{-\frac{\pi}{c} y_{\nu\nu} t}$$

$$< C_8 \prod_{\nu=1}^{n} y_{\nu\nu}^{\frac{1-n}{2}} \sum_{t=1}^{\infty} e^{-\frac{\pi}{2c} y_{\nu\nu} t}$$

since

$$t^{\frac{n-1}{2}} e^{-\frac{\pi}{c} y_{\nu\nu} t} = y_{\nu\nu}^{-\frac{n-1}{2}} \{(y_{\nu\nu}t)^{\frac{n-1}{2}} e^{-\frac{\pi}{2c} y_{\nu\nu} t}\} e^{-\frac{\pi}{2c} y_{\nu\nu} t}$$

and the factor in the curly brackets is bounded. Using the inequality

$$\sum_{t=1}^{\infty} e^{-tx} = \frac{e^{-x}}{1-e^{-x}} \leqq \frac{1}{x} e^{-\frac{1}{2}x} \qquad (x > 0)$$

we get further

$$\sum_{T>0} e^{-\pi\sigma(YT)} < c_9 \prod_{\nu=1}^{n} y_{\nu\nu}^{-\frac{n+1}{2}} e^{-\frac{\pi}{4c} y_{\nu\nu}}$$

$$\leqq c_9 |Y|^{-\frac{n+1}{2}} e^{-\frac{\pi}{4c}\sigma(Y)}$$

$$\leqq c_9 |Y|^{-\frac{n+1}{2}} e^{-\frac{\pi n}{4c}\sqrt[n]{|Y|}} .$$

Thus the integrand in $\eta(s,u)$ is majorized by

$$c_6 c_9 |Y|^{\mathcal{R}\!e\, s - k - \frac{\kappa}{n} - \frac{n+1}{2}} e^{-\frac{\pi n}{4c}\sqrt[n]{|Y|}} .$$

The existence of the integral $\eta(s,u)$ follows now for

$$\mathcal{R}\!e\, s > k + \frac{\kappa}{n} + \frac{n+1}{2}$$

because the integral

$$\int_{\mathcal{R}_n} e^{-a|Y|^{\alpha}} |Y|^{s} dv = \frac{n+1}{2} v_n \frac{1}{\alpha} \Gamma(\tfrac{s}{\alpha}) a^{-\frac{s}{\alpha}}$$

exists for $a,\alpha,s > 0$. We obtain the last formula by applying Lemma 2 of §10 with $\phi(y) = e^{-ay^{\alpha}} y^{s}$.

We evaluate $\eta(s,u)$. The following operations are justified for sufficiently large values of $\mathcal{R}\!e\, s$.

$$\eta(s,u) = \sum_{T>0} a(T) \int_{\mathcal{R}} |Y|^{s} \hat{u}(Y) P_k e^{-2\pi\sigma(YT)} dv$$

$$= (-2\pi)^n \sum_{T>0} a(T)|T| \int_{\mathcal{R}} |Y|^{s-k} \hat{u}(Y) \hat{M}_n |Y|^{k+1} e^{-2\pi\sigma(YT)} dv .$$

We replace $T$ by $T[U]$, where $T$ runs through a complete set of representatives of the equivalence classes $\{T\} > 0$ and $U$ over all unimodular matrices. Keeping in mind that a given $T > 0$ will then appear $\varepsilon(T)$ times we get

$$\eta(s,u) = (-2\pi)^n \sum_{\{T\}>0} \frac{a(T)|T|}{\varepsilon(T)} \sum_U \int_{\mathcal{R}} |Y|^{s-k}\hat{u}(Y)\hat{M}_n|Y|^{k+1}e^{-2\pi\sigma(Y[U']T)}dv$$

$$= (-2\pi)^n \sum_{\{T\}>0} \frac{a(T)|T|}{\varepsilon(T)} \sum_U \int_{\mathcal{R}[U']} |Y|^{s-k}\hat{u}(Y)\hat{M}_n|Y|^{k+1}e^{-2\pi\sigma(YT)}dv.$$

Here we used the invariance of $u(Y)$ and $\hat{M}_n$ under $Y \mapsto Y[U']$. Since $\bigcup_U \mathcal{R}[U']$ covers the space of all positive matrices twice, we obtain

$$\eta(s,u) = 2(-2\pi)^n \sum_{\{T\}>0} \frac{a(T)|T|}{\varepsilon(T)} \int_{Y>0} |Y|^{s-k}\hat{u}(Y)\hat{M}_n|Y|^{k+1}e^{-2\pi\sigma(YT)}dv.$$

Since (§6) $\hat{M}_n$ is the adjoint of $M_n$ and since by the definition of a Größen-character the partial derivatives of $u(Y)$ of a given degree $h$ do not grow faster to infinity than a function of type

$$C(\sigma(Y))^{\kappa}|Y|^{-\frac{\kappa+h}{n}} \qquad (C,\kappa > 0),$$

we have

$$\eta(s,u) = 2(-2\pi)^n \sum_{\{T\}>0} \frac{a(T)|T|}{\varepsilon(T)} \int_{Y>0} e^{-2\pi\sigma(YT)}|Y|^{k+1}M_n|Y|^{s-k}\hat{u}(Y)dv.$$

Now we have to find a suitable expression for $M_n|Y|^{s-k}\hat{u}(Y)$. We recall that $u(Y)$ is an eigenfunction of $L$ so that

$$\sigma(Y \tfrac{\partial}{\partial Y})^h \hat{u}(Y) = \lambda_h \hat{u}(Y) \qquad (1 \leqq h \leqq n),$$

and in particular $\lambda_1 = 0$. In §7 we used already an operator identity of the form

$$|Y|^{k-s} M_n |Y|^{s-k} = s^n + q(s, \sigma(Y \tfrac{\partial}{\partial Y}), \ldots, \sigma(Y \tfrac{\partial}{\partial Y})^n),$$

where $q(x_0, x_1, \ldots, x_n)$ is a polynomial with constant coefficients whose degree in $s$ is less than $n$. We apply this identity to $\hat{u}(Y)$ and get

$$|Y|^{k-s} M_n |Y|^{s-k} \hat{u}(Y) = (s^n + q(s, 0, \lambda_2, \ldots, \lambda_n))\hat{u}(Y)$$

or

$$M_n |Y|^{s-k} \hat{u}(Y) = \chi(s, \hat{u}) |Y|^{s-k} \hat{u}(Y),$$

where $\chi(s, \hat{u}) = s^n + \ldots$ is a polynomial in $s$ of degree $n$ with constant coefficients depending on $\hat{u}$. We can thus write

$$\eta(s,u) = 2(-2\pi)^n \chi(s,\hat{u}) \sum_{\{T\}>0} \frac{a(T)|T|}{\varepsilon(T)} \int_{Y>0} e^{-2\pi\sigma(YT)} |Y|^{s+1} \hat{u}(Y)\, dv.$$

It follows from the main result of §7 (replacing $X$, $s$, $u$ by $\frac{1}{2\pi} T^{-1}$, $s+1$, $\hat{u}$, respectively) that

$$\int_{Y>0} e^{-2\pi\sigma(YT)} |Y|^{s+1} \hat{u}(Y)\, dv =$$

$$= \pi^{\frac{1}{4}n(n-1)} (2\pi)^{-n(s+1)} \prod_{\nu=1}^{n} \Gamma(s+1-\alpha_\nu) |T|^{-s-1} u(T)$$

$$= \pi^{\frac{1}{4}n(n-1)} (2\pi)^{-n(s+1)} f(s) \prod_{\nu=1}^{n} \Gamma(s-\alpha_\nu) |T|^{-s-1} u(T),$$

where

$$f(s) = (s-\alpha_1)(s-\alpha_2) \ldots (s-\alpha_n),$$

$$|Y|^{\frac{1}{2}(n-1)-s} M_n |Y|^{s-\frac{1}{2}(n-1)} \hat{\Omega}(Y) = f(s)\hat{\Omega}(Y)$$

(p. 94). Comparing the last formula with the equation which defines $\chi(s,\hat{\Omega})$, we get

$$f(s) = \chi(s+k-\frac{n-1}{2}, \hat{\Omega})$$

and so finally

$$\eta(s,u) = 2(-1)^n \pi^{\frac{1}{2}n(n-1)} \chi(s,\Omega)\chi(s+k-\frac{n-1}{2}, \Omega)(2\pi)^{-ns} \prod_{\nu=1}^{n} \Gamma(s-\alpha_\nu) D(s,u),$$

where the Dirichlet series

$$D(s,u) = \sum_{\{T\}>0} \frac{a(T)u(T)}{\varepsilon(T)|T|^s}$$

generalizes the series $D(s)$ introduced earlier.

The functional equation for $\eta(s,u)$ or, more precisely, for $D(s,u)$ can now be proved very simply. Indeed

$$\eta(s,u) = \int_{\mathcal{R}} h(Y)|Y|^s \hat{\Omega}(Y)dv$$

$$= \int_{\substack{\mathcal{R} \\ |Y|\geq 1}} \{h(Y)|Y|^s \Omega(Y) + h(\hat{Y})|Y|^{-s} u(Y)\}dv$$

$$= \int_{\substack{\mathcal{R} \\ |Y|\geq 1}} h(Y)\{|Y|^s \Omega(Y) + (-1)^{\frac{1}{2}nk}|Y|^{k-s} u(Y)\}dv,$$

and this representation shows that $\eta(s,u)$ is an entire function of $s$ and that

$$\eta(k-s,u) = (-1)^{\frac{1}{2}nk}\eta(s,\hat{u}).$$

Introduce now the function

$$\xi(s,u) = (2\pi)^{-ns}\prod_{\nu=1}^{n}\Gamma(s-\alpha_\nu)D(s,u).$$

We claim, and this is the main result of this section, that $\xi(s,u)$ satisfies the same functional equation

$$\xi(k-s,u) = (-1)^{\frac{1}{2}nk}\xi(s,\hat{u}).$$

Since

$$\eta(s,u) = 2(-1)^n\pi^{\frac{1}{2}n(n-1)}\chi(s,\hat{u})\chi(s+k-\tfrac{n-1}{2},\hat{u})\xi(s,u),$$

we have only to check that $\chi(s,\hat{u})\chi(s+k-\frac{n-1}{2},\hat{u})$ is invariant under the simultaneous substitutions $s \mapsto k-s$, $u \mapsto \hat{u}$. The defining relation

$$M_n|Y|^{s-k}\hat{u}(Y) = \chi(s,\hat{u})|Y|^{s-k}\hat{u}(Y)$$

of $\chi(s,\hat{u})$ transforms under the substitutions $Y \mapsto \hat{Y}$, $s \mapsto 2k-\frac{n-1}{2}-s$ into

$$\hat{M}_n|Y|^{s-k+\frac{1}{2}(n-1)}u(Y) = \chi(2k-\tfrac{n-1}{2}-s,\hat{u})|Y|^{s-k+\frac{1}{2}(n-1)}u(Y),$$

or, using the formula for $\hat{M}_n$,

$$M_n|Y|^{s-k}u(Y) = (-1)^n\chi(2k-\tfrac{n-1}{2}-s,\hat{u})|Y|^{s-k}u(Y),$$

so that

$$\chi(s,u) = (-1)^n \chi(2k - \tfrac{n-1}{2} - s\ ,\hat{u})$$

and

$$\chi(k-s,u)\chi(2k-s-\tfrac{n-1}{2},u) = \chi(s+k-\tfrac{n-1}{2},\hat{u})\chi(s,\hat{u}),$$

which is what we wanted to prove.

Together with $\eta(s,u)$ also

$$\chi(s,\hat{u})D(s,u) = \prod_{\nu=1}^{n} (s+\tfrac{n-1}{2}-k-\alpha_\nu)D(s,u)$$

is an entire function of $s$. Since the roots $\alpha_1,\alpha_2,\ldots,\alpha_n$ depend on $\hat{u}$, we shall write $\alpha_\nu = \alpha_\nu(\hat{u})$ and correspondingly $\hat{\alpha}_\nu = \alpha_\nu(u)$. The identity

$$f(s) = \prod_{\nu=1}^{n} (s-\alpha_\nu) = \chi(s+k-\tfrac{n-1}{2},\hat{u})$$

yields

$$\chi(s,u) = \prod_{\nu=1}^{n} (s-k+\tfrac{n-1}{2}-\hat{\alpha}_\nu)$$

and

$$(-1)^n\chi(2k-\tfrac{n-1}{2}-s\ ,\hat{u}) = \prod_{\nu=1}^{n} (s-k+\alpha_\nu).$$

But the last two polynomials are identical, hence we can order the roots in such a way that

$$\alpha_\nu + \hat{\alpha}_\nu = \frac{n-1}{2} \qquad (1 \leqq \nu \leqq n),$$

which is in accordance with the relation

$$\sum_{\nu=1}^{n} \alpha_\nu = \sum_{\nu=1}^{n} \hat{\alpha}_\nu = \frac{n-1}{4}$$

(p. 94).

## §16. *Zeta functions attached to quadratic forms*

Let $S = S^{(m)}$ be a positive real matrix and $u(Y)$ a Größencharacter in the sense of §10 defined for $Y = Y^{(n)} > 0$, where $m > n$. Denote by $w_0(X)$ a homogeneous polynomial of degree $2nk$ in the entries of $X = X^{(m,n)}$ which verifies

$$w_0(XV) = |V|^{2k} w_0(X) \qquad \text{for all} \quad V = V^{(n)},$$

where $k$ is an integer $\geqq 0$. We introduce the Dirichlet series

$$\phi_0(s, S; w_0, u) = \sum_{\{G\}} \frac{w_0(QG)u(S[G])}{|S[G]|^{s+k}} \qquad (S = Q'Q, \quad Q = Q^{(m)}),$$

where the summation indicates that $G = G^{(m,n)}$ runs through a complete set of non-associated integral matrices of rank $n$, calling $G$ and $G_1$ associated if $G_1 = GU$, with a unimodular $U = U^{(n)}$.

The interest in the analytical properties of the functions defined by these Dirichlet series was prompted by the following number-theoretical problem [15]. In the Minkowski domain $\mathcal{R} = \mathcal{R}_n$ of reduced positive matrices let $\mathfrak{A}$ denote a subcone, i.e., a subset such that if $Y$ belongs to $\mathfrak{A}$ then $\lambda Y \in \mathfrak{A}$ for all $\lambda > 0$. On the other hand denote by $\mathfrak{L}$ a subset of the set of all real matrices $X = X^{(m,n)}$ such that if $X \in \mathfrak{L}$, then $XV \in \mathfrak{L}$ for any nonsingular $V = V^{(n)}$. Under suitable measurability assumptions on $\mathfrak{A}$ and $\mathfrak{L}$ one wants to find an asymptotic estimate for the number $a_t(\mathfrak{A}, \mathfrak{L})$ of all integral matrices $G = G^{(m,n)} \in \mathfrak{L}$ such that

$$S[G] = T \quad \text{with} \quad T \in \mathfrak{A} \text{ and } |T| = t.$$

Let $u$ be a function defined on the space of all positive matrices $Y = Y^{(n)}$ such that

$$u(Y) = \begin{cases} 1 & \text{for } Y \in \mathfrak{A}, \\ 0 & \text{for } Y \in \mathfrak{R}, \quad Y \notin \mathfrak{A}, \end{cases}$$

$$u(Y[U]) = u(Y) \quad \text{for all unimodular } U,$$

and let $w_0$ be a function defined on the set of all real matrices $X = X^{(m,n)}$ such that

$$\frac{w_0(QG)}{|S[G]|^k} = \begin{cases} 1 & \text{if } G \in \mathfrak{L} \\ 0 & \text{otherwise.} \end{cases} \qquad (S = Q'Q)$$

The Dirichlet series formed with these functions is then obviously

$$\phi_0(s,S;w_0,u) = \sum_{t=1}^{\infty} \frac{a_t(\mathfrak{A},\mathfrak{L})}{t^s}.$$

If we approximate $u$ by Größen-characters and $w_0$ by spherical harmonics, we are led to Dirichlet series of the above sort. The number-theoretical problem can be couched also in terms of lattices.

Returning to the Dirichlet series introduced at the beginning, it can be shown that $\phi_0$ is a meromorphic function of $s \in \mathbb{C}$ and satisfies a functional equation of the Riemann type if $w_0(X)$ is a harmonic function, i.e.,

$$\sigma\left(\frac{\partial}{\partial X'}\frac{\partial}{\partial X}\right)w_0(X) = 0 \qquad \text{with } X = (x_{\mu\nu}), \quad \frac{\partial}{\partial X} = \left(\frac{\partial}{\partial x_{\mu\nu}}\right).$$

In order to prove that $\phi_0$ is meromorphic also for the more general polynomials $w_0(X)$ described above, it is reasonable to enlarge the set of admitted polynomials $w_0$ and to consider all polynomials $w_0(X)$

which satisfy $w_o(XV) = w_o(X)$ for all $V = V^{(n)}$ with $V'V = E$. Now it is known from the theory of algebraic invariants that such a polynomial is a polynomial in the entries of $XX'$ so that we can write $w_o(X) = w(XX')$. Instead of $\phi_o$ we then get a more general function $\phi(s,S;w,u)$, which is not a pure Dirichlet series any more but has the desired properties: $\phi$ satisfies a functional equation of the Riemann type and it is meromorphic in $\mathbb{C}$. The quotient

$$\phi(s,S;w,u) : \phi_o(s,S;w_o,u)$$

is a polynomial in $s$ with constant coefficients if $\phi_o$ is defined at all, i.e., if $w_o$ is of the above given type.

We start our investigation with the theta-series

$$\theta(Y,S;w) = \sum_G w(Y[G'Q'])e^{-\pi\sigma(YS[G])},$$

where $G$ runs through all integral matrices $G = G^{(m,n)}$. Writing $Y = R'R$ we also have

$$w(Y[G'Q']) = w(QGR'RG'Q') = w_o(QGR').$$

It can be shown [16, pp. 3-8] that

$$\theta(\hat{Y},\hat{S};w) = |S|^{\frac{1}{2}n}|Y|^{\frac{1}{2}m}\theta(Y,S;\hat{w}),$$

where $\hat{Y} = Y^{-1}$, $\hat{S} = S^{-1}$ and

$$\hat{w}(XX') = \int w(TT')e^{-\pi\sigma(T+iX)'(T+iX)}[dT]$$

$$= e^{-\frac{\Delta}{4\pi}} w(-XX')$$

with

$$\Delta = \sigma(\frac{\partial}{\partial X'} \frac{\partial}{\partial X}),$$

and the domain of integration is the space of all real matrices $T = T^{(m,n)}$.

For a given $w(XX')$ there exists a differential operator $L(X \frac{\partial}{\partial X'})$, which is a polynomial in the entries of $X \frac{\partial}{\partial X'}$ with constant coefficients, such that

$$L(X \frac{\partial}{\partial X'})e^{-\pi\sigma(X'X)} = w(XX')e^{-\pi\sigma(X'X)}$$

[16, pp. 8-9]. Conversely it is obvious that if $L(X \frac{\partial}{\partial X'})$ is given, then the polynomial $w_0(X)$ defined by

$$L(X \frac{\partial}{\partial X'})e^{-\pi\sigma(X'X)} = w_0(X)e^{-\pi\sigma(X'X)}$$

satisfies $w_0(XV) = w_0(X)$ for all orthogonal $V$, i.e., $w_0(X) = w(XX')$.

With the differential operator

$$P_k = |Y|^{-k}\hat{M}_n|Y|^k M_n,$$

which already in §15 helped us to eliminate disturbing terms, we form

$$\Theta(Y,S;w) = P_{\frac{1}{2}m}\theta(Y,S;w).$$

Using

$$\hat{P}_k = |Y|^k P_k |Y|^{-k}$$

we have again

$$\Theta(\hat{Y},\hat{S};w) = |S|^{\frac{1}{2}n}|Y|^{\frac{1}{2}m}\Theta(Y,S;\hat{w}).$$

We claim that in the series

$$\Theta(Y,S;w) = \sum_{G} P_{\frac{1}{2}m} w(Y[G'Q'])e^{-\pi\sigma(YS[G])}$$

the terms with $r$ = rank $G < n$ vanish. Indeed, if $G = G^{(m,n)} = (G_1^{(m,r)},0)U$, where $r$ = rank $G$ and $U = U^{(n)}$ is a non-singular matrix, then

$$G' = U'\begin{pmatrix} G_1' \\ 0 \end{pmatrix} .$$

Furthermore

$$\sigma(YS[G]) = \sigma(YG'Q'QG) = \sigma(QGYG'Q') = \sigma(Y[G'Q']),$$

where

$$Y[G'Q'] = Y\left[U'\begin{pmatrix} G_1' \\ 0 \end{pmatrix}Q'\right] = Y\left[U'\begin{pmatrix} G_1'Q' \\ 0 \end{pmatrix}\right] .$$

Since $M_n = |Y|\,|\frac{\partial}{\partial Y}|$, it is sufficient to prove that

$$|Y|\,|\tfrac{\partial}{\partial Y}|w(Y[G'Q'])e^{-\pi\sigma(YS[G])}$$

$$= |Y|\,|\tfrac{\partial}{\partial Y}|w(Y\left[U'\begin{pmatrix} G_1'Q' \\ 0 \end{pmatrix}\right])e^{-\pi\sigma(Y\left[U'\begin{pmatrix} G_1'Q' \\ 0 \end{pmatrix}\right])}$$

is zero. Since $P_{\frac{1}{2}m}$ and $M_n$ are invariant operators, we have only to consider the expression obtained by the substitution $Y \mapsto Y[U'^{-1}]$:

$$|Y|\,|\tfrac{\partial}{\partial Y}|w\left(Y\begin{bmatrix} G_1'Q' \\ 0 \end{bmatrix}\right)e^{-\pi\sigma\left(Y\begin{bmatrix} G_1'Q' \\ 0 \end{bmatrix}\right)} =$$

$$= |Y|\,|\tfrac{\partial}{\partial Y}|\,\omega(Y_1[G_1Q'])e^{-\pi\sigma(Y_1[G_1Q'])}$$

where $Y_1$ is defined by

$$Y = \begin{pmatrix} Y_1^{(r)} & * \\ * & * \end{pmatrix}.$$

Now $|\frac{\partial}{\partial Y}|f(Y_1) = 0$ if $r < n$ for an arbitrary function $f(Y_1)$, since $f(Y_1)$ is then independent of the entries $y_{n1}, y_{n2}, \ldots, y_{nn}$. This proves our claim.

Let $\mathcal{R}$ be again the Minkowski domain of reduced positive matrices, $dv = |Y|^{-\frac{1}{2}(n+1)}[dY]$, and introduce the transform

$$\Xi(s,S;\omega,u) = \int_{\mathcal{R}} \Theta(Y,S;\omega)|Y|^s\hat{u}(Y)dv \qquad (\hat{u}(Y) = u(\hat{Y})).$$

Performing the transformation already considered in §15, we obtain the expression

$$\Xi(s,S;\omega,u) = \int_{\substack{\mathcal{R} \\ |Y| \geqq 1}} \{\Theta(Y,S;\omega)|Y|^s\hat{u}(Y) + |S|^{-\frac{1}{2}n}\Theta(Y,\hat{S};\hat{\omega})|Y|^{\frac{1}{2}m-s}u(Y)\}dv,$$

which proves that $\Xi(s,S;\omega,u)$ is a holomorphic function of $s \in \mathbb{C}$ and satisfies the functional equation

$$|S|^{\frac{1}{2}n}\Xi(\tfrac{m}{2}-s,S;\omega,u) = |\hat{S}|^{\frac{1}{2}n}\Xi(s,\hat{S};\hat{\omega},\hat{u}).$$

We compute $\Xi(s,S;\omega,u)$. We have

$$\Xi(s,S;\omega,u) = \sum_{\substack{G \\ \text{rank } G = n}} \int_{\mathcal{R}} |Y|^s\hat{u}(Y)P_{\frac{1}{2}m}\omega(Y[G'Q'])e^{-\pi\sigma(YS[G])}dv$$

$$= \sum_{\{G\}} \sum_{U} \int_{\mathcal{R}} |Y|^s\hat{u}(Y)P_{\frac{1}{2}m}\omega(Y[U'G'Q'])e^{-\pi\sigma(S[G]Y[U'])}dv,$$

where the first summation indicates that $G$ runs through a complete set of representatives of the cosets $G\Gamma_n$ modulo the group $\Gamma_n$ of unimodular matrices $U = U^{(n)}$, with rank $G = n$, and where in the second summation $U$ runs through $\Gamma_n$. Since $\bigcup_{U\in\Gamma_n} \mathcal{R}[U']$ covers the space of positive matrices twice, it follows that

$$\Xi(s,S;w,u) = \sum_{\{G\}} \sum_{U} \int_{\mathcal{R}[U']} |Y|^s \hat{u}(Y) P_{\frac{1}{2}m} w(Y[G'Q']) e^{-\pi\sigma(S[G]Y)} dv$$

$$= 2 \sum_{\{G\}} \int_{Y>0} |Y|^s \hat{u}(Y) P_{\frac{1}{2}m} w(Y[G'Q']) e^{-\pi\sigma(S[G]Y)} dv$$

$$= 2 \sum_{\{G\}} \int_{Y>0} |Y|^{s-\frac{1}{2}m} \hat{u}(Y) \hat{M}_n |Y|^{\frac{1}{2}m} M_n w(Y[G'Q']) e^{-\pi\sigma(S[G]Y)} dv$$

$$= 2 \sum_{\{G\}} \int_{Y>0} \{M_n |Y|^{s-\frac{1}{2}m} \hat{u}(Y)\} |Y|^{\frac{1}{2}m} \{M_n w(Y[G'Q']) e^{-\pi\sigma(S[G]Y)}\} dv .$$

As in §15 we have

$$M_n |Y|^{s-\frac{1}{2}m} \hat{u}(Y) = \chi(s,\hat{u}) |Y|^{s-\frac{1}{2}m} \hat{u}(Y),$$

where $\chi(s,\hat{u}) = s^n + \ldots$ is a polynomial in $s$ of degree $n$ with constant coefficients depending on $\hat{u}$. We can proceed with

$$\Xi(s,S;w,u) = 2\chi(s,\hat{u}) \sum_{\{G\}} \int_{Y>0} |Y|^s \hat{u}(Y) M_n w(Y[G'Q']) e^{-\pi\sigma(S[G]Y)} dv$$

$$= 2\chi(s,\hat{u}) \sum_{\{G\}} \int_{Y>0} e^{-\pi\sigma(S[G]Y)} w(Y[G'Q']) \hat{M}_n |Y|^s \hat{u}(Y) dv.$$

The above relation for $\hat{u}$ together with

$$M_n = (-1)^n |Y|^{-\frac{1}{2}(n-1)} \hat{M}_n |Y|^{\frac{1}{2}(n-1)}$$

yields

$$\hat{M}_n |Y|^{s-\frac{1}{2}(m-n+1)} \hat{u}(Y) = (-1)^n \chi(s,\hat{u}) |Y|^{s-\frac{1}{2}(m-n+1)} \hat{u}(Y)$$

or

$$\hat{M}_n |Y|^s \hat{u}(Y) = (-1)^n \chi(s + \frac{m-n+1}{2}, \hat{u}) |Y|^s \hat{u}(Y),$$

and therefore

$$E(s,S;w,u) =$$

$$= 2(-1)^n \chi(s,\hat{u}) \chi(s + \frac{m-n+1}{2}, \hat{u}) \sum_{\{G\}} \int_{Y>0} e^{-\pi\sigma(S[G]Y)} |Y|^s w(Y[G'Q']) \hat{u}(Y) dv.$$

Next we eliminate $w$ from under the integral sign. Setting $Y = RR'$ and replacing $X$ by $XR$ in the relation which associates with $w(XX')$ the differential operator $L(X \frac{\partial}{\partial X^{\mathsf{T}}})$, we obtain

$$L(X \frac{\partial}{\partial X^{\mathsf{T}}}) e^{-\pi\sigma(Y[X'])} = w(Y[X']) e^{-\pi\sigma(Y[X'])}.$$

If we set $X = QG$ in the general term of the above series, we have $\sigma(S[G]Y) = \sigma(G'Q'QGY) = \sigma(Y[G'Q']) = \sigma(Y[X'])$ and the integral becomes

$$J(s,X;w,\hat{u}) = \int_{Y>0} e^{-\pi\sigma(Y[X'])} |Y|^s w(Y[X']) \hat{u}(Y) dv$$

$$= L(X \frac{\partial}{\partial X^{\mathsf{T}}}) J(s,X;1,\hat{u}),$$

where, in the notation of §7, we have

$$J(s,X;1,\hat{u}) = \int_{Y>0} e^{-\pi\sigma(Y[X'])}|Y|^{s}\hat{u}(Y)dv =$$

$$= J_s(\tfrac{1}{\pi}(X'X)^{-1},\hat{u}) = \pi^{\frac{1}{4}n(n-1)-ns}\prod_{\nu=1}^{n}\Gamma(s-\alpha_\nu)|X'X|^{-s}u(X'X),$$

and $\alpha_1,\alpha_2,\ldots,\alpha_n$ are the roots of the polynomial

$$f(s) = \chi(s+\frac{m-n+1}{2},\hat{u}) = \prod_{\nu=1}^{n}(s-\alpha_\nu)$$

defined by

$$M_n|Y|^{s-\frac{1}{2}(n-1)}\hat{u}(Y) = f(s)|Y|^{s-\frac{1}{2}(n-1)}\hat{u}(Y)$$

or

$$\hat{M}_n|Y|^{s}\hat{u}(Y) = (-1)^n f(s)|Y|^{s}\hat{u}(Y)$$

(p. 94).

If we substitute back $QG$, with a variable $G$, instead of $X$, then $\frac{\partial}{\partial X}$ becomes $Q'^{-1}\frac{\partial}{\partial G}$ and $X\frac{\partial}{\partial X'}$ will become $QG\frac{\partial}{\partial G'}Q^{-1}$, so that

$$\int_{Y>0} e^{-\pi\sigma(S[G]Y)}|Y|^{s}w(Y[G'Q'])\hat{u}(Y)dv =$$

$$= J(s,QG;w,\hat{u}) = L(QG\frac{\partial}{\partial G'}Q^{-1})J(s,QG;1,\hat{u}) =$$

$$= \pi^{\frac{1}{4}n(n-1)-ns}\prod_{\nu=1}^{n}\Gamma(s-\alpha_\nu)L(QG\frac{\partial}{\partial G'}Q^{-1})\left\{\frac{u(S[G])}{|S[G]|^{s}}\right\}$$

and finally

$\Xi(s,S;w,u) =$

$$= (-1)^n 2\pi^{\frac{1}{4}n(n-1)} \chi(s,\hat{u})\chi(s+\frac{m-n+1}{2},\hat{u})\pi^{-ns}\prod_{\nu=1}^{n}\Gamma(s-\alpha_\nu)\cdot\phi(s,S;w,u),$$

where

$$\phi(s,S;w,u) = \sum_{\{G\}} L(QG\frac{\partial}{\partial G^T}Q')\left\{\frac{u(S[G])}{|S[G]|^s}\right\}.$$

We saw at the end of §15 (with $2k$ instead of $m$) that the function $\chi(s,\hat{u})\chi(s+\frac{m-n+1}{2},\hat{u})$ remains invariant if we replace $s$ by $\frac{1}{2}m-s$ and $u$ by $\hat{u}$. It follows therefore from the functional equation of $\Xi(s,S;w,u)$ that the function

$$\xi(s,S;u,v) = \pi^{-ns}\prod_{\nu=1}^{n}\Gamma(s-\alpha_\nu)|S|^{\frac{1}{4}n}\cdot\phi(s,S;w,u)$$

satisfies the functional equation

$$\xi(\tfrac{m}{2}-s,S;w,u) = \xi(s,\hat{S};\hat{w},\hat{u}).$$

Together with $\Xi(s,S;w,u)$, also

$$\chi(s,\hat{u})\phi(s,S;w,u) = \prod_{\nu=1}^{n}(s-\frac{m-n+1}{2}-\alpha_\nu)\phi(s,S;w,u)$$

is an entire function of $s\in\mathbb{C}$.

We return to the special case considered at the beginning, when $w(XX') = w_0(X)$, where

$$w_0(XV) = |V|^{2k}w_0(X) \qquad \text{for all} \quad V = V^{(n)}.$$

Then setting $Y = RR'$ we have

$$w(Y[G'Q']) = w_0(QGR) = |R|^{2k} w_0(QG) = |Y|^k w_0(QG)$$

and so

$$J(s, QG; w, \hat{u}) = w_0(QG) \int_{Y>0} e^{-\pi\sigma(S[G]Y)} |Y|^{s+k} \hat{u}(Y)\, dv$$

$$= w_0(QG) J(s+k, QG; 1, \hat{u})$$

$$= \pi^{\frac{1}{4}n(n-1)-n(s+k)} \prod_{\nu=1}^{n} \Gamma(s+k-\alpha_\nu) \frac{w_0(QG)u(S[G])}{|S[G]|^{s+k}},$$

hence

$$\Xi(s, S; w, u) =$$

$$= (-1)^n 2\pi^{\frac{1}{4}n(n-1)} \chi(s, \hat{u}) \chi(s + \frac{m-n+1}{2}, \hat{u}) \pi^{-n(s+k)} \prod_{\nu=1}^{n} \Gamma(s+k-\alpha_\nu) \phi_0(s, S; w_0, u),$$

where $\phi_0(s, S; w_0, u)$ is the Dirichlet series introduced at the beginning

$$\phi_0(s, S; w_0, u) = \sum_{\{G\}} \frac{w_0(QG)u(S[G])}{|S[G]|^{s+k}} \qquad (S = Q'Q).$$

Comparing the two expressions for $\Xi(s, S; w, u)$ we obtain

$$\phi(s, S; w, u) = \pi^{-nk} \prod_{\nu=1}^{n} \frac{\Gamma(s+k-\alpha_\nu)}{\Gamma(s-\alpha_\nu)} \cdot \phi_0(s, S; w_0, u)$$

$$= \pi^{-nk} \prod_{\nu=0}^{k-1} f(s+\nu) \cdot \phi_0(s, S; w_0, u),$$

so that the quotient $\phi/\phi_o$ is indeed a polynomial as we asserted at the beginning.

For number-theoretical applications the computation of the residues of $\phi_o = \phi_o(s,S;w_o,u)$ has a special significance. We will not solve the general problem here, but we can show, using a method of C. L. Siegel [30], that there exists a positive constant $\delta$ such that all functions $\phi_o$ are regular in the half-plane $\mathcal{R}e\, s > \frac{1}{2}m - \delta$ with the exception of the point $s = \frac{1}{2}m$ where $\phi_o$ is either regular or has a pole of order 1, provided that

$$|u(Y)| < C\left(\frac{\sigma(Y)}{\sqrt[n]{|Y|}}\right)^{\kappa} \qquad \text{with } \kappa < \frac{n}{2}\,.$$

We can choose for instance $\delta = \frac{1}{2} - \frac{\kappa}{n}$. We shall obtain an explicit formula for the residue

$$\lim_{s\to\frac{1}{2}m} (s - \tfrac{m}{2})\phi_o(s,S;w_o,u)$$

of $\phi_o$ at $s = \frac{1}{2}m$.

We use the previous notation and introduce

$$f(Y,S;w) = \sum_{\text{rank } G = n} w(Y[G'Q'])e^{-\pi\sigma(S[G]Y)},$$

$$g(Y,S;w) = \sum_{\text{rank } G < n} w(Y[G'Q'])e^{-\pi\sigma(S[G]Y)},$$

$$h(Y,S;w) = \sum_{G\neq 0} w(Y[G'Q'])e^{-\pi\sigma(S[G]Y)},$$

so that

$$f(Y,S;w) = |S|^{-\frac{1}{2}n}|Y|^{-\frac{1}{2}m}h(\hat{Y},\hat{S};\hat{w}) - g(Y,S;w) + |S|^{-\frac{1}{2}n}|Y|^{-\frac{1}{2}m}\hat{w}(0).$$

By direct computation of

$$\eta(s,S;w,u) = \int_{\mathcal{R}} f(Y,S;w)|Y|^s \hat{u}(Y)dv$$

we get

$$\eta(s,S;w,u) = 2\pi^{\frac{1}{2}n(n-1)-ns} \prod_{\nu=1}^{n} \Gamma(s-\alpha_\nu)\phi(s,S;w,u).$$

Instead of decomposing the domain of integration $\mathcal{R} = \mathcal{R}_n$ by the hypersurface $|Y| = 1$ into $\mathcal{R} - \mathcal{R}'$ and $\mathcal{R}'$, where

$$\mathcal{R}' = \mathcal{R}_n' = \{Y \in \mathcal{R}_n \mid |Y| < 1\},$$

we separate off from $\mathcal{R}$ the subset

$$\mathcal{R}'' = \mathcal{R}_n'' = \{Y = (y_{\mu\nu}) \in \mathcal{R}_n \mid y_{nn} < 1\}.$$

The "theta transformation formula" then yields

$$\eta(s,S;w,u) = I_1 + I_2 - I_3 - I_4 + I_5,$$

where

$$I_1 = \int_{\mathcal{R}-\mathcal{R}''} f(Y,S;w)|Y|^s \hat{u}(Y)dv,$$

$$I_2 = |S|^{-\frac{1}{2}n} \int_{\mathcal{R}''} h(\hat{Y},\hat{S};\hat{w})|Y|^{s-\frac{1}{2}m} \hat{u}(Y)dv,$$

$$I_3 = \int_{\mathcal{R}''} g(Y,S;w)|Y|^s \hat{u}(Y)dv,$$

$$I_4 = \hat{w}(0)|S|^{-\frac{1}{2}n} \int_{\mathcal{R}'-\mathcal{R}''} |Y|^{s-\frac{1}{2}m} \hat{u}(Y)dv,$$

$$I_5 = \hat{w}(0)|S|^{-\frac{1}{2}n} \int_{\mathcal{R}'} |Y|^{s-\frac{1}{2}m} \hat{u}(Y)dv.$$

In our deduction the estimation of $I_3$ will be the most difficult part.

If we write $Y = (y_{\mu\nu})$, $y_\nu = y_{\nu\nu}$ $(1 \leqq \nu \leqq n)$, $Y_0 = (\delta_{\mu\nu} y_\nu)$, then, according to reduction theory (§9), $Y \in \mathcal{R}$ implies the inequalities

$$-y_\mu \leqq 2y_{\mu\nu} \leqq y_\mu \leqq y_\nu \qquad (1 \leqq \mu < \nu \leqq n),$$

$$|Y| \leqq y_1 y_2 \cdots y_n \leqq c_1 |Y|,$$

$$c_2 Y_0 < Y < c_3 Y_0, \qquad c_2 Y_0^{-1} < Y^{-1} < c_3 Y_0^{-1},$$

where $c_1$, $c_2$, $c_3$ are constants which depend on $n$ only. Therefore there exist positive constants $C$ and $\kappa$ such that the Größen-character $\hat{u}(Y)$ satisfies the inequality

$$|\hat{u}(Y)| < C y_n^{\kappa} (y_1 y_2 \cdots y_n)^{-\frac{\kappa}{n}} \qquad \text{for} \quad Y \in \mathcal{R}.$$

We assume that

$$0 < \kappa < \frac{n}{2}.$$

If we denote by $p$ the degree of the polynomial $w(W)$, then

$$\frac{w(W)}{(1 + \sigma(W))^p}$$

is bounded for $W \geqq 0$. Thus there exists a constant $C_1 > 0$ such that

$$|w(Y[G'Q'])| < C_1 (1 + \sigma(S[G]Y))^p,$$

and analogously

$$|\hat{w}(\hat{Y}[G'Q^{-1}])| < C_2 (1 + \sigma(\hat{S}[G]\hat{Y}))^p$$

with another constant $C_2 > 0$.

We shall denote by $c_4, c_5, c_6, \ldots$ positive constants which depend on $n$, $S$, $C$, $\kappa$, $C_1$, $C_2$, $p$. First of all we prove that the integrals $I_\nu$, $\nu = 1,2,3,4$, are holomorphic functions of $s$ in the half-plane $\mathcal{R}e\, s > \frac{m}{2} - \delta$ $(\delta = \frac{1}{2} - \frac{\kappa}{n})$. Without loss of generality we may assume that $s$ is real.

1. We assume that $G = (g_{\mu\nu})$ has rank $n$. Then

$$\frac{\pi}{2}\,\sigma(S[G]Y) > 3c_4\sigma(G'GY_o) \geqq c_4\sigma(G'GY_o) + 2c_4\sigma(Y)$$

because

$$\sigma(G'GY_o) = \sum_{\mu=1}^{m}\sum_{\nu=1}^{n} g_{\mu\nu}^2 y_\nu \geqq \sum_{\nu=1}^{n} y_\nu = \sigma(Y).$$

Therefore

$$|w(Y[G'Q'])\, e^{-\pi\sigma(S[G]Y)} < c_5 e^{-\frac{1}{2}\pi\sigma(S[G]Y)}$$

$$< c_5 e^{-2c_4\sigma(Y) - c_4\sigma(G'GY_o)}$$

and

$$|f(Y,S;w)| < c_5 e^{-2c_4\sigma(Y)} \sum_G e^{-c_4\sigma(G'GY_o)}$$

$$= c_5 e^{-2c_4\sigma(Y)} \prod_{\nu=1}^{n} \{\sum_{g=-\infty}^{\infty} e^{-c_4 g^2 y_\nu}\}^m.$$

Observing that for any $\delta > 0$ there exists $C = C(\delta) > 0$ such that for $y > 0$ we have

$$\sum_{g=-\infty}^{\infty} e^{-\pi g^2 y} = \frac{1}{\sqrt{y}} \sum_{g=-\infty}^{\infty} e^{-\pi g^2 \frac{1}{y}} < C\,\frac{1}{\sqrt{y}}\, e^{\delta y}$$

and so

$$\sum_{g=-\infty}^{\infty} e^{-c_4 g^2 y_\nu} < c_6 \frac{1}{\sqrt{y}} e^{\frac{c_4 y_\nu}{m}} ,$$

we obtain

$$|f(Y,S;w)| < c_7 e^{-c_4 \sigma(Y)} (y_1 y_2 \cdots y_n)^{-\frac{1}{2}m} .$$

Since $\mathcal{R} - \mathcal{R}''$ is a subset of

$$\vartheta_1 = \{Y \mid -y_\mu \leqq 2y_{\mu\nu} \leqq y_\mu \leqq y_\nu \quad (\mu < \nu), \quad y_n \geqq 1\},$$

it follows that

$$|I_1| < c_8 \int_{\vartheta_1} e^{-c_4 y_n} (y_1 y_2 \cdots y_n)^{s - \frac{m}{2} - \frac{n+1}{2} - \frac{\kappa}{n}} y_n^{\kappa} [dY]$$

$$< c_9 \int_{\substack{y_1 \leqq \cdots \leqq y_n \\ y_n \geqq 1}} e^{-c_4 y_n} (y_1 y_2 \cdots y_n)^{s - \frac{m}{2} - \frac{n+1}{2} - \frac{\kappa}{n}} y_n^{\kappa} y_1^{n-1} y_2^{n-2} \cdots y_{n-1} dy_1 dy_2 \cdots dy_n .$$

With the help of the substitution

$$\theta_\mu = \frac{y_\mu}{y_{\mu+1}} \quad (1 \leqq \mu < n), \qquad y = y_n,$$

this can be rewritten as

$$|I_1| < c_9 \int_1^\infty e^{-c_4 y} y^{n(s - \frac{m}{2}) - 1} dy \prod_{\nu=1}^{n-1} \int_0^1 \theta^{\nu(s - \frac{m}{2} + \frac{n-\nu}{2} - \frac{\kappa}{n}) - 1} d\theta .$$

These integrals exist for $s > \frac{m}{2} - \delta$ if $\delta = \frac{1}{2} - \frac{\kappa}{n}$ .

2. We assume $G = (g_{\mu\nu}) \neq 0$. Then

$$\frac{\pi}{2}\sigma(\hat{S}[G]\hat{Y}) > 2c_4\sigma(G'GY_0^{-1}) = c_4\sigma(G'GY_0^{-1}) + c_4 \sum_{\mu,\nu} g_{\mu\nu}^2 y_\nu^{-1}$$

$$> c_4\sigma(G'G) + c_4 y_n^{-1},$$

provided that $Y \in \mathcal{R}''$, when in particular $y_1 \leqq y_2 \leqq \ldots \leqq y_n < 1$. Therefore

$$|\hat{w}(\hat{Y}[G'Q^{-1}])| e^{-\pi\sigma(\hat{S}[G]\hat{Y})} < c_5 e^{-\frac{1}{2}\pi\sigma(\hat{S}[G]\hat{Y})} < e^{-c_4 y_n^{-1} - c_4\sigma(G'G)}$$

and

$$|h(\hat{Y},\hat{S};\hat{w})| < c_5 e^{-c_4 y_n^{-1}} \sum_{G\neq 0} e^{-c_4\sigma(G'G)} = c_6 e^{-c_4 y_n^{-1}} \qquad (Y \in \mathcal{R}'').$$

Since $\mathcal{R}''$ is a subset of

$$\vartheta_2 = \{Y \mid -y_\mu \leqq 2y_{\mu\nu} \leqq y_\mu \leqq y_\nu < 1 \quad (\mu \neq \nu)\},$$

we get

$$|I_2| < c_7 \int_{\vartheta_2} e^{-c_4 y_n^{-1}} (y_1 y_2 \ldots y_n)^{s - \frac{m}{2} - \frac{n+1}{2} - \frac{\kappa}{n}} y_n^{\kappa} [dY]$$

$$< c_8 \int_{y_1 \leqq y_2 \leqq \ldots \leqq y_n < 1} e^{-c_4 y_n^{-1}} (y_1 y_2 \ldots y_n)^{s - \frac{m}{2} - \frac{n+1}{2} - \frac{\kappa}{n}} y_n^{\kappa} y_1^{n-1} y_2^{n-2} \ldots y_{n-1} dy_1 dy_2 \ldots dy_n$$

$$= c_8 \int_0^1 e^{-c_4 y^{-1}} y^{n(s - \frac{m}{2}) - 1} dy \prod_{\nu=1}^{n-1} \int_0^1 \theta^{\nu(s - \frac{m}{2} + \frac{n-\nu}{2} - \frac{\kappa}{n}) - 1} d\theta.$$

These integrals exist again for $s > \frac{m}{2} - \delta$ if $\delta = \frac{1}{2} - \frac{\kappa}{n}$.

3. We decompose $g(Y,S;w)$ as follows:

$$g(Y,S;w) = \sum_{h=0}^{n-1} g_h(Y,S;w),$$

$$g_h(Y,S;w) = \sum_{\text{rank } G = h} w(Y[G'Q'])e^{-\pi\sigma(S[G]Y)}.$$

The matrices $G$ of rank $h > 0$ can be represented uniquely as products $G = BP'$, where $P = P^{(n,h)} = (\mathfrak{p}_1, \mathfrak{p}_2, \ldots, \mathfrak{p}_h)$ runs over a complete set $\{P\}$ of primitive matrices, not right-associated with respect to the unimodular group, and $B = B^{(m,h)} = (\mathfrak{b}_1, \mathfrak{b}_2, \ldots, \mathfrak{b}_h)$ over all integral matrices of rank $h$. We can assume that $Y[P]$ is reduced, i.e., that $Y[P] \in \mathcal{R}_h$. Obviously

$$g_h(Y,S;w) = \sum_{\{P\}} g_P ,$$

where

$$g_P = \sum_{B} w(Y[PB'Q'])e^{-\pi\sigma(S[B]\cdot Y[P])}.$$

From

$$\frac{\pi}{2}\sigma(S[B]Y[P]) > c_4\sigma(B'BY[P]) > c_2c_4 \sum_{\nu=1}^{h} Y[\mathfrak{p}_\nu]\mathfrak{b}_\nu'\mathfrak{b}_\nu$$

$$> c_2^2c_4 \sum_{\nu=1}^{h} Y_0[\mathfrak{p}_\nu]\mathfrak{b}_\nu'\mathfrak{b}_\nu$$

we obtain with $c_5 = c_2^2c_4$ the estimate

$$|w(Y[PB'Q'])|e^{-\pi\sigma(S[B]Y[P])} < c_6 e^{-\frac{1}{2}\pi\sigma(S[B]Y[P])}$$

$$< c_6 e^{-c_5 \sum_{\nu=1}^{h} Y_0[\mathfrak{g}_\nu]\mathfrak{b}'_\nu \mathfrak{b}_\nu}.$$

If $\mathfrak{b}$ denotes a column-vector with $m$ integral components, we obtain from the inequality already used in 1.,

$$\sum_{\mathfrak{b}\neq 0} e^{-c_5 Y_0[\mathfrak{g}_\nu]\mathfrak{b}'\mathfrak{b}} < e^{-\frac{2}{3}c_5 Y_0[\mathfrak{g}_\nu]} \sum_{\mathfrak{b}} e^{-\frac{1}{3}c_5 Y_0[\mathfrak{g}_\nu]\mathfrak{b}'\mathfrak{b}}$$

$$= e^{-\frac{2}{3}c_5 Y_0[\mathfrak{g}_\nu]} \left( \sum_{g=-\infty}^{\infty} e^{-\frac{1}{3}c_5 Y_0[\mathfrak{g}_\nu]g^2} \right)^m$$

$$< c_7 e^{-\frac{2}{3}c_5 Y_0[\mathfrak{g}_\nu]} \left( (Y_0[\mathfrak{g}_\nu])^{-\frac{1}{2}} e^{\frac{1}{3m}c_5 Y_0[\mathfrak{g}_\nu]} \right)^m$$

$$= c_7 (Y_0[\mathfrak{g}_\nu])^{-\frac{m}{2}} e^{-\frac{1}{3}c_5 Y_0[\mathfrak{g}_\nu]}.$$

Consequently with $c_8 = \frac{1}{3}c_5$ we have

$$|g_P| < c_6 \sum_B e^{-c_5 \sum_{\nu=1}^{h} Y_0[\mathfrak{g}_\nu]\mathfrak{b}'_\nu \mathfrak{b}_\nu}$$

$$\leqq c_6 \prod_{\nu=1}^{h} \sum_{\mathfrak{b}\neq 0} e^{-c_5 Y_0[\mathfrak{g}_\nu]\mathfrak{b}'\mathfrak{b}}$$

$$< c_9 \prod_{\nu=1}^{h} \{(Y_0[\mathfrak{g}_\nu])^{-\frac{m}{2}} e^{-c_8 Y_0[\mathfrak{g}_\nu]}\}$$

and

$$|g_h(Y,S;w)| < c_9 \sum_{\{P\}} \prod_{\nu=1}^{h} \{(Y_o[\mathfrak{p}_\nu])^{-\frac{m}{2}} e^{-c_8 Y_o[\mathfrak{p}_\nu]}\}.$$

Let $q$ be an integer, $1 \leqq q \leqq n$. Introducing

$$j_q = \sum_{\mathfrak{p}} (Y_o[\mathfrak{p}])^{-\frac{m}{2}} e^{-c_8 Y_o[\mathfrak{p}]},$$

where $\mathfrak{p}' = (p_1, p_2, \ldots, p_n)$ runs through all integral row-vectors with $p_q \neq 0$, we can write

$$|g_h(Y,S;w)| < h!c_9 \sum_{1 \leqq q_1 < \ldots < q_h \leqq n} j_{q_1} j_{q_2} \cdots j_{q_h}.$$

For $k = 1,2,3,\ldots$ and $0 < y_1 \leqq y_2 \leqq \ldots \leqq y_n < 1$ the number of integral solutions $(p_1, p_2, \ldots, p_n)$ of the inequality

$$k \leqq Y_o[\mathfrak{p}] = y_1 p_1^2 + \ldots + y_n p_n^2 < k+1$$

is less than $c_{10} k^{\frac{n}{2}} (y_1 \ldots y_n)^{-\frac{1}{2}}$. Denote by $j_q^*$ the partial series of $j_q$ which contains the terms of $j_q$ with the property $Y_o[\mathfrak{p}] \geqq 1$, and set $j_q^{**} = j_q - j_q^*$. Clearly

$$j_q^* < c_{10}(y_1 \ldots y_n)^{-\frac{1}{2}} \sum_{k=1}^{\infty} k^{\frac{n-m}{2}} e^{-c_8 k} = c_{11}(y_1 y_2 \ldots y_n)^{-\frac{1}{2}}.$$

In order to estimate $j_q^{**}$, we write

$$k_t = \left[ \frac{-\log y_t}{\log 2} \right] \qquad (1 \leqq t \leqq n), \qquad k_{n+1} = 0.$$

The condition $Y \in \mathfrak{R}''$ implies the relations

$$k_{t+1} \leqq k_t, \qquad 2^{-k_t-1} < y_t \leqq 2^{-k_t} \qquad (1 \leqq t \leqq n).$$

Let $k \geqq 0$ be a given integer and determine $t = t_k$ in $0 \leqq t \leqq n$ so that

$$2^{-k}y_1^{-1} \geqq \dots \geqq 2^{-k}y_t^{-1} > 1 \geqq 2^{-k}y_{t+1}^{-1} \geqq \dots \geqq 2^{-k}y_n^{-1}.$$

If $t = 0$, then there are no terms to the left of 1, and if $t = n$, then there are no terms to the right of 1. We assume that

$$2^{-k-1} \leqq Y_o[\mathfrak{g}] = y_1p_1^2 + \dots + y_np_n^2 < 2^{-k}$$

has at least one integral solution $\mathfrak{g}'$ with $p_q \neq 0$. Then $2^{-k_q-1} < y_q < 2^{-k}$ so that $k_q + 1 > k$ or $k_q \geqq k$. Since $p_\nu^2 < 2^{-k}y_\nu^{-1}$, we obtain $p_\nu^2 < 1$ for $\nu > t = t_k$, i.e., $p_{t+1} = \dots = p_n = 0$. Consequently $q \leqq t$ and the number of all integral solutions is less than

$$c_{12} \prod_{\nu=1}^{t} (2^{-k}y_\nu^{-1})^{\frac{1}{2}} = c_{12}\, 2^{-k\frac{t}{2}} (y_1y_2\cdots y_t)^{-\frac{1}{2}}.$$

Therefore

$$j_q^{**} < c_{12} \sum_{k=0}^{k_q} 2^{(k+1)\frac{m}{2} - k\frac{t}{2}} (y_1y_2\cdots y_t)^{-\frac{1}{2}}$$

$$= c_{13} \sum_{k=0}^{k_q} 2^{k_q\frac{m-t}{2}} (y_1y_2\cdots y_t)^{-\frac{1}{2}}$$

$$\leqq c_{13} \sum_{k=0}^{k_q} y_q^{-\frac{m-t}{2}} (y_1y_2\cdots y_t)^{-\frac{1}{2}}$$

$$\leq c_{13} \sum_{k=0}^{k_q} y_q^{-\frac{m-q}{2}} (y_1 y_2 \cdots y_q)^{-\frac{1}{2}}$$

$$= c_{13}(k_q + 1) y_q^{\frac{q-m}{2}} (y_1 y_2 \cdots y_q)^{-\frac{1}{2}}$$

$$< c_{14}(y_1 y_2 \cdots y_q)^{-\frac{1}{2}} y_q^{\frac{q-m}{2}} \log \frac{2}{y_q} .$$

The same estimate is valid also for $j_q^*$ and $j_q$ with an appropriate $c_{14}$. Thus we obtain

$$|g_h(Y,S;w)| < c_{15} \sum_{1 \leq q_1 < \cdots < q_h \leq n} \prod_{\nu=1}^{h} \{(y_1 y_2 \cdots y_{q_\nu})^{-\frac{1}{2}} y_{q_\nu}^{-\frac{m-q_\nu}{2}} \log \frac{2}{y_{q_\nu}}\}$$

$$\leq c_{15} \sum_{1 \leq q_1 < \cdots < q_h \leq n} \prod_{\nu=1}^{h} \{(y_1 y_2 \cdots y_\nu)^{-\frac{1}{2}} y_\nu^{-\frac{m-\nu}{2}} \log \frac{2}{y_\nu}\}$$

$$= c_{15} \binom{n}{h} \prod_{\nu=1}^{h} \{y_\nu^{\nu - \frac{m+h+1}{2}} \log \frac{2}{y_\nu}\}$$

$$\leq c_{16} \prod_{\nu=1}^{n-1} \{y_\nu^{\nu - \frac{m+n}{2}} \log \frac{2}{y_\nu}\} ,$$

and this estimate holds trivially for $h = 0$.

It follows that

$$|g(Y,S;w)| < c_{17} \prod_{\nu=1}^{n-1} \{y_\nu^{\nu - \frac{m+n}{2}} \log \frac{2}{y_\nu}\}$$

and finally

$$|I_3| < c_{18} \int\limits_{0<y_1<\dots<y_n<1} \prod_{\nu=1}^{n-1} \{y_\nu^{\nu-\frac{m+n}{2}} \log \frac{2}{y_\nu}\} \prod_{\nu=1}^{n} \{y_\nu^{s+n-\nu-\frac{\kappa}{n}-\frac{n+1}{2}}\} y_n^\kappa dy_1 \dots dy_n$$

$$\leqq c_{18} \left\{ \int_0^1 y^{s-\frac{m}{2}+\frac{1}{2}-\frac{\kappa}{n}} \log \frac{2}{y} \frac{dy}{y} \right\}^{n-1} \int_0^1 y^{s-\frac{n}{2}+\frac{1}{2}-\frac{\kappa}{n}+\kappa} \frac{dy}{y},$$

and these integrals exist for $s > \frac{m}{2} - \delta$ if $\delta = \frac{1}{2} - \frac{\kappa}{n}$.

4. Since $\mathfrak{R}' - \mathfrak{R}''$ is contained in the point set defined by

$$-y_\mu \leqq 2y_{\mu\nu} \leqq y_\mu < y_\nu \quad (1 \leqq \mu < \nu \leqq n), \quad 1 \leqq y_n, \quad y_1 y_2 \cdots y_n < c_1,$$

we obtain

$$|I_4| < c_4 \int\limits_{\substack{0<y_1<\dots<y_{n-1} \\ 1<y_n,\ y_1 y_2 \cdots y_n < c_1}} y_n^\kappa \prod_{\nu=1}^{n} \{y_\nu^{s-\frac{m}{2}-\frac{n+1}{2}-\frac{\kappa}{n}+n-\nu} dy_\nu\}.$$

The substitution

$$\theta_\nu = \frac{y_\nu}{y_{\nu+1}} \quad (1 \leqq \nu \leqq n-2), \quad y = y_n, \quad t = y_1 y_2 \cdots y_n$$

with

$$\frac{\partial(y_1,\dots,y_n)}{\partial(\theta_1,\dots,\theta_{n-2},t,y)} = \frac{y_2 y_3 \cdots y_{n-2} y_{n-1}^2}{(n-1)t}$$

yields

$$|I_4| < \frac{c_4}{n-1} \int_0^{c_1} t^{s - \frac{m}{2} + \frac{1}{2} - \frac{\kappa}{n}} \frac{dt}{t} \int_1^{\infty} y^{-\frac{n}{2} + \kappa} \frac{dy}{y} \prod_{\nu=1}^{n-2} \int_0^1 \theta^{\frac{\nu(n-\nu-1)}{2}} \frac{d\theta}{\theta},$$

and these integrals exist if $s > \frac{m}{2} - \delta, \quad \delta = \frac{1}{2} - \frac{\kappa}{n} > 0.$

5. Let us set

$$Y = yY_1, \quad y > 0, \quad |Y_1| = 1, \quad dv_1 = \frac{y\ dv}{n\ dy}.$$

Then

$$I_5 = \hat{\omega}(0)|S|^{-\frac{n}{2}} n \int_0^1 y^{n(s - \frac{m}{2})} \frac{dy}{y} \int_{\substack{Y_1 \in \mathcal{R} \\ |Y_1|=1}} \hat{u}(Y_1) dv_1$$

$$= |S|^{-\frac{n}{2}} \frac{\hat{\omega}(0)}{s - \frac{m}{2}} \int_{\substack{Y_1 \in \mathcal{R} \\ |Y_1|=1}} \hat{u}(Y_1) dv_1 ,$$

provided that this integral exists. For $s = \frac{m}{2} + \frac{n+1}{2}$ we obtain the identity

$$\hat{\omega}(0)|S|^{-\frac{n}{2}} \int_{\mathcal{R}'} \hat{u}(Y)[dY] = \hat{\omega}(0)|S|^{-\frac{n}{2}} \frac{2}{n+1} \int_{\substack{Y_1 \in \mathcal{R} \\ |Y_1|=1}} \hat{u}(Y_1) dv_1,$$

hence

$$I_5 = |S|^{-\frac{n}{2}} \hat{\omega}(0) \frac{n+1}{2s - m} \int_{\mathcal{R}'} \hat{u}(Y)[dY].$$

It remains to prove that this integral exists for $\kappa < \frac{n}{2}$.

If we write

$$Y = \begin{pmatrix} X & \mathfrak{y} \\ \mathfrak{y}' & y \end{pmatrix} \in \mathcal{R}',$$

then

$$X \in \mathcal{R}_{n-1}, \qquad \frac{1}{c_1}|X|y \leqq \frac{1}{c_1} y_1 y_2 \cdots y_n \leqq |Y|,$$

$$|X| \leqq y^{n-1} \leqq \left(\frac{c_1|Y|}{|X|}\right)^{n-1} < \frac{c_1^{n-1}}{|X|^{n-1}}$$

and so

$$|X| < c_4 = c_1^{1-\frac{1}{n}}, \qquad 0 < y < \frac{c_1}{|X|}.$$

We estimate

$$\left|\int_{\mathcal{R}'} \hat{u}(Y)[dY]\right| < c_5 \int_{\substack{X\in\mathcal{R}_{n-1} \\ |X|<c_4 \\ 0<y<\frac{c_1}{|X|}}} y^{\kappa-\frac{\kappa}{n}} |X|^{-\frac{\kappa}{n}+1} [dX]dy$$

$$= c_6 \int_{\substack{X\in\mathcal{R}_{n-1} \\ |X|<c}} |X|^{-\kappa}[dX] = c_7 \int_{\mathcal{R}'_{n-1}} |X|^{-\kappa}[dX].$$

The last integral exists for $\kappa < \frac{n}{2}$ because for $s > 0$ we have

$$\int_{\mathcal{R}'} |Y|^{s-\frac{n+1}{2}} [dY] = \int_{\mathcal{R}'} |Y|^s dv < e \int_{\mathcal{R}} e^{-|Y|} |Y|^s dv = e^{\frac{n+1}{2}} v_n \Gamma(s)$$

by virtue of the formula on p. 213 with $a = \alpha = 1$.

Since the map $Y_1 \mapsto Y_1^{-1}$ leaves $dv_1$ invariant and commutes with the unimodular group, we have

$$\int_{\substack{Y_1 \in \mathcal{R} \\ |Y_1|=1}} \hat{u}(Y_1) dv_1 = \int_{\substack{Y_1 \in \mathcal{R} \\ |Y_1|=1}} u(Y_1) dv_1 .$$

For functions $u(Y)$, $u^*(Y)$ which are defined on $Y > 0$, homogeneous of degree 0 and invariant under the unimodular group we define a scalar product by

$$(u,u^*) = \frac{1}{V_n} \int_{\substack{Y \in \mathcal{R} \\ |Y|=t}} u(Y) \overline{u^*(Y)} |Y| \frac{dv}{d|Y|} .$$

The integral does not depend on $t$. We choose the constant $V_n$ so that $(1,1) = 1$, i.e., $V_n$ is given by

$$V_n = \int_{\substack{Y \in \mathcal{R} \\ |Y|=1}} |Y| \frac{dv}{d|Y|} .$$

It follows that

$$V_n \int_0^\infty f(t) dt = \int_{\mathcal{R}} f(|Y|) |Y| dv$$

for an arbitrary function $f(t)$. The special choice

$$f(t) = \begin{cases} t^{\frac{1}{2}(n-1)} & \text{if } 0 \leqq t \leqq 1, \\ 0 & \text{if } t > 1, \end{cases}$$

yields the well-known value

$$\frac{2}{n+1} V_n = \int_{\mathcal{R}'} [dY] = \frac{2}{n+1} \prod_{\nu=2}^{n} \xi(\nu), \qquad \xi(s) = \pi^{-\frac{1}{2}s}\Gamma(\tfrac{s}{2})\zeta(s),$$

or

$$V_n = \zeta(2)\zeta(3) \ldots \zeta(n) \prod_{\nu=1}^{n} \{\pi^{-\frac{1}{2}\nu}\Gamma(\tfrac{\nu}{2})\}.$$

With the scalar product notation we have

$$\int_{\substack{Y_1 \in \mathcal{R} \\ |Y_1|=1}} u(Y_1)dv_1 = V_n(u,1).$$

We shall interpret now $\hat{\omega}(0)$ in a similar way. For polynomials $\phi(X)$, $\phi^*(X)$ in the entries of $X = X^{(m,n)}$ it is meaningful to introduce scalar products on the unit sphere and on the Graßmannian manifold:

$$[\phi,\phi^*] = \frac{\Gamma\left(\frac{mn}{2}\right)}{\pi^{\frac{1}{2}mn}} \int_{\sigma(X'X)=1} \phi(X)\,\overline{\phi^*(X)}\,\frac{[dX]}{d\sigma(X'X)},$$

$$\langle\phi,\phi^*\rangle = \frac{\prod_{\nu=0}^{n-1} \Gamma\left(\frac{m-\nu}{2}\right)}{\pi^{\frac{1}{2}mn - \frac{1}{4}n(n-1)}} \int_{X'X=E} \phi(X)\,\overline{\phi^*(X)}\,\frac{[dX]}{[d(X'X)]}.$$

In case of $n = 1$ the two formulae coincide. The constant factors in front of the integrals are chosen so that $[1,1] = \langle 1,1\rangle = 1$ (see [17], formula (46)).

Assume now that

$$\phi(XV) = |V|^{2k}\phi(X), \qquad \phi^*(XV) = |V|^{2k^*}\phi^*(X).$$

Then

$$[\phi,\phi^*] = \frac{\Gamma\left(\frac{mn}{2}\right)}{\pi^{\frac{1}{2}mn}} \int\limits_{\sigma(X'X)=y} \phi(X)\ \overline{\phi^*(X)}\ (\sigma(X'X))^{1-\frac{1}{2}mn-n(k+k^*)}\ \frac{[dX]}{d\sigma(X'X)}$$

and

$$\langle\phi,\phi^*\rangle = \frac{\prod\limits_{\nu=0}^{n-1}\Gamma\left(\frac{m-\nu}{2}\right)}{\pi^{\frac{1}{2}mn-\frac{1}{4}n(n-1)}} \int\limits_{X'X=Y} \phi(X)\ \overline{\phi^*(X)}\ |X'X|^{\frac{1}{2}(n+1)-\frac{1}{2}m-k-k^*}\ \frac{[dX]}{[d(X'X)]}$$

for arbitrary $y > 0$ and $Y > 0$ because these integrals do not depend on $y$ and $Y$, as one can show by substituting $X\sqrt{y}$ and $X\sqrt{Y}$ for $X$, and they coincide respectively with the definitions of the scalar products in case of $y = 1$ and $Y = E$. For given functions $f(y)$, $F(Y)$ we obtain

$$[\phi,\phi^*]\int\limits_0^\infty f(y)dy =$$

$$= \frac{\Gamma\left(\frac{mn}{2}\right)}{\pi^{\frac{1}{2}mn}} \int \phi(X)\ \overline{\phi^*(X)}\ f(\sigma(X'X))(\sigma(X'X))^{1-\frac{1}{2}mn-n(k+k^*)}[dX]$$

and

$$\langle\phi,\phi^*\rangle \int\limits_{Y>0} F(Y)[dY] =$$

$$= \frac{\prod_{\nu=0}^{n-1} \Gamma\left(\frac{m-\nu}{2}\right)}{\pi^{\frac{1}{2}mn - \frac{1}{4}n(n-1)}} \int \phi(X)\ \overline{\phi^*(X)}\ F(X'X)|X'X|^{\frac{1}{2}(n+1-m)-k-k^*}[dX].$$

In both cases we have to integrate over the whole $X$-space.

If we choose

$$f(y) = y^{n(k+k^*)+\frac{1}{2}mn-1}\ e^{-y},$$

$$F(Y) = |Y|^{\frac{1}{2}(m-n-1)+k+k^*}\ e^{-\sigma(Y)},$$

and observe that

$$\int\limits_0^\infty f(y)dy = \Gamma\left(n(k+k^*) + \frac{mn}{2}\right),$$

$$\int\limits_{Y>0} F(Y)[dY] = \pi^{\frac{1}{4}n(n-1)} \prod_{\nu=0}^{n-1} \Gamma\left(\frac{m-\nu}{2} + k + k^*\right),$$

we get

$$[\phi,\phi^*]\Gamma(n(k+k^*) + \frac{mn}{2}) \frac{\pi^{\frac{1}{2}mn}}{\Gamma\left(\frac{mn}{2}\right)} =$$

$$= \langle\phi,\phi^*\rangle\ \pi^{\frac{1}{4}n(n-1)} \prod_{\nu=0}^{n-1} \frac{\Gamma\left(\frac{m-\nu}{2} + k + k^*\right)}{\Gamma\left(\frac{m-\nu}{2}\right)}\ \pi^{\frac{1}{2}mn - \frac{1}{4}n(n-1)}$$

or

$$[\phi,\phi^*] = \frac{\Gamma\left(\frac{mn}{2}\right)}{\Gamma\left(\frac{mn}{2}+n(k+k^*)\right)} \prod_{\nu=0}^{n-1} \frac{\Gamma\left(\frac{m-\nu}{2}+k+k^*\right)}{\Gamma\left(\frac{m-\nu}{2}\right)} <\phi,\phi^*> .$$

In particular for $\phi^* = 1$ we have

$$[\phi,1] = \frac{\Gamma\left(\frac{mn}{2}\right)}{\Gamma\left(\frac{mn}{2}+nk\right)} \prod_{\nu=0}^{n-1} \frac{\Gamma\left(\frac{m-\nu}{2}+k\right)}{\Gamma\left(\frac{m-\nu}{2}\right)} <\phi,1> .$$

We proceed with the computation of

$$\hat{\omega}(0) = \int \omega(TT')e^{-\pi\sigma(T'T)}[dT]$$

$$= \int_0^\infty e^{-\pi t}\left\{ \int_{\sigma(T'T)=t} \omega(TT') \frac{[dT]}{d\sigma(T'T)} \right\} dt.$$

If $\omega(TT')$ is a homogeneous polynomial of degree $2h$, then the substitution $T = \sqrt{t}\, X$ yields

$$\hat{\omega}(0) = \int_0^\infty e^{-\pi t}\, t^{h+\frac{1}{2}mn-1}\, dt \int_{\sigma(X'X)=1} \omega(XX') \frac{[dX]}{d\sigma(X'X)}$$

$$= \frac{\Gamma\left(\frac{mn}{2}+h\right)}{\pi^{\frac{1}{2}mn+h}} \frac{\pi^{\frac{1}{2}mn}}{\Gamma\left(\frac{mn}{2}\right)} [\omega,1] = \frac{\Gamma\left(\frac{mn}{2}+h\right)}{\Gamma\left(\frac{mn}{2}\right)\pi^h} [\omega,1].$$

Let us now assume, like at the beginning of this section, that $\omega(XX') = \omega_o(X)$ is a homogeneous polynomial of degree $2nk$ and that

$$\omega_o(XV) = |V|^{2k}\, \omega_o(X) \qquad \text{for all } V = V^{(n)}.$$

Using the preceding formula with $h = nk$ and the formula expressing $[\phi,1]$ in terms of $<\phi,1>$ we obtain

$$\hat{w}(0) = \pi^{-nk} \prod_{\nu=0}^{n-1} \frac{\Gamma\left(\frac{m-\nu}{2}+k\right)}{\Gamma\left(\frac{m-\nu}{2}\right)} <w_o,1>$$

and so our above result yields

$$\lim_{s\to\frac{1}{2}m} (s-\tfrac{m}{2})\eta(s,S;w,u) = |S|^{-\frac{1}{2}n}\, \hat{w}(0) \int\limits_{\substack{Y_1\in\mathcal{R}\\ |Y_1|=1}} u(Y_1)dv_1$$

$$= |S|^{-\frac{1}{2}n}\, \pi^{-nk} \prod_{\nu=0}^{n-1} \left\{ \frac{\Gamma\left(\frac{m-\nu}{2}+k\right)\Gamma\left(\frac{\nu+1}{2}\right)}{\Gamma\left(\frac{m-\nu}{2}\right)\pi^{\frac{1}{2}(\nu+1)}} \right\} \prod_{\nu=2}^{n} \zeta(\nu)<w_o,1>(u,1).$$

As we have seen on p. 230

$$\phi(s,S;w,u) = \pi^{-nk} \prod_{\nu=1}^{n} \frac{\Gamma(s+k-\alpha_\nu)}{\Gamma(s-\alpha_\nu)} \phi_o(s,S;w_o,u),$$

hence

$$\eta(s,S;w,u) = 2\pi^{\frac{1}{2}n(n-1)-n(s+k)} \prod_{\nu=1}^{n} \Gamma(s+k-\alpha_\nu)\ \phi_o(s,S;w_o,u)$$

and

$$\lim_{s\to\frac{1}{2}m} (s-\tfrac{m}{2})\phi_o(s,S;w_o,u) =$$

$$= \frac{1}{2}\pi^{\frac{1}{2}n(m-n)}|S|^{-\frac{1}{2}n} \prod_{\nu=1}^{n} \left\{ \frac{\Gamma\left(\frac{m}{2}+k-\frac{\nu-1}{2}\right)\Gamma\left(\frac{\nu}{2}\right)}{\Gamma\left(\frac{m}{2}+k-\alpha_\nu\right)\Gamma\left(\frac{m+1-\nu}{2}\right)} \right\} \prod_{\nu=2}^{n} \zeta(\nu)<w_o,1>(u,1).$$

The function $\phi_o(s,S;w_o,u)$ and the scalar product $\langle w_o,1\rangle$ remain unchanged if we replace $w_o(X)$ and $k$ by $w_o(X)|X'X|^l$ and $k+l$ respectively. Therefore

$$\prod_{\nu=1}^{n} \frac{\Gamma\left(\frac{m}{2}+k+l-\frac{\nu-1}{2}\right)}{\Gamma\left(\frac{m}{2}+k+l-\alpha_\nu\right)} = \prod_{\nu=1}^{n} \frac{\Gamma\left(\frac{m}{2}+k-\frac{\nu-1}{2}\right)}{\Gamma\left(\frac{m}{2}+k-\alpha_\nu\right)} \qquad \text{for} \quad l \in \mathbb{N},$$

provided that $\langle w_o,1\rangle(u,1) \neq 0$ and $\kappa < \frac{1}{2}n$. Choosing in particular $w_o = 1$ we obtain

$$\prod_{\nu=1}^{n} \frac{\Gamma\left(\frac{m}{2}+l-\frac{\nu-1}{2}\right)}{\Gamma\left(\frac{m}{2}+l-\alpha_\nu\right)} = \prod_{\nu=1}^{n} \frac{\Gamma\left(\frac{m}{2}-\frac{\nu-1}{2}\right)}{\Gamma\left(\frac{m}{2}-\alpha_\nu\right)} \qquad \text{for} \quad l \in \mathbb{N},$$

hence

$$\prod_{\nu=1}^{n} \frac{\Gamma\left(\frac{m}{2}+l+1-\frac{\nu-1}{2}\right)}{\Gamma\left(\frac{m}{2}+l-\frac{\nu-1}{2}\right)} = \prod_{\nu=1}^{n} \frac{\Gamma\left(\frac{m}{2}+l+1-\alpha_\nu\right)}{\Gamma\left(\frac{m}{2}+l-\alpha_\nu\right)}$$

or

$$\prod_{\nu=1}^{n} \left(l+\frac{m}{2}-\frac{\nu-1}{2}\right) = \prod_{\nu=1}^{n} \left(l+\frac{m}{2}-\alpha_\nu\right) \qquad \text{for} \quad l \in \mathbb{N}$$

if $(u,1) \neq 0$ and $\kappa < \frac{1}{2}n$. This implies

$$\alpha_\nu = \frac{\nu-1}{2} \qquad \text{for} \quad \nu = 1,2,\ldots,n$$

if we arrange $\alpha_1,\alpha_2,\ldots,\alpha_n$ in an increasing order, and leads to the following simplification:

$$\lim_{s\to\frac{1}{2}m} (s-\tfrac{m}{2})\phi_0(s,S;w_0,u) =$$

$$= \frac{1}{2}\,\pi^{\frac{1}{2}n(m-n)}|S|^{-\frac{1}{2}n} \prod_{\nu=1}^{n} \frac{\Gamma\left(\frac{\nu}{2}\right)}{\Gamma\left(\frac{m+1-\nu}{2}\right)} \prod_{\nu=2}^{n} \zeta(\nu)\langle w_0,1\rangle(u,1).$$

We summarize what we proved:

*Theorem. The zeta function* $\phi_0(s,S;w_0,u)$ *defined by*

$$\phi_0(s,S;w_0,u) = \sum_{\{G\}} \frac{w_0(QG)u(S[G])}{|S[G]|^{s+k}} \qquad (S = Q'Q)$$

*has the following property:*

$$\phi_0(s,S;w_0,u) - \pi^{\frac{1}{2}n(m-n)}|S|^{-\frac{1}{2}n} \prod_{\nu=1}^{n} \frac{\Gamma\left(\frac{\nu}{2}\right)}{\Gamma\left(\frac{m+1-\nu}{2}\right)} \prod_{\nu=2}^{n} \zeta(\nu)\, \frac{\langle w_0,1\rangle(u,1)}{2s-m}$$

*is holomorphic in the half-plane* $\mathcal{R}e\, s > \frac{m}{2} - \delta$ *provided that the polynomial* $w_0(X)$ *and the Größen-character* $u(Y)$ *satisfy the conditions*

$$w_0(XV) = |V|^{2k}\, w_0(X) \qquad \textit{for all } V$$

*and*

$$|u(Y)| < C\left(\frac{\sigma(Y)}{\sqrt[n]{|Y|}}\right)^{\kappa}, \qquad \kappa < \frac{n}{2},$$

*and we choose* $\delta = \frac{1}{2} - \frac{\kappa}{n}$.

## §17. Selberg's zeta functions

In this section we shall give a number theoretical approach to the theory of Eisenstein series of the group of unimodular matrices. When we speak of symmetric matrices, we should keep in mind that such matrices define quadratic forms.

*Definition. A sequence of matrices* $\overset{n}{Y}, \overset{n-1}{Y}, \ldots, \overset{1}{Y}$ *is called a descending chain if*

1. $\overset{h}{Y}$ *is a real positive h-rowed symmetric matrix for* $h = 1,2,\ldots,n$,

2. *each link* $\overset{h}{Y}$ $(1 \leq h < n)$ *can be represented by the preceding link* $\overset{h+1}{Y}$, *i.e., there exists an integral matrix* $G_h = G_h^{(h+1,h)}$ *such that*

$$\overset{h}{Y} = \overset{h+1}{Y}[G_h].$$

Of course $G_h$ must be of rank $h$. The descending chain is called *primitive* if all the matrices $G_h$ $(1 \leq h < n)$ are primitive, i.e., can be completed by an additional column to unimodular matrices.

Denote by $\Gamma_n$ the group of $n$-rowed unimodular matrices and, for later use, by $B_n$ the group of all triangular unimodular matrices, i.e., matrices $U \in \Gamma_n$ of the form

$$U = \begin{pmatrix} * & \cdots & * \\ & \ddots & \vdots \\ 0 & & * \end{pmatrix}.$$

If we replace the links $\overset{h}{Y}$ of a descending chain by equivalent matrices $\overset{h}{Y}{}^* = \overset{h}{Y}[U_h]$, where $U_h \in \Gamma_h$, then we get a descending chain

$\overset{n}{Y}{}^*, \overset{n-1}{Y}{}^*, \ldots, \overset{1}{Y}{}^*$ since obviously $\overset{h}{Y}{}^* > 0$ and $\overset{h}{Y}{}^* = \overset{h+1}{Y}{}^*[G_h^*]$ with $G_h^* = U_{h+1}^{-1} G_h U_h$ for $1 \leqq h < n$. A descending chain is determined by the matrices $\overset{n}{Y}, G_{n-1}, G_{n-2}, \ldots, G_1$ since $\overset{h}{Y} = \overset{n}{Y}[G_{n-1} \ldots G_h]$. We shall consider descending chains with variable first links $\overset{n}{Y}$ and fixed $G_{n-1}, G_{n-2}, \ldots, G_1$. We shall say that two such descending chains determined by $\overset{n}{Y}, G_{n-1}, G_{n-2}, \ldots, G_1$ and $\overset{n}{Y}{}^*, G_{n-1}^*, G_{n-2}^*, \ldots, G_1^*$, respectively, are equivalent if there exist $U_h \in \Gamma_h$ with $1 \leqq h \leqq n$ such that $G_h^* = U_{h+1}^{-1} G_h U_h$ for $1 \leqq h < n$ and $\overset{n}{Y}{}^* = \overset{n}{Y}[U_n]$. We shall denote by $\langle \overset{n}{Y}, \ldots, \overset{1}{Y} \rangle$ the class of descending chains equivalent to $\overset{n}{Y}, \ldots, \overset{1}{Y}$. If one descending chain in the equivalence class $\langle \overset{n}{Y}, \ldots, \overset{1}{Y} \rangle$ is primitive, then all the descending chains in the equivalence class are primitive; we shall say that such an equivalence class is primitive.

We determine special representatives of the classes $\langle \overset{n}{Y}, \ldots, \overset{1}{Y} \rangle$, where $\overset{n}{Y}$ belongs to a given equivalence class $\langle Y \rangle = \{Y[U] \mid U \in \Gamma_n\}$. This can be done by means of

*Lemma* 1. *Let* $G_h = G_h^{(h+1,h)}$ $(1 \leqq h < n)$ *be a given sequence of integral matrices. There exist* $U_h \in \Gamma_h$ *with* $1 \leqq h \leqq n$ *such that*

$$U_{h+1}^{-1} G_h U_h = \begin{pmatrix} D_h \\ 0 \end{pmatrix}$$

*for* $1 \leqq h < n$, *where*

$$D_h = \begin{pmatrix} * & \cdots & * \\ & \ddots & \vdots \\ 0 & & * \end{pmatrix}$$

*is an integral upper triangular matrix.*

*Proof* by induction on $n$. For $n = 2$ the lemma is obvious, so we can assume that $n > 2$ and that the lemma is true for $n-1$ instead of $n$. First of all, it is possible to choose $U_h$ so that the last line of $U_{h+1}^{-1}G_h$ vanishes for $1 \leqq h < n$, and this property is not lost if we multiply the matrix from the right by $U_h$. Thus we may assume from the outset that $G_h$ has the form

$$G_h = \begin{pmatrix} H_{h-1} & * \\ 0 & 0 \end{pmatrix},$$

with an integral matrix $H_{h-1} = H^{(h,h-1)}$. By virtue of the induction hypothesis there exist $V_{h-1} \in \Gamma_{h-1}$ with $2 \leqq h \leqq n$ such that $V_h^{-1}H_{h-1}V_{h-1}$ has an upper triangular integral matrix in the first $h-1$ rows and zeros in the last row. Then setting

$$U_h = \begin{pmatrix} V_{h-1} & * \\ 0 & 1 \end{pmatrix}$$

we have $U_h \in \Gamma_h$ for $1 \leqq h \leqq n$ and

$$U_{h+1}^{-1}G_hU_h = \begin{pmatrix} V_h^{-1} & * \\ 0 & 1 \end{pmatrix}\begin{pmatrix} H_{h-1} & * \\ 0 & 0 \end{pmatrix}\begin{pmatrix} V_{h-1} & * \\ 0 & 1 \end{pmatrix}$$

$$= \begin{pmatrix} V_h^{-1}H_{h-1}V_h & * \\ 0 & 0 \end{pmatrix},$$

where the last matrix has the desired form.

We denote by $\Delta_n$ the set of all $n$-rowed integral upper triangular matrices of rank $n$. Observe that $\Delta_n$ is not a group and that $\Delta_n B_n = \Delta_n$.

*Lemma* 2. *For each left coset* $UB_n$, $D_{n-1}B_{n-1}$, ..., $D_1B_1$ *in* $\Gamma_n/B_n$, $\Delta_{n-1}/B_{n-1}$, ..., $\Delta_1/B_1$, *respectively, pick a fixed representative. Then there is a one-to-one correspondence between the classes* $\langle\overset{n}{Y},\dots,\overset{1}{Y}\rangle$ *of descending chains with a given* $\langle\overset{n}{Y}\rangle = \langle Y\rangle$ *and the sequences* $U$, $D_{n-1}$, ..., $D_1$ *of the given system of representatives. The correspondence is given by*

$$\overset{n}{Y} = Y[U], \qquad \overset{h}{Y} = Y\left[U\begin{pmatrix}D_{n-1}\\0\end{pmatrix}\cdots\begin{pmatrix}D_h\\0\end{pmatrix}\right] \qquad (1 \leqq h < n).$$

*Proof*. It follows from Lemma 1 that each class $\langle\overset{n}{Y},\dots,\overset{1}{Y}\rangle$ has a representative $\overset{n}{Y},\dots,\overset{1}{Y}$ where

$$\overset{n}{Y} = Y[U^*], \qquad \overset{h}{Y} = \overset{h+1}{Y}\begin{bmatrix}D_h^*\\0\end{bmatrix},$$

where $U^*\in\Gamma_n$ and $D_h^*\in\Delta_h$ for $1 \leqq h < n$. The type of the matrix $\begin{pmatrix}D_h^*\\0\end{pmatrix}$ is invariant with respect to the transformation

$$\begin{pmatrix}D_h^*\\0\end{pmatrix} \mapsto V_{h+1}^{-1}\begin{pmatrix}D_h^*\\0\end{pmatrix}V_h \qquad \text{with } V_h\in B_h, \qquad h = 1,2,\dots,n.$$

We set

$$U = U^*V_n, \qquad \begin{pmatrix}D_h\\0\end{pmatrix} = V_{h+1}^{-1}\begin{pmatrix}D_h^*\\0\end{pmatrix}V_h \qquad (1 \leqq h < n),$$

and first choose $V_n$ so that $U$ is the given representative of the coset $UB_n = U^*B_n$; next we choose $V_{n-1}$ so that $D_{n-1}$ is the given representative of the coset $D_{n-1}B_{n-1}$; and so on. It follows that each class $\langle\overset{n}{Y},\dots,\overset{1}{Y}\rangle$ contains a representative $\overset{n}{Y},\dots,\overset{1}{Y}$ of the type described in Lemma 2.

It remains to prove that to different sequences $U, D_{n-1}, \ldots, D_1$ of given representatives there correspond different classes $\langle \overset{n}{Y}, \ldots, \overset{1}{Y} \rangle$. Let

$$\overset{n}{Y} = Y[U], \qquad \overset{h}{Y} = Y\left[U\begin{pmatrix} D_{n-1} \\ 0 \end{pmatrix} \cdots \begin{pmatrix} D_h \\ 0 \end{pmatrix}\right] \qquad (1 \leqq h < n),$$

$$\overset{n}{Y}{}^* = Y[U^*], \qquad \overset{h}{Y}{}^* = Y\left[U^*\begin{pmatrix} D^*_{n-1} \\ 0 \end{pmatrix} \cdots \begin{pmatrix} D^*_h \\ 0 \end{pmatrix}\right] \qquad (1 \leqq h < n),$$

where $U, D_{n-1}, \ldots, D_1$ and $U^*, D^*_{n-1}, \ldots, D^*_1$ are two sequences of given representatives, and assume that

$$\langle \overset{n}{Y}, \ldots, \overset{1}{Y} \rangle = \langle \overset{n}{Y}{}^*, \ldots, \overset{1}{Y}{}^* \rangle .$$

We have to prove that $U^* = U$ and $D^*_h = D_h$ for $1 \leqq h < n$. The assumption means that

$$\overset{h}{Y}{}^* = \overset{h}{Y}[U_h] \quad \text{with} \quad U_h \in \Gamma_h \quad \text{for} \quad h = 1, 2, \ldots, n,$$

i.e.,

$$Y[U^*] = Y[UU_n],$$

$$Y\left[U^*\begin{pmatrix} D^*_{n-1} \\ 0 \end{pmatrix} \cdots \begin{pmatrix} D^*_h \\ 0 \end{pmatrix}\right] = Y\left[U\begin{pmatrix} D_{n-1} \\ 0 \end{pmatrix} \cdots \begin{pmatrix} D_h \\ 0 \end{pmatrix} U_h\right] \qquad (1 \leqq h < n).$$

By the definition of an equivalence, these identities are required to hold for a variable $Y$ and therefore they yield

$$U^* = \pm UU_n,$$

$$U^*\begin{pmatrix} D^*_{n-1} \\ 0 \end{pmatrix} \cdots \begin{pmatrix} D^*_h \\ 0 \end{pmatrix} = \pm U\begin{pmatrix} D_{n-1} \\ 0 \end{pmatrix} \cdots \begin{pmatrix} D_h \\ 0 \end{pmatrix} U_h \qquad (1 \leqq h < n),$$

i.e.,

$$U^* = \pm U U_n, \qquad U_{h+1}\begin{pmatrix} D_h^* \\ 0 \end{pmatrix} = \pm \begin{pmatrix} D_h \\ 0 \end{pmatrix} U_h \qquad (1 \leqq h < n).$$

The relations for $h = 1,2,\ldots,n-1$ show successively that $U_2 \in B_2$, $U_3 \in B_3$, ..., $U_n \in B_n$. Hence $U^* B_n = U B_n$ and therefore $U^* = U$. It now follows that $U_n = \pm E^{(n)}$ and $D_{n-1}^* = \pm D_{n-1} U_{n-1}$. Consequently again $D_{n-1}^* B_{n-1} = D_{n-1} B_{n-1}$ so that $D_{n-1}^* = D_{n-1}$ and $U_{n-1} = \pm E^{(n-1)}$. Proceeding in this way we see that $U^* = U$ and $D_h^* = D_h$ for $1 \leqq h < n$, q.e.d.

The integral upper triangular matrices

$$D_h = (d_{\mu\nu}^{(h)}) \quad : \quad \begin{array}{ll} 0 \leqq d_{\mu\nu}^{(h)} < d_{\mu\mu}^{(h)} & \text{for } 1 \leqq \mu < \nu \leqq h, \\ d_{\mu\nu}^{(h)} = 0 & \text{for } 1 \leqq \nu < \mu \leqq h \end{array}$$

form a complete set of representatives of all cosets $D_h B_h$ with an upper triangular matrix $D_h$ of rank $h$. Two different matrices of this special type belong to different cosets. The number of cosets $D_h B_h$ with given $d_{\mu\mu}^{(h)}$ $(1 \leqq \mu \leqq h)$ is obviously $\prod_{\mu=1}^{h} (d_{\mu\mu}^{(h)})^{h-\mu}$ .

The class $<\overset{n}{Y},\ldots,\overset{1}{Y}>$ is primitive if and only if $\begin{pmatrix} D_h \\ 0 \end{pmatrix}$ is primitive, i.e., $D_h$ unimodular for $1 \leqq h < n$. Since $\Delta_h \cap \Gamma_h = B_h$, we can in this case choose $D_h = E^{(h)}$ as the special representatives and obtain

*Lemma* 3. *In each left coset* $UB_n$ *belonging to* $\Gamma_n/B_n$ *choose a fixed representative. Then there is a one-to-one correspondence between the primitive classes* $<\overset{n}{Y},\ldots,\overset{1}{Y}>$ *with given* $<\overset{n}{Y}> = <Y>$ *and the given set of coset representatives* $U: \Gamma_n/B_n$. *The correspondence is given by*

$$\overset{n}{Y} = Y[U], \qquad \overset{h}{Y} = Y\left[U\begin{pmatrix} E^{(h)} \\ 0 \end{pmatrix}\right] \qquad (1 \leqq h < n).$$

We now introduce Selberg's zeta functions. Denote by $z_1, z_2, \ldots, z_{n-1}$ a system of $n-1$ complex variables and by $s_1, s_2, \ldots, s_n$ a system of $n$ complex variables, the two being related by the equations

$$z_h = s_{h+1} - s_h + \frac{1}{2} \qquad h = 1,2,\ldots,n-1.$$

If $Y$ is a variable positive matrix we set $Y = \overset{n}{Y}$ and define

$$\zeta^*(Y;s_1,\ldots,s_n) = \sum_{\substack{\langle \overset{n}{Y},\ldots,\overset{1}{Y}\rangle \\ \text{primitive}}} |\overset{n-1}{Y}|^{-z_{n-1}} |\overset{n-2}{Y}|^{-z_{n-2}} \ldots |\overset{1}{Y}|^{-z_1},$$

$$\zeta(Y;s_1,\ldots,s_n) = \sum_{\langle \overset{n}{Y},\ldots,\overset{1}{Y}\rangle} |\overset{n-1}{Y}|^{-z_{n-1}} |\overset{n-2}{Y}|^{-z_{n-2}} \ldots |\overset{1}{Y}|^{-z_1}.$$

Lemma 3 yields

$$\zeta^*(Y;s_1,\ldots,s_n) = \sum_{U:\Gamma_n/B_n} \prod_{h=1}^{n-1} \left| Y\left[U\begin{pmatrix} E^{(h)} \\ 0 \end{pmatrix}\right]\right|^{s_h - s_{h+1} - \frac{1}{2}}$$

$$= |Y|^{-s_n - \frac{1}{4}(n-1)} \sum_{U:\Gamma_n/B_n} f_s(Y[U]),$$

where

$$f_s(Y) = \prod_{h=1}^{n} \left| Y\begin{bmatrix} E^{(h)} \\ 0 \end{bmatrix} \right|^{s_h - s_{h+1} - \frac{1}{2}}$$

$$= |Y|^{s_n + \frac{1}{2}(n-1)} \prod_{h=1}^{n-1} \left| Y\begin{bmatrix} E^{(h)} \\ 0 \end{bmatrix} \right|^{s_h - s_{h+1} - \frac{1}{2}}$$

is the function introduced in §6. Up to a factor which is a power of $|Y|$, this is the series $u_{1,2,\ldots,n-1}(Y)$ we introduced in §10, where we proved in particular that it converges for $\mathcal{Re}\, z_h > 1$ $(1 \leqq h < n)$.

We shall now show that there is a close connection between $\zeta$ and $\zeta^*$. Using the special representatives $D_h$ described above after the proof of Lemma 2, we get

$$\begin{pmatrix} D_{n-1} \\ 0 \end{pmatrix}\begin{pmatrix} D_{n-2} \\ 0 \end{pmatrix} \cdots \begin{pmatrix} D_h \\ 0 \end{pmatrix} = \begin{pmatrix} T_h \\ 0^{(n-h,h)} \end{pmatrix}$$

with $T_h = T_h^{(h)} = (t_{\mu\nu}^{(h)})$, $t_{\mu\nu}^{(h)} = 0$ for $\mu > \nu$ and

$$t_{\nu\nu}^{(h)} = \prod_{j=h}^{n-1} d_{\nu\nu}^{(j)} \qquad \text{for } 1 \leqq \nu \leqq h.$$

It follows, in the notation of Lemma 2, that

$$\overset{h}{|Y|} = \left| Y\left[ U\begin{pmatrix} T_h \\ 0 \end{pmatrix} \right] \right| = \left| Y\left[ U\begin{pmatrix} E^{(h)} \\ 0 \end{pmatrix} \right] \right| \cdot |T_h|^2$$

$$= \left| Y\left[ U\begin{pmatrix} E^{(h)} \\ 0 \end{pmatrix} \right] \right| \prod_{\nu=1}^{h} \prod_{j=h}^{n-1} (d_{\nu\nu}^{(j)})^2,$$

hence

$$\prod_{h=1}^{n-1} |\overset{h}{Y}|^{-z_h} = \prod_{h=1}^{n-1} \left\{ \left| Y\left[ U\begin{pmatrix} E^{(h)} \\ 0 \end{pmatrix} \right] \right|^{-z_h} \prod_{\nu=1}^{h} \prod_{j=h}^{n-1} (d_{\nu\nu}^{(j)})^{-2z_h} \right\}$$

and

$$\zeta(Y;s_1,\dots,s_n) = \zeta^*(Y;s_1,\dots,s_n) \sum_{D_{n-1},\dots,D_1} \prod_{h=1}^{n-1} \prod_{\nu=1}^{h} \prod_{j=h}^{n-1} (d_{\nu\nu}^{(j)})^{-2z_h}$$

$$= \zeta^*(Y;s_1,\dots,s_n) \sum_{\substack{d_\mu^{(h)} \geqq 1 \\ 1 \leqq h < n \\ 1 \leqq \mu \leqq h}} \left\{ \prod_{j=1}^{n-1} \prod_{\nu=1}^{j} (d_\nu^{(j)})^{j-\nu} \right\} \left\{ \prod_{h=1}^{n-1} \prod_{\nu=1}^{h} \prod_{j=h}^{n-1} (d_\nu^{(j)})^{-2z_h} \right\}$$

$$= \zeta^*(Y;s_1,\dots,s_n) \sum_{\substack{d_\mu^{(h)} \geqq 1 \\ 1 \leqq h < n \\ 1 \leqq \mu \leqq h}} \prod_{j=1}^{n-1} \prod_{\nu=1}^{j} (d_\nu^{(j)})^{-2(z_\nu + z_{\nu+1} + \dots + z_j) + j - \nu}$$

$$= \zeta^*(Y;s_1,\dots,s_n) \prod_{1 \leqq \nu \leqq j < n} \zeta(2(z_\nu + z_{\nu+1} + \dots + z_j) - j + \nu),$$

where

$$\zeta(s) = \sum_{n=1}^{\infty} n^{-s}$$

is Riemann's zeta function. Considering that

$$2(z_\nu + z_{\nu+1} + \dots + z_j) - j + \nu = 2(s_{j+1} - s_\nu) + 1,$$

we obtain the announced relation

$$\zeta(Y;s_1,\ldots,s_n) = \zeta^*(Y;s_1,\ldots,s_n) \prod_{1\leq i<j\leq n} \zeta(2(s_j - s_i)+1).$$

Let us write $s = (s_1,s_2,\ldots,s_n)$ and $\tilde{s} = (-s_n,-s_{n-1},\ldots,-s_1)$. In §6 we proved that

$$f_s(Y^{-1}) = f_{\tilde{s}}(Y[\tilde{U}]),$$

where

$$\tilde{U} = \begin{pmatrix} 0 & & 1 \\ & \cdot^{\cdot^{1}} & \\ 1 & & 0 \end{pmatrix},$$

so that in particular

$$f_s(Y^{-1}[U]) = f_{\tilde{s}}(Y[U'^{-1}\tilde{U}]).$$

Now let $U$ run through a complete set of representatives of the cosets belonging to $\Gamma_n/B_n$. Then $\Gamma_n = \bigcup_U UB_n$ and by transposition $\Gamma_n = \bigcup_U U'^{-1}B_n' = \bigcup_U U'^{-1}\tilde{U}\tilde{U}B_n'$. But $\tilde{U}\in\Gamma_n$ and $\tilde{U}B_n'\tilde{U}^{-1} = B_n$ so that $\Gamma_n = \bigcup_U U'^{-1}\tilde{U}B_n$ and we see that together with $U$ also $U'^{-1}\tilde{U}$ runs through a complete set of representatives belonging to $\Gamma_n/B_n$. So we get

$$\zeta^*(Y^{-1};s) = |Y|^{s_n+\frac{1}{2}(n-1)} \sum_{U:\Gamma_n/B_n} f_s(Y^{-1}[U])$$

$$= |Y|^{s_n+\frac{1}{2}(n-1)} \sum_{U:\Gamma_n/B_n} f_{\tilde{s}}(Y[U'^{-1}\tilde{U}]) = |Y|^{s_n-s_1+\frac{1}{2}(n-1)} \zeta^*(Y;\tilde{s})$$

and thus also

$$\zeta(Y^{-1};s) = |Y|^{s_n - s_1 + \frac{1}{2}(n-1)} \zeta(Y;\tilde{s})$$

since the factor

$$Z(s) = \prod_{1 \leq i < j \leq n} \zeta(2(s_j - s_i) + 1)$$

satisfies $Z(\tilde{s}) = Z(s)$. Furthermore the function

$$\phi(s) = \prod_{1 \leq \nu < \mu \leq n} \{\pi^{-(s_\mu - s_\nu + \frac{1}{2})} (s_\mu - s_\nu + \tfrac{1}{2})(s_\nu - s_\mu + \tfrac{1}{2}) \Gamma(s_\mu - s_\nu + \tfrac{1}{2})\}$$

$$= \{\pi^{\frac{1}{2}n(n-1) + (n+1)\sum_{\nu=1}^{n} s_\nu} \prod_{\substack{\mu,\nu=1 \\ \mu \neq \nu}}^{n} (s_\mu - s_\nu + \tfrac{1}{2})\} \, \pi^{-2\sum_{\nu=1}^{n} \nu s_\nu} \prod_{1 \leq \nu < \mu \leq n} \Gamma(s_\mu - s_\nu + \tfrac{1}{2})$$

satisfies

$$\phi(\tilde{s}) = \phi(s).$$

The main result will be:

*Theorem* 1. (1) *The function*

$$\prod_{1 \leq \nu < \mu \leq n} (s_\mu - s_\nu - \tfrac{1}{2}) \zeta(Y; s_1, \ldots, s_n)$$

*is holomorphic in* $\mathbb{C}^n$.

(2) *The meromorphic function*

$$\pi^{-2\sum_{\nu=1}^{n} \nu s_\nu} \prod_{1 \leq \nu < \mu \leq n} \Gamma(s_\mu - s_\nu + \tfrac{1}{2}) |Y|^{s_n} \zeta(Y; s_1, \ldots, s_n)$$

*is invariant under the group of all cyclic permutations of the variables* $s_1, s_2, \ldots, s_n$.

*Proof.* In the case $n = 2$ the theorem asserts well-known properties of Epstein's zeta function (cf. Hecke, Math. Werke). The main point is here the transcription to the variables $s_1$, $s_2$. For $n = 2$ we have

$$\zeta(Y;s) = \frac{1}{2} \sum_{(x,y)\neq(0,0)} \left(Y\begin{bmatrix} x \\ y \end{bmatrix}\right)^{-z}, \qquad z = s_2 - s_1 + \frac{1}{2},$$

where $Y = Y^{(2)} > 0$. The theta series

$$\theta(y,Y) = \sum_{a,b} e^{-\pi Y \left[\begin{smallmatrix} a \\ b \end{smallmatrix}\right] y}$$

satisfies the relation

$$\theta(y^{-1},Y^{-1}) = y|Y|^{\frac{1}{2}}\theta(y,Y),$$

a special case of the theta-relation on p. 222 if we set $m = 2$, $n = 1$, $w = 1$, $G = \begin{bmatrix} a \\ b \end{bmatrix}$, replace $S$ by $Y$ and $Y$ by $y$. The Mellin transform of $\theta_1(y,Y) = \theta(y,Y) - 1$ is equal to

$$\xi(z,Y) = \int_0^\infty \theta_1(y,Y)y^{z-1}dy$$

$$= \sum_{(a,b)\neq(0,0)} \int_0^\infty e^{-\pi Y \left[\begin{smallmatrix} a \\ b \end{smallmatrix}\right] y} y^{z-1}dy = \pi^{-z}\Gamma(z) \sum_{(a,b)\neq(0,0)} \left(Y\begin{bmatrix} a \\ b \end{bmatrix}\right)^{-z}$$

and

$$\xi(z,Y) = \int_1^\infty \{\theta_1(y,Y)y^{z-1} + \theta_1(y^{-1},Y)y^{-z-1}\}dy.$$

The above theta-relation yields

$$\theta_1(y^{-1},Y) = y|Y|^{-\frac{1}{2}}\theta_1(y,Y^{-1}) + y|Y|^{-\frac{1}{2}} - 1$$

hence

$$\xi(z,Y) = \int_1^\infty \{\theta_1(y,Y)y^z + |Y|^{-\frac{1}{2}}\theta_1(y,Y^{-1})y^{1-z}\}\frac{dy}{y} + |Y|^{-\frac{1}{2}}\frac{1}{z-1} - \frac{1}{z},$$

which shows that

$$|Y|^{-\frac{1}{2}}\xi(1-z,Y^{-1}) = \xi(z,Y).$$

If we set

$$Y = \begin{pmatrix} \alpha & \beta \\ \beta & \gamma \end{pmatrix},$$

then

$$Y^{-1} = |Y|^{-1}\begin{pmatrix} \gamma & -\beta \\ -\beta & \alpha \end{pmatrix}$$

and

$$Y^{-1}\begin{bmatrix} a \\ b \end{bmatrix} = (\gamma a^2 - 2\beta ab + \alpha b^2)|Y|^{-1}.$$

Substituting $-b$ for $b$ we see that

$$Y^{-1}\begin{bmatrix} a \\ -b \end{bmatrix} = |Y|^{-1}Y\begin{bmatrix} b \\ a \end{bmatrix}.$$

Writing

$$D(s,Y) = \sum_{(a,b)\neq(0,0)}\left(Y\begin{bmatrix} a \\ b \end{bmatrix}\right)^{-s} = 2\zeta(Y;s)$$

we obtain therefore

$$D(z,Y^{-1}) = |Y|^{z}D(z,Y).$$

Using $\xi(z,Y) = \pi^{-z}\Gamma(z)D(z,Y)$ and the above equation for $\xi(z,Y)$, we get

$$\xi(z,Y) = |Y|^{\frac{1}{2}-z}\xi(1-z,Y).$$

If we write $s_2 - s_1 + \frac{1}{2} = z$, then the permutation $(s_1,s_2) \mapsto (s_2,s_1)$ corresponds to $z \mapsto 1-z$. The first assertion of Theorem 1 is clearly equivalent to saying that $(z-1)D(z,Y)$ is a holomorphic function of $(s_1,s_2) \in \mathbb{C}^2$. But $(z-1)z\xi(z,Y)$ is an entire function of $z$, and

$$\begin{aligned}(z-1)z\,\xi(z,Y) &= \pi^{-z}(z-1)z\,\Gamma(z)D(z,Y)\\ &= \pi^{-z}\,\Gamma(z+1)(z-1)D(z,Y),\end{aligned}$$

so that $(z-1)D(z,Y)$ is an entire function of $z$.

The second assertion of Theorem 1 amounts to saying that

$$\pi^{-2s_1-4s_2}\,\Gamma(z)|Y|^{s_2}D(z,Y)$$

is invariant under the permutation of $s_1$ and $s_2$, or under $z \mapsto 1-z$. But

$$-2s_1 - 4s_2 = -z - 3(s_1+s_2) + \frac{1}{2}$$

and $-3(s_1+s_2)+\frac{1}{2}$ is obviously invariant under $s_1 \leftrightarrow s_2$, so that we have only to prove that

$$\pi^{-z}\,\Gamma(z)|Y|^{s_2}D(z,Y) = \pi^{-(1-z)}\Gamma(1-z)|Y|^{s_1}D(1-z,Y),$$

i.e., that

$$\pi^{-z}\Gamma(z)D(z,Y) = |Y|^{\frac{1}{2}-z}\pi^{-(1-z)}\Gamma(1-z)D(1-z,Y),$$

and this is precisely the above functional equation for $D(z,Y)$.

We can thus assume that $n \geqq 3$, and for technical reasons we shall replace $Y^{(n)}$ by $S^{(n+1)}$ and $s = (s_1,\dots,s_n)$ by $(s,s_{n+1}) = (s_1,\dots,s_{n+1})$ so that from now on we assume that $n \geqq 2$. We introduce the theta-series

$$\theta(Y,S) = \sum_G e^{-\pi\sigma(S[G]Y)},$$

where $S = S^{(n+1)} > 0$, $Y = Y^{(n)} > 0$, and the relation on p. 222 gives now

$$\theta(\hat{Y},\hat{S}) = |S|^{\frac{1}{2}n}|Y|^{\frac{1}{2}(n+1)}\theta(Y,S) \qquad (\hat{Y} = Y^{-1},\ \hat{S} = S^{-1}).$$

With the help of the differential operator

$$\mathrm{P}_k = |Y|^{-k}\hat{\mathrm{M}}_n|Y|^k\mathrm{M}_n$$

introduced in §15, we form, similarly as in §16, the function

$$\Theta(Y,S) = \mathrm{P}_{\frac{1}{2}(n+1)}\theta(Y,S).$$

Using

$$\hat{\mathrm{P}}_k = |Y|^k\mathrm{P}_k|Y|^{-k}$$

we have again

$$\Theta(\hat{Y},\hat{S}) = |S|^{\frac{1}{2}n}|Y|^{\frac{1}{2}(n+1)}\Theta(Y,S).$$

In the series

$$\Theta(Y,S) = \sum_G \mathrm{P}_{\frac{1}{2}(n+1)}\, e^{-\pi\sigma(S[G]Y)}$$

it is sufficient to extend the summation over all integral matrices $G = G^{(n+1,n)}$ whose rank is equal to $n$, because

$$\left|\frac{\partial}{\partial Y}\right| e^{-\pi\sigma(S[G]Y)} = |-\pi S[G]| e^{-\pi\sigma(S[G]Y)}$$

vanishes if rank $G < n$.

We denote again by $\mathcal{R}$ the Minkowski domain of positive reduced matrices, write $dv = |Y|^{-\frac{1}{2}(n+1)}[dY]$, and compute the function

$$\xi(S;s,s_{n+1}) = \int_{\mathcal{R}} \Theta(Y,S)|Y|^{s_{n+1}-s_1+\frac{1}{2}n}\zeta(Y,\tilde{s})dv$$

$$= Z(s)\sum_G \int_{\mathcal{R}} |Y|^{s_{n+1}-s_1+\frac{1}{2}n}\zeta^*(Y,\tilde{s})P_{\frac{1}{2}(n+1)}e^{-\pi\sigma(S[G]Y)}dv$$

$$= (-\pi)^n Z(s)\sum_G |S[G]| \cdot$$

$$\cdot \int_{\mathcal{R}} |Y|^{s_{n+1}-\frac{1}{2}(n+1)} \sum_{U:\Gamma_n/B_n} f_{\tilde{s}}(Y[U])\hat{M}_n|Y|^{\frac{1}{2}(n+3)}e^{-\pi\sigma(S[G]Y)}dv.$$

Using the relation

$$|Y|^t f_s(Y) = f_{s+t}(Y),$$

where $(s+t)_\nu = s_\nu + t$, we get

$$\xi(S;s,s_{n+1}) = (-\pi)^n Z(s)\sum_{U:\Gamma_n/B_n}\sum_G |S[G]| \cdot$$

$$\cdot \int_{\mathcal{R}} f_{\tilde{s}+s_{n+1}-\frac{1}{2}(n+1)}(Y[U])\hat{M}_n|Y|^{\frac{1}{2}(n+3)}e^{-\pi\sigma(S[G]Y)}dv.$$

We replace $G$ by $GW'U'$, where the new $G$ runs through a complete set of representatives of the cosets $G\mathrm{B}_n'$, and $W$ runs through the matrices in $\mathrm{B}_n$, while $U'$ is fixed. Since $f_s(Y)$ is invariant under $Y \mapsto Y[W]$, $W \in \mathrm{B}_n$, we get

$$\xi(S;s,s_{n+1}) = (-\pi)^n Z(s) \sum_{G/\mathrm{B}_n'} |S[G]| \sum_{U:\Gamma_n/\mathrm{B}_n} \sum_{W\in\mathrm{B}_n}$$

$$\int_{\mathcal{R}} f_{\tilde{s}+s_{n+1}-\frac{1}{4}(n+1)}(Y[UW])\hat{\mathrm{M}}_n |Y|^{\frac{1}{2}(n+3)} e^{-\pi\sigma(S[G]Y[UW])} dv.$$

Taking into account that $UW$ gives all unimodular matrices, that $\bigcup_U \mathcal{R}[U]$ covers the space of all positive matrices twice, and that $\hat{\mathrm{M}}_n$ is the adjoint of $\mathrm{M}_n$, we obtain

$$\xi(S;s,s_{n+1}) = (-\pi)^n Z(s) \sum_{G/\mathrm{B}_n'} |S[G]| \sum_{U\in\Gamma_n}$$

$$\int_{\mathcal{R}[U]} f_{\tilde{s}+s_{n+1}-\frac{1}{4}(n+1)}(Y)\hat{\mathrm{M}}_n |Y|^{\frac{1}{2}(n+3)} e^{-\pi\sigma(S[G]Y)} dv$$

$$= 2(-\pi)^n Z(s) \sum_{G/\mathrm{B}_n'} |S[G]| \int_{Y>0} f_{\tilde{s}+s_{n+1}-\frac{1}{4}(n+1)}(Y)\hat{\mathrm{M}}_n |Y|^{\frac{1}{2}(n+3)} e^{-\pi\sigma(S[G]Y)} dv$$

$$= 2(-\pi)^n Z(s) \sum_{G/\mathrm{B}'} |S[G]| \int_{Y>0} e^{-\pi\sigma(S[G]Y)} |Y|^{\frac{1}{2}(n+3)} \mathrm{M}_n f_{\tilde{s}+s_{n+1}-\frac{1}{4}(n+1)}(Y) dv.$$

By virtue of the formula

$$M_n f_s(Y) = \prod_{\nu=1}^{n} (s_\nu + \frac{n-1}{4}) f_s(Y),$$

proved at the end of §6, of the formula

$$\int_{Y>0} e^{-\sigma(YX^{-1})} f_s(Y) dv = \pi^{\frac{1}{4}n(n-1)} \prod_{\nu=1}^{n} \Gamma(s_\nu - \frac{n-1}{4}) \cdot f_s(X),$$

also proved in §6, and the above formula for $|Y|^t f_s(Y)$ with $t = \frac{n+3}{2}$, we have

$$\xi(S; s, s_{n+1}) = 2(-\pi)^n \prod_{\nu=1}^{n} (s_{n+1} - s_\nu - \tfrac{1}{2}) Z(s) \sum_{G/B_n'} |S[G]| \cdot$$

$$\cdot \int_{Y>0} e^{-\pi\sigma(S[G]Y)} f_{\tilde{s}+s_{n+1}+1+\frac{1}{4}(n+1)}(Y) dv$$

$$= 2\pi^{n+\frac{1}{4}n(n-1)} \prod_{\nu=1}^{n} \{(s_\nu - s_{n+1} + \tfrac{1}{2}) \Gamma(s_{n+1} - s_\nu + \tfrac{3}{2})\} Z(s) \cdot$$

$$\cdot \sum_{G/B_n'} |S[G]| f_{\tilde{s}+s_{n+1}+1+\frac{1}{4}(n+1)}(\tfrac{1}{\pi}(S[G])^{-1}).$$

Using again the formula relating $f_s$ and $f_{\tilde{s}}$, the fact that $f_s$ is homogeneous of degree $\sum_{\nu=1}^{n} s_\nu$ and $|Y| f_s(Y) = f_{s+1}(Y)$, we get

$\xi(S;s,s_{n+1}) =$

$$= 2\prod_{\nu=1}^{n}\{\pi^{-(s_{n+1}-s_\nu+\frac{1}{2})}(s_{n+1}-s_\nu+\tfrac{1}{2})(s_\nu-s_{n+1}+\tfrac{1}{2})\Gamma(s_{n+1}-s_\nu+\tfrac{1}{2})\}Z(s)\,\cdot$$

$$\cdot\sum_{G/B_n'} f_{s-s_{n+1}-\frac{1}{2}(n+1)}(S[G\tilde{U}]).$$

Now $\bigcup_G GB_n' = \bigcup_G G\tilde{U}\tilde{U}B_n'\tilde{U}^{-1} = \bigcup_G G\tilde{U}B_n$, so that $G\tilde{U}$ runs through a complete set of representatives of the cosets $GB_n$, and thus

$$\xi(S;s,s_{n+1}) = \frac{2\phi(s,s_{n+1})}{\phi(s)}\,Z(s)\sum_{G/B_n} f_{s-s_{n+1}-\frac{1}{2}(n+1)}(S[G]),$$

where $\phi$ is the function introduced before the statement of Theorem 1.

Next we prove

$$Z(s)\sum_{G/B_n} f_{s-s_{n+1}-\frac{1}{2}(n+1)}(S[G]) = \zeta(S;s,s_{n+1}).$$

Let us write $G = U\begin{pmatrix} D_n \\ 0 \end{pmatrix}$ with $U\in\Gamma_{n+1}$ and an upper triangular matrix $D_n = D_n^{(n)}$. The summation over $G/B_n$ is equivalent to the summation over $U: \Gamma_{n+1}/B_{n+1}$ and $D_n: \Delta_n/B_n$, where $\Delta_n$ denotes again the set of $n$-rowed integral upper triangular matrices of rank $n$. For the $D_n$ we can take the special representatives

$$D_n = (d_{\mu\nu}):\quad \begin{matrix} 0 \leqq d_{\mu\nu} < d_{\mu\mu} = d_\mu & \text{for } 1 \leqq \mu < \nu \leqq n, \\ d_{\mu\nu} = 0 & \text{for } 1 \leqq \nu < \mu \leqq n. \end{matrix}$$

This yields

$$\sum_{G/B_n} f_{s-s_{n+1}-\frac{1}{2}(n+1)}(S[G]) = \sum_{U:\Gamma_{n+1}/B_{n+1}} \sum_{D_n} \prod_{h=1}^{n} \left| S\left[U\begin{pmatrix} D_n \\ 0 \end{pmatrix}\begin{pmatrix} E^{(h)} \\ 0 \end{pmatrix}\right]\right|^{-z_h}$$

$$= \sum_{U:\Gamma_{n+1}/B_{n+1}} \sum_{D_n} \left\{\prod_{h=1}^{n} \left| S\left[U\begin{pmatrix} E^{(h)} \\ 0 \end{pmatrix}\right]\right|^{-z_h}\right\}\left\{\prod_{h=1}^{n} (d_1\dots d_h)^{-2z_h}\right\}$$

$$= \zeta^*(S;s,s_{n+1}) \sum_{d_1,\dots,d_n=1}^{\infty} \left\{\prod_{\nu=1}^{n} d_\nu^{n-\nu}\right\}\left\{\prod_{h=1}^{n} (d_1 d_2\dots d_h)^{-2z_h}\right\}$$

$$= \zeta^*(S;s,s_{n+1}) \sum_{d_1,\dots,d_n=1}^{\infty} \prod_{\nu=1}^{n} d_\nu^{-2(z_\nu+\dots+z_n)+n-\nu}$$

$$= \zeta^*(S;s,s_{n+1}) \prod_{\nu=1}^{n} \zeta(2(z_\nu+\dots+z_n)-n+\nu)$$

$$= \zeta^*(S;s,s_{n+1}) \prod_{\nu=1}^{n} \zeta(2(s_{n+1}-s_\nu)+1),$$

from where our assertion follows.

We have thus proved the relation

$$\xi(S;s,s_{n+1}) = \frac{2\phi(s,s_{n+1})}{\phi(s)} \zeta(S;s,s_{n+1}).$$

From the representation

$$\xi(S;s,s_{n+1}) =$$

$$= \int\limits_{\substack{\mathcal{R}\\|Y|\geq 1}} \{\Theta(Y,S)|Y|^{s_{n+1}-s_1+\frac{1}{2}n}\zeta(Y,\tilde{s}) + \Theta(\hat{Y},S)|Y|^{s_1-s_{n+1}-\frac{1}{2}n}\zeta(\hat{Y},\tilde{s})\}dv$$

$$= \int\limits_{\substack{\mathcal{R}\\|Y|\geq 1}} \{\Theta(Y,S)|Y|^{s_{n+1}-s_1+\frac{1}{2}n}\zeta(Y,\tilde{s}) + |S|^{-\frac{1}{2}n}\Theta(Y,\hat{S})|Y|^{s_n-s_{n+1}+\frac{1}{2}n}\zeta(Y,s)\}dv$$

we see that $\phi(s,s_{n+1})\zeta(S;s,s_{n+1})$ is holomorphic in the domain

$$\mathcal{L} =$$

$$= \{(s_1,s_2,\ldots,s_{n+1}) \mid \mathcal{Re}\,(s_{\nu+1}-s_\nu+\tfrac{1}{2}) > 1 \quad (1 \leqq \nu < n), \quad s_{n+1} \text{ arbitrary}\},$$

and moreover that the functional equation

$$|S|^{\frac{1}{2}n}\xi(S;s,s_{n+1}) = \xi(\hat{S};\tilde{s},-s_{n+1})$$

is satisfied. The right-hand side of this equation can be rewritten as

$$\xi(\hat{S};\tilde{s},-s_{n+1}) = \frac{2\phi(\tilde{s},-s_{n+1})}{\phi(s)}\zeta(\hat{S};\tilde{s},-s_{n+1})$$

$$= \frac{2\phi(s_{n+1},s_1,\ldots,s_n)}{\phi(s_1,\ldots,s_n)}|S|^{-s_{n+1}+s_n+\frac{1}{2}n}\zeta(S;s_{n+1},s_1,\ldots,s_n),$$

provided that $\mathcal{Re}\,(s_{\nu+1}-s_\nu+\frac{1}{2}) > 1 \quad (1 \leqq \nu < n)$ and $\mathcal{Re}\,(s_1-s_{n+1}+\frac{1}{2}) > 1$. For the left-hand side we get

$$|S|^{\frac{1}{2}n}\xi(S;s,s_{n+1}) = \frac{2\phi(s_1,\ldots,s_{n+1})}{\phi(s_1,\ldots,s_n)}\,|S|^{\frac{1}{2}n}\zeta(S;s_1,\ldots,s_{n+1}).$$

This shows that the function

$$\eta(S;s_1,\ldots,s_{n+1}) = \phi(s_1,\ldots,s_{n+1})|S|^{s_{n+1}}\zeta(S;s_1,\ldots,s_{n+1})$$

is invariant with respect to the cyclic permutation

$$\pi\colon (s_1,s_2,\ldots,s_n,s_{n+1}) \mapsto (s_{n+1},s_1,\ldots,s_{n-1},s_n).$$

Therefore $\eta(S;s_1,\ldots,s_{n+1})$ is holomorphic in the domain

$$\vartheta = \bigcup_{\nu=0}^{n} \pi^{\nu}(\mathcal{L}).$$

The projection of $\vartheta \subset \mathbb{C}^{n+1}$ in $\mathbb{R}^{n+1}$ is a domain $\mathcal{Y}$, which consists of the points $(\mathcal{R}e\, s_1,\ \mathcal{R}e\, s_2,\ldots,\ \mathcal{R}e\, s_{n+1})$. Clearly

$$\vartheta = \mathcal{Y} + i\,\mathbb{R}^{n+1} \subset \mathbb{C}^{n+1};$$

a domain of this type is called a tube domain. It is known [4; Theorem 2.5.10, p. 41] that the envelope of holomorphy $\vartheta^*$ of $\vartheta$ is the tube domain

$$\vartheta^* = \mathcal{Y}^* + i\,\mathbb{R}^{n+1},$$

where $\mathcal{Y}^*$ denotes the convex hull of $\mathcal{Y}$ in the sense of euclidean geometry. From the defining inequalities of $\mathcal{L}$ follows that $\mathcal{Y}$ contains a straight line parallel to the $(n+1)^{\text{th}}$ axis of $\mathbb{R}^{n+1}$. Since $\mathcal{Y}$ is invariant under the group generated by $\pi$, it follows that $\mathcal{Y}$ contains a system of straight lines $l_1,l_2,\ldots,l_{n+1}$ such that $l_\nu$ is parallel to the $\nu^{\text{th}}$ axis of $\mathbb{R}^{n+1}$. The convex hull of the system

$l_1, l_2, \ldots, l_{n+1}$ is already the whole space $\mathbb{R}^{n+1}$. This proves that $\mathcal{Y}^* = \mathbb{R}^{n+1}$ and $\vartheta^* = \mathbb{C}^{n+1}$, i.e., that $\eta(S; s_1, \ldots, s_{n+1})$ is holomorphic in $\mathbb{C}^{n+1}$. Now Theorem 1 in the changed notation is obvious because

$$(\phi(s_1, s_2, \ldots, s_{n+1}))^{-1} \prod_{1 \leq \nu < \mu \leq n+1} (s_\mu - s_\nu - \tfrac{1}{2})$$

is holomorphic in $\mathbb{C}^{n+1}$ and

$$(\phi(s_1, s_2, \ldots, s_{n+1}))^{-1} \, \pi^{-2 \sum_{\nu=1}^{n+1} \nu s_\nu} \prod_{1 \leq \nu < \mu \leq n+1} \Gamma(s_\mu - s_\nu + \tfrac{1}{2})$$

is invariant under all permutations of the variables.

Selberg proceeded by another method and proved moreover the invariance of $\eta(Y; s)$ under the group of all permutations of the variables. The following proof for the transposition of the first two variables will be along his lines. Using the concept of a shortened (ordinary or primitive) descending chain, whose definition is self-explanatory, we can write

$$\zeta(Y; s) = \sum_{\langle \overset{n}{Y}, \ldots, \overset{2}{Y} \rangle} \left\{ \prod_{\nu=2}^{n-1} |\overset{\nu}{Y}|^{-z_\nu} \right\} \overset{2}{\zeta}(\overset{2}{Y}, z_1), \qquad \langle Y \rangle = \langle \overset{n}{Y} \rangle,$$

where $\overset{2}{\zeta}(Y; z_1) = \overset{2}{\zeta}(Y; s_1, s_2)$. We consider this function in dependence of $z_1 = s_2 - s_1 + \frac{1}{2}$. In order to prove the functional equation, and moreover to estimate this zeta function, we use the representation - obtained from the one used in the proof of the preceding theorem for the case $n = 2$ by a slight change of notation -

$$\eta(\overset{2}{Y},z) = \int_0^\infty \theta_1(\overset{2}{Y},y)y^{z-1}dy = \pi^{-z}\Gamma(z)\zeta(\overset{2}{Y};z)$$

$$= \int_1^\infty \{\theta_1(\overset{2}{Y},y)y^z + |\overset{2}{Y}|^{-\frac{1}{2}}\theta_1(\overset{2}{Y}^{-1},y)y^{1-z}\}\frac{dy}{y} - \frac{1}{2}\left(\frac{1}{z} + \frac{1}{1-z}|Y|^{-\frac{1}{2}}\right),$$

with

$$2\theta_1(\overset{2}{Y},y) + 1 = \sum_{a,b=-\infty}^{\infty} e^{-\pi \overset{2}{Y}\left[\begin{smallmatrix}a\\b\end{smallmatrix}\right] y}.$$

As we have seen, from here it follows that

$$\eta(\overset{2}{Y},z) = |\overset{2}{Y}|^{-\frac{1}{2}}\eta(\overset{2}{Y}^{-1},1-z) = |\overset{2}{Y}|^{\frac{1}{2}-z}\eta(\overset{2}{Y},1-z).$$

Now let us assume

(1) $\qquad |z_1| > 1, \qquad |z_1-1| > 1, \qquad -2 < \mathfrak{Re}\, z_1 < 3.$

Then we can estimate

$$|\eta(\overset{2}{Y},z_1)| < \int_0^\infty \theta_1(\overset{2}{Y},y)y^2dy + |\overset{2}{Y}|^{-\frac{1}{2}}\int_0^\infty \theta_1(\overset{2}{Y}^{-1},y)y^2dy + \frac{1}{2}(1+|\overset{2}{Y}|^{-\frac{1}{2}})$$

$$= \eta(\overset{2}{Y},3) + \eta(\overset{2}{Y}^{-1},3)|\overset{2}{Y}|^{-\frac{1}{2}} + \frac{1}{2}(1+|\overset{2}{Y}|^{-\frac{1}{2}})$$

$$= \eta(\overset{2}{Y},3) + \eta(\overset{2}{Y},3)|\overset{2}{Y}|^{\frac{5}{2}} + \frac{1}{2}(1+|\overset{2}{Y}|^{-\frac{1}{2}}).$$

The series

$$(2) \qquad \pi^{-z_1}\Gamma(z_1)\zeta(Y,s) = \sum_{\substack{n \quad\; 2 \\ <Y,\ldots,Y>}} \left\{ \prod_{\nu=2}^{n-1} |\overset{\nu}{Y}|^{-z_\nu} \right\} \eta(\overset{2}{Y}, z_1)$$

converges for $\mathcal{R}e\, z_\nu > 1$ $(1 \leqq \nu < n)$: this was proved in §10, as well as the convergence of

$$\sum_{\substack{n \quad\; 2 \\ <Y,\ldots,Y>}} \prod_{\nu=2}^{n-1} |\overset{\nu}{Y}|^{-z_\nu} \quad \text{for} \quad \mathcal{R}e\, z_\nu > 1 \;\; (2 < \nu < n), \quad \mathcal{R}e\, z_2 > \tfrac{3}{2},$$

because this series differs from

$$\sum_{\substack{n \quad\; 2 \\ <Y,\ldots,Y> \\ \text{primitive}}} \prod_{\nu=2}^{n-1} |\overset{\nu}{Y}|^{-z_\nu} = u_{2,3,\ldots,n-1}(Y)|Y|^{-\frac{1}{n}\sum_{\nu=2}^{n-1} \nu z_\nu}$$

only by a product of values of Riemann's zeta function, and we have seen in §10 that the series $u_{2,3,\ldots,n-1}(Y)$ formed with the exponents $z_2, z_3, \ldots, z_{n-1}$ converges in the given domain. From the estimate of $\eta(\overset{2}{Y}, z_1)$ we proved to hold in the domain defined by (1) we conclude that the series (2) converges in the domain

$$\vartheta_1 = \{|z_1| > 1, \;\; |z_1 - 1| > 1, \;\; -2 < \mathcal{R}e\, z_1 < 3, \;\; \mathcal{R}e\, z_2 > \tfrac{7}{2},$$

$$\mathcal{R}e\, z_\nu > 1 \;\; (2 < \nu < n)\}.$$

This shows that the functional equation

$$\sum_{\substack{n \\ \langle Y,\ldots,Y\rangle}}{}_{2}\left\{\prod_{\nu=3}^{n-1} |\overset{\nu}{Y}|^{-z_\nu}\right\} |\overset{2}{Y}|^{-z_2} \eta(\overset{2}{Y},z_1) =$$

$$= \sum_{\substack{n \\ \langle Y,\ldots,Y\rangle}}{}_{2}\left\{\prod_{\nu=3}^{n-1} |\overset{\nu}{Y}|^{-z_\nu}\right\} |\overset{2}{Y}|^{\frac{1}{2}-z_1-z_2} \eta(\overset{2}{Y},1-z_1)$$

holds in the subdomain $\vartheta_2$ of $\vartheta_1$ obtained by the restriction $\mathcal{R}e\, z_2 > 6$, because in $\vartheta_2$ both series are convergent. This means that

$$\pi^{-z_1}\Gamma(z_1)\zeta(Y,s) = \pi^{-(s_2-s_1+\frac{1}{2})}\Gamma(s_2-s_1+\tfrac{1}{2})\zeta(Y,s)$$

is invariant under the transformation

$$z_1 \mapsto 1-z_1, \quad z_2 \mapsto z_1+z_2-\tfrac{1}{2}, \quad z_\nu \mapsto z_\nu \quad (2<\nu<n), \quad s_n \mapsto s_n,$$

which is precisely the transposition of $s_1$ and $s_2$. It is now obvious that also

$$\prod_{1\leq\nu<\mu\leq n} \{\pi^{-(s_\mu-s_\nu+\frac{1}{2})}\Gamma(s_\mu-s_\nu+\tfrac{1}{2})\}|Y|^{s_n}\zeta(Y,s)$$

is invariant under the transposition of $s_1$ and $s_2$. We proved already the invariance under the group consisting of all cyclic permutations. These two results yield:

*Theorem 2. The meromorphic function*

$$\pi^{-2\sum_{\nu=1}^{n}\nu s_\nu} \prod_{1\leq\nu<\mu\leq n} \Gamma(s_\mu-s_\nu+\tfrac{1}{2})|Y|^{s_n}\zeta(Y;s)$$

*is invariant under all permutations of the variables* $s_1,s_2,\ldots,s_n$.

To conclude this section we investigate a general type of zeta functions related to the Größen-characters $u_{k_1,k_2,\ldots,k_r}(Y)$ considered in §10. A sequence $\overset{k_{r+1}}{Y}, \overset{k_r}{Y}, \ldots, \overset{k_1}{Y}$ of positive matrices is a *descending chain* if each link $\overset{k_\nu}{Y}$ $(1 \leqq \nu \leqq r)$ can be represented by the preceding link $\overset{k_{\nu+1}}{Y}$ $(k_{r+1} = n)$, i.e.,

$$\overset{k_\nu}{Y} = \overset{k_{\nu+1}}{Y}[G_\nu] \text{ with integral } G_\nu = G_\nu^{(k_{\nu+1},k_\nu)} \text{ of rank } k_\nu .$$

Define $\langle Y\rangle = \{Y[U] \mid U \in \Gamma\}$, where $\Gamma = \Gamma_n$ is the unimodular group. Then the equivalence class $\langle \overset{k_{r+1}}{Y}, \ldots, \overset{k_1}{Y}\rangle$ is by definition uniquely determined by the sequence $\langle \overset{k_{r+1}}{Y}\rangle, \ldots, \langle \overset{k_1}{Y}\rangle$. Again a descending chain will be called primitive if the matrices $G_\nu$ $(1 \leqq \nu \leqq r)$ are primitive. We write $\langle\ \rangle_p$ instead of $\langle\ \rangle$ to indicate that the class consists of primitive chains.

The general zeta functions are now defined by

$$\zeta^*(Y;s,h) = \sum_{\langle \overset{k_{r+1}}{Y}, \ldots, \overset{k_1}{Y}\rangle_p} \prod_{\nu=1}^{r} |\overset{k_\nu}{Y}|^{-s_\nu},$$

$$\zeta(Y;s,h) = \sum_{\langle \overset{k_{r+1}}{Y}, \ldots, \overset{k_1}{Y}\rangle} \prod_{\nu=1}^{r} |\overset{k_\nu}{Y}|^{-s_\nu},$$

where $\langle \overset{k_{r+1}}{Y}\rangle = \langle Y\rangle$ and

$$s = (s_1,s_2,\ldots,s_{r+1}), \qquad h = (h_1,h_2,\ldots,h_{r+1}).$$

These sequences determine

$$k_\nu = \sum_{\mu=1}^{\nu} h_\mu \quad (1 \leqq \nu \leqq r+1), \qquad s_\nu = s_{\nu+1} - s_\nu + \frac{h_{\nu+1} + h_\nu}{4} \quad (1 \leqq \nu \leqq r).$$

Similarly to the case $r = n-1$ we can prove the relation

$$\zeta(Y;s,h) = \zeta^*(Y;s,h)Z(s,h),$$

where

$$Z(s,h) = \prod_{1 \leqq \nu < \mu \leqq r+1} \prod_{\kappa=0}^{h_\nu - 1} \zeta(2(s_\mu - s_\nu) + \frac{h_\mu + h_\nu}{2} - \kappa).$$

A proof can be based on Lemma 5 below.

Define the set of matrices

$$\Delta_{h_1,h_2,\ldots,h_\kappa} = \{(Q_{\mu\nu}) \ (1 \leqq \mu,\nu \leqq \kappa) \mid Q_{\mu\nu} = Q_{\mu\nu}^{(h_\mu,h_\nu)} \text{ integral}, \ |Q_{\nu\nu}| \neq 0, \ Q_{\mu\nu} = 0 \ (\mu > \nu)\}.$$

The intersection $\Gamma_{h_1,h_2,\ldots,h_\kappa} = \Delta_{h_1,h_2,\ldots,h_\kappa} \cap \Gamma_{k_\kappa}$ is a subgroup of $\Gamma_{k_\kappa}$. The notation harmonizes with that of §10.

*Lemma* 4. *A given sequence of integral matrices* $G_\nu = G_\nu^{(k_{\nu+1},k_\nu)}$ *with* rank $G_\nu = k_\nu < k_{\nu+1}$ $(1 \leqq \nu \leqq r)$ *can be transformed into*

$$U_{\nu+1}^{-1} G_\nu U_\nu = \begin{pmatrix} D_\nu \\ 0 \end{pmatrix} \qquad (1 \leqq \nu \leqq r),$$

*with* $U_\nu \in \Gamma_{k_\nu}$ $(1 \leqq \nu \leqq r+1)$, *so that* $U_{r+1}, D_r, \ldots, D_1$ *belong to a given complete set of representatives of cosets in*

$$\Gamma/\Gamma_{h_1,h_2,\dots,h_{r+1}},\ \Delta_{h_1,h_2,\dots,h_r}/\Gamma_{h_1,h_2,\dots,h_r},\ \dots,\ \Delta_{h_1}/\Gamma_{h_1},$$

*respectively.*

*Proof.* It is easy to show by induction on $r$ that there exist unimodular matrices $\tilde{U}_\nu \in \Gamma_{k_\nu}$ $(1 \leqq \nu \leqq r+1)$ such that

$$\tilde{U}_{\nu+1}^{-1} G_\nu \tilde{U}_\nu = \begin{pmatrix} \tilde{D}_\nu \\ 0 \end{pmatrix} \qquad (1 \leqq \nu \leqq r),$$

with $\tilde{D}_\nu \in \Delta_{1,1,\dots,1}$, where the subscript 1 figures $k_\nu$ times. It is only necessary to observe that for a given $G_r$ there exist always $\tilde{U}_{r+1} \in \Gamma_{k_{r+1}}$ and $\tilde{D}_r \in \Delta_{1,1,\dots,1}$ $(\subset \Delta_{k_r})$

$$\tilde{U}_{r+1}^{-1} G_r = \begin{pmatrix} \tilde{D}_r \\ 0 \end{pmatrix} .$$

It remains only to prove the existence of $V_\nu \in \Gamma_{h_1,h_2,\dots,h_\nu}$ $(1 \leqq \nu \leqq r+1)$ such that

$$\tilde{U}_{r+1} V_{r+1} = U_{r+1}, \qquad V_{\nu+1}^{-1} \begin{pmatrix} \tilde{D}_\nu \\ 0 \end{pmatrix} V_\nu = \begin{pmatrix} D_\nu \\ 0 \end{pmatrix} \qquad (1 \leqq \nu \leqq r),$$

where $U_{r+1}, D_r, \dots, D_1$ are coset representatives as stated in the lemma. First we choose a suitable $V_{r+1}$. Clearly

$$V_{r+1}^{-1} \begin{pmatrix} \tilde{D}_r \\ 0 \end{pmatrix} = \begin{pmatrix} D_r^* \\ 0 \end{pmatrix}, \qquad D_r^* \in \Delta_{h_1,h_2,\dots,h_r}.$$

For an appropriate $V_r$ we have $D_r^* V_r = D_r$. Next

$$V_r^{-1} \begin{pmatrix} \tilde{D}_{r-1} \\ 0 \end{pmatrix} = \begin{pmatrix} D_{r-1}^* \\ 0 \end{pmatrix}, \qquad D_{r-1}^* \in \Delta_{h_1,h_2,\dots,h_{r-1}},$$

and we determine $V_{r-1}$ so that $D^{*}_{r-1}V_{r-1} = D_{r-1}$. Proceeding in this fashion we prove the lemma.

Special representatives of the equivalence classes are given by

*Lemma* 5. *There is a one-to-one correspondence between the classes* $<\overset{k_{r+1}}{Y},\ldots,\overset{k_1}{Y}>$ *with given* $<\overset{k_{r+1}}{Y}> = <Y>$ *and the sequences* $U, D_r, \ldots, D_1$ *of given representatives of the cosets in*

$$\Gamma/\Gamma_{h_1,h_2,\ldots,h_{r+1}},\ \Delta_{h_1,h_2,\ldots,h_r}/\Gamma_{h_1,h_2,\ldots,h_r},\ \ldots,\ \Delta_{h_1}/\Gamma_{h_1},$$

*respectively. The correspondence is given by*

$$\overset{k_{r+1}}{Y} = Y[U], \qquad \overset{k_\nu}{Y} = Y\left[U\begin{pmatrix}D_r\\0\end{pmatrix}\cdots\begin{pmatrix}D_\nu\\0\end{pmatrix}\right] \qquad (1 \leqq \nu \leqq r).$$

We omit the proof because it is similar to that of Lemma 2.

It is not hard now to prove the relation between $\zeta^{*}(Y;s,h)$ and $\zeta(Y;s,h)$ with the help of the special representatives

$$D_\nu = \begin{pmatrix} d_1 & d_{12} & \cdots & d_{1k_\nu} \\ & d_2 & \cdots & d_{2k_\nu} \\ & & \ddots & \vdots \\ 0 & & & d_{k_\nu} \end{pmatrix} : \quad 0 \leqq d_{\mu\kappa} < d_\mu \qquad (1 \leqq \mu < \kappa \leqq k_\nu).$$

We do not go into details.

Observing that for primitive classes $<\overset{k_{r+1}}{Y},\ldots,\overset{k_1}{Y}>_p$ necessarily $D_\nu = E^{(k_\nu)}$, we obtain

$$\zeta^{*}(Y;s,h) = u_{k_1,k_2,\ldots,k_r}(Y)\,|Y|^{-\frac{1}{n}\sum_{\nu=1}^{r} k_\nu s_\nu}.$$

This proves the convergence of the series which defines the zeta function for

$$(3) \qquad \mathcal{R}e\, z_\nu > \tfrac{1}{2}(h_{\nu+1}+h_\nu) \qquad (1 \leqq \nu \leqq r).$$

In order to investigate a transformation formula for the zeta functions with respect to the map $Y \mapsto Y^{-1}$, we introduce $\tilde{U} = (\delta_{n+1-\mu,\nu})$ and define an involution $s,h \mapsto \tilde{s},\tilde{h}$ by

$$\tilde{s}_\nu = -s_{r+2-\nu}, \qquad \tilde{h}_\nu = h_{r+2-\nu} \qquad (1 \leqq \nu \leqq r+1).$$

For

$$\tilde{k}_\nu = \sum_{\mu=1}^{\nu} \tilde{h}_\mu \quad (1 \leqq \nu \leqq r+1), \qquad \tilde{z}_\nu = \tilde{s}_{\nu+1} - \tilde{s}_\nu + \frac{\tilde{h}_{\nu+1}+\tilde{h}_\nu}{4} \quad (1 \leqq \nu \leqq r)$$

we get the relations

$$\tilde{k}_\nu = k_{r+1} - k_{r+1-\nu} \quad (1 \leqq \nu \leqq r+1), \qquad \tilde{z}_\nu = z_{r+1-\nu} \quad (1 \leqq \nu \leqq r)$$

with $k_o = 0$. It follows immediately that the conditions for convergence (3) are invariant with respect to the involution $s,h \mapsto \tilde{s},\tilde{h}$. We can prove the formula

$$\prod_{\nu=1}^{r} |Y_{\tilde{k}_\nu}|^{-\tilde{z}_\nu} = |Y|^{-s_{r+1}-\tilde{s}_{r+1}-\frac{1}{4}(k_r+\tilde{k}_r)} \prod_{\nu=1}^{r} |Y^{-1}[\tilde{U}]_{k_\nu}|^{-z_\nu}$$

essentially in the same way as in the case $r = n-1$ and obtain

$$\zeta^*(Y;\tilde{s},\tilde{h}) = \sum_{U:\Gamma/\Gamma_{\tilde{h}_1,\tilde{h}_2,\ldots,\tilde{h}_{r+1}}} \prod_{\nu=1}^{r} |Y[U]_{\tilde{k}_\nu}|^{-\tilde{z}_\nu}$$

$$= |Y|^{-s_{r+1}-\tilde{s}_{r+1}-\frac{1}{4}(k_r+\tilde{k}_r)} \sum_{U:\Gamma/\Gamma_{\tilde{h}_1,\tilde{h}_2,\ldots,\tilde{h}_{r+1}}} \prod_{\nu=1}^{r} |Y^{-1}[U'^{-1}\tilde{U}]_{k_\nu}|^{-z_\nu}.$$

Observing that the matrices $U'^{-1}\tilde{U}$ yield a complete set of coset representatives of $\Gamma/\Gamma_{h_1,h_2,\dots,h_{r+1}}$, because of $\tilde{U}\Gamma'_{\tilde{h}_1,\tilde{h}_2,\dots,\tilde{h}_{r+1}}\tilde{U} = \Gamma_{h_1,h_2,\dots,h_{r+1}}$, we obtain

$$\zeta^*(Y;\tilde{s},\tilde{h}) = |Y|^{-s_{r+1}-\tilde{s}_{r+1}-\frac{1}{2}(k_r+\tilde{k}_r)}\zeta^*(Y^{-1};s,h)$$

and

$$|Y|^{s_{r+1}+\frac{1}{2}k_r}Z(\tilde{s},\tilde{h})\zeta(Y;s,h) = |Y|^{-\tilde{s}_{r+1}-\frac{1}{2}\tilde{k}_r}Z(s,h)\zeta(Y^{-1};\tilde{s},\tilde{h}).$$

We prove that $\zeta(Y;s,h)$ is meromorphic by forming successive residues of $\zeta(Y;s)$ $(r = n-1)$ at points contained in the boundary of the domain of convergence each time. Of course this domain changes at every step because the number of variables decreases. For the proof an estimate of $\zeta(Y;s_1,s_2,h_1,h_2)$ is needed. With $S^{(m)}$, $z$, $n$, $m-n$ instead of $Y^{(n)}$, $z_1 = s_2 - s_1 + \frac{1}{2}(h_2 + h_1)$, $h_1$, $h_2$, respectively, this zeta function becomes

$$\phi_n(S,z) = \sum_{\{G\}} |S[G]|^{-z}$$

where $G = G^{(m,n)}$ runs over a complete set of integral matrices of rank $n$ which are not associated pairwise. In §16 we proved for

$$\Xi_n(S,z) = 2\varepsilon_n(z)\varepsilon_n(\tfrac{m}{2}-z)\prod_{\nu=0}^{n-1}\{\pi^{-z+\frac{1}{2}\nu}\Gamma(z-\tfrac{\nu}{2})\}\phi_n(S,z)$$

the representation

$$\Xi_n(S,z) = \int_{\substack{\mathcal{R}\\ |Y|>1}} \{\Theta(Y,S)|Y|^z + |S|^{-\frac{1}{2}n}\Theta(Y,S^{-1})|Y|^{\frac{1}{2}m-z}\}dv$$

where

$$\Theta(Y,S) = \sum_{G} P_{\frac{1}{2}m}\, e^{-\pi\sigma(S[G]Y)} \qquad (\text{rank } G = n).$$

As we have seen in §15 (p. 211), there exists a positive constant $C_1$ which as well as further on $C_2$, $C_3$ and $C_4$ depends only on $m$ such that

$$|P_{\frac{1}{2}m}\, e^{-\pi\sigma(S[G]Y)}| < C_1 e^{-\frac{1}{2}\pi\sigma(S[G]Y)}.$$

We obtain

$$|\Theta(Y,S)| < C_1 \theta_1(Y,\tfrac{1}{2}S)$$

where

$$\theta_1(Y,S) = \sum_{G} e^{-\pi\sigma(S[G]Y)} \qquad (\text{rank } G = n)$$

and furthermore

$$\int_{\mathcal{R}} \theta_1(Y,S)|Y|^{z} dv = 2 \prod_{\nu=0}^{n-1} \{\pi^{-z+\frac{1}{2}\nu}\Gamma(z-\tfrac{\nu}{2})\}\phi_n(S,z)$$

for $\mathcal{R}e\, z > \frac{1}{2}m$ because of

$$\theta_1(Y,S) < C|Y|^{-\frac{1}{2}m} e^{-\varepsilon\sigma(Y)} \qquad \text{for } Y \in \mathcal{R}$$

with certain positive constants $C = C(S)$ and $\varepsilon = \varepsilon(S)$.

We introduce the point set

$$\vartheta(m) = \{z \mid -\tfrac{1}{2} \leq \mathcal{R}e\, z \leq \tfrac{m+1}{2},\ |\mathcal{I}m\, z| \leq m,\ |z-\tfrac{\nu}{2}| \geq \tfrac{1}{5}\ (0 \leq \nu \leq m)\}$$

which is invariant with respect to $z \mapsto \frac{m}{2} - z$. Then we obtain for $z \in \vartheta(m)$ the estimate

$$|\Xi_n(S,z)| < C_1 \int\limits_{\substack{\mathcal{R}\\ |Y|>1}} \{\theta_1(Y,\tfrac{1}{2}S) + |S|^{-\frac{1}{2}n}\theta_1(Y,\tfrac{1}{2}S^{-1})\}|Y|^{\frac{1}{2}(m+1)}dv$$

$$< C_1 \int\limits_{\mathcal{R}} \{\theta_1(Y,\tfrac{1}{2}S) + |S|^{-\frac{1}{2}n}\theta_1(Y,\tfrac{1}{2}S^{-1})\}|Y|^{\frac{1}{2}(m+1)}dv$$

$$= 2C_1 \prod_{\nu=0}^{n-1} \{\pi^{-\frac{1}{2}(m+1-\nu)}\Gamma(\tfrac{m+1-\nu}{2})\}\{\phi_n(\tfrac{1}{2}S,\tfrac{m+1}{2}) + |S|^{-\frac{1}{2}n}\phi_n(\tfrac{1}{2}S^{-1},\tfrac{m+1}{2})\}$$

$$= C_2\{\phi_n(S,\tfrac{m+1}{2}) + |S|^{-\frac{1}{2}n}\phi_n(S^{-1},\tfrac{m+1}{2})\}.$$

The transformation formula of $\zeta(Y;s,h)$ with respect to $Y,s,h \mapsto Y^{-1},\tilde{s},\tilde{h}$ yields in our special case

$$\prod_{\nu=0}^{m-n-1} \zeta(2z-\nu)\cdot\phi_n(S^{-1},z) = \prod_{\nu=0}^{n-1} \zeta(2z-\nu)\cdot|S|^z\phi_{m-n}(S,z).$$

We apply this formula with $z = \frac{1}{2}(m+1)$ and obtain

$$|\Xi_n(S,z)| < C_3\{\phi_n(S,\tfrac{m+1}{2}) + |S|^{\frac{1}{2}(m+1-n)}\phi_{m-n}(S,\tfrac{m+1}{2})\}.$$

Denote by $\delta = \delta(m) > 0$ a lower bound of the absolute value of

$$2\varepsilon_n(z)\varepsilon_n(\tfrac{m}{2}-z)\prod_{\nu=0}^{n-1}\{\pi^{-z+\frac{1}{2}\nu}\Gamma(z-\tfrac{\nu}{2})\} \qquad \text{for } z\in\vartheta(m),$$

then finally

$$|\phi_n(S,z)| < C_4\{\phi_n(S,\tfrac{m+1}{2}) + |S|^{\frac{1}{2}(m+1-n)}\phi_{m-n}(S,\tfrac{m+1}{2})\} \qquad \text{for } z\in\vartheta(m)$$

with $C_4 = C_3\delta^{-1} = C_4(m) > 0$.

*Theorem* 3. *For a given integer* $\rho$ *in* $1 \leqq \rho \leqq r$ *the function*

$$(z_\rho - \tfrac{1}{2}(h_{\rho+1} + h_\rho))\zeta^*(Y;s,h)$$

*is holomorphic in the domain defined by*

$$|z_\rho - \tfrac{1}{2}(h_{\rho+1} + h_\rho)| < \tfrac{1}{4}, \qquad \mathcal{Re}\, z_\nu > n \quad (1 \leqq \nu \leqq r, \quad \nu \neq \rho).$$

*Its value at* $z_\rho = \frac{1}{2}(h_{\rho+1} + h_\rho)$ *is, up to a positive constant factor, equal to* $\zeta^*(Y;\hat{s},\hat{h})|Y|^{-\frac{1}{2}h_r^*}$, *where* $h_r^* = 0$ *or* $h_r$ *for* $\rho < r$ *or* $\rho = r$ *respectively and*

$$\hat{s} = (\hat{s}_1,\hat{s}_2,\ldots,\hat{s}_r), \qquad \hat{h} = (\hat{h}_1,\hat{h}_2,\ldots,\hat{h}_r),$$

$$\hat{s}_\nu = \begin{cases} s_\nu & \text{for } 1 \leqq \nu < \rho, \\ s_\rho + \frac{1}{4}h_{\rho+1} & \text{for } \nu = \rho, \\ s_{\nu+1} & \text{for } \rho < \nu \leqq r, \end{cases} \qquad \hat{h}_\nu = \begin{cases} h_\nu & \text{for } 1 \leqq \nu < \rho, \\ h_\rho + h_{\rho+1} & \text{for } \nu = \rho, \\ h_{\nu+1} & \text{for } \rho < \nu \leqq r. \end{cases}$$

*The function* $\zeta(Y;s,h)$ *is meromorphic in* $\mathbb{C}^{r+1}$.

*Proof.* We rewrite $\zeta^*(Y;s,h)$ as follows:

$$\zeta^*(Y;s,h) = \sum_p \left\{ \prod_{\substack{\nu=1 \\ \nu\neq\rho}}^{r} |\overset{k_\nu}{Y}|^{-z_\nu} \right\} \sum_{\langle \overset{k_\rho}{Y} \rangle} |\overset{k_\rho}{Y}|^{-z_\rho},$$

where the letter $p$ indicates that the summation is extended over all $\langle \overset{k_{r+1}}{Y}, \ldots, \overset{k_{\rho+1}}{Y}, \overset{k_{\rho-1}}{Y}, \ldots, \overset{k_1}{Y} \rangle_p$. In the interior sum $\langle \overset{k_\rho}{Y} \rangle$ ranges over all classes such that there exist primitive matrices $G_\rho$ and $G_{\rho-1}$ with

$$\overset{k_\rho}{Y} = \overset{k_{\rho+1}}{Y}[G_\rho], \qquad \overset{k_{\rho-1}}{Y} = \overset{k_\rho}{Y}[G_{\rho-1}].$$

Since ${}^{k_{\rho-1}}Y = {}^{k_{\rho+1}}Y\,[Q]$ holds with the primitive matrix $Q = G_\rho G_{\rho-1}$, we can choose ${}^{k_{\rho+1}}Y$ in $< {}^{k_{\rho+1}}Y >$ so that

$$(4) \qquad {}^{k_{\rho-1}}Y = {}^{k_{\rho+1}}Y \begin{bmatrix} E \\ 0 \end{bmatrix}$$

is valid. It follows that

$$G_\rho G_{\rho-1} = \varepsilon \begin{pmatrix} E \\ 0 \end{pmatrix}$$

with $\varepsilon = \pm 1$, and we can assume that $\varepsilon = 1$ because $G_{\rho-1}$ can be replaced by $\varepsilon G_{\rho-1}$. We can furthermore achieve

$$G_{\rho-1} = \begin{pmatrix} E \\ 0 \end{pmatrix}$$

by a transformation of the type $G_\rho, G_{\rho-1} \mapsto G_\rho U, U^{-1} G_{\rho-1}$ with $U \in \Gamma_{k_\rho}$. The solutions of

$$G_\rho \begin{pmatrix} E^{(k_{\rho-1})} \\ 0 \end{pmatrix} = \begin{pmatrix} E^{(k_{\rho-1})} \\ 0 \end{pmatrix}$$

have the form

$$G_\rho = \begin{pmatrix} E^{(k_{\rho-1})} & A \\ 0 & P \end{pmatrix}.$$

Multiplication of $G_\rho$ from the right-hand side by a matrix of type

$$\begin{pmatrix} E^{(k_{\rho-1})} & * \\ 0 & * \end{pmatrix} \quad (\in \Gamma_{k_\rho})$$

does not destroy the above relation. We can choose this matrix so that $A$ is replaced by 0 and $P$ by a given representative of the class $\{P\}$ of matrices associated with $P$. So we obtain

$$G_\rho = \begin{pmatrix} E^{(k_{\rho-1})} & 0 \\ 0 & P \end{pmatrix} \qquad \text{with primitive} \quad P = P^{(h_\rho + h_{\rho+1}, h_\rho)}$$

and the summation over $<\overset{k_\rho}{Y}>$ is equivalent to the summation over $\{P\}$. By (4) we have

$$\overset{k_{\rho+1}}{Y} = \begin{pmatrix} \overset{k_{\rho-1}}{Y} & 0 \\ 0 & S \end{pmatrix} \begin{bmatrix} E & X \\ 0 & E \end{bmatrix}$$

with appropriate matrices $S = S^{(h_\rho + h_{\rho+1})} > 0$ and $X = X^{(k_{\rho-1}, h_\rho + h_{\rho+1})}$. Choose $W \in \Gamma_{h_\rho + h_{\rho+1}}$ so that

$$P = W \begin{pmatrix} E^{(h_\rho)} \\ 0 \end{pmatrix},$$

then

$$G_\rho = \begin{pmatrix} E & 0 \\ 0 & W \end{pmatrix} \begin{pmatrix} E^{(k_\rho)} \\ 0 \end{pmatrix}$$

and

$$\overset{k_\rho}{Y} = \overset{k_{\rho+1}}{Y}[G_\rho] = \begin{pmatrix} \overset{k_{\rho-1}}{Y} & 0 \\ 0 & S \end{pmatrix} \begin{bmatrix} E & XW \\ 0 & W \end{bmatrix}_{k_\rho}$$

$$= \begin{pmatrix} \overset{k_{\rho-1}}{Y} & 0 \\ 0 & S[W] \end{pmatrix} \begin{bmatrix} E & XW \\ 0 & E \end{bmatrix}_{k_\rho} = \begin{pmatrix} \overset{k_{\rho-1}}{Y} & 0 \\ 0 & S[W]_{h_\rho} \end{pmatrix} \begin{bmatrix} E^{(k_{\rho-1})} & * \\ 0 & E^{(h_\rho)} \end{bmatrix},$$

hence

$$|{}^{k_\rho}Y| = |{}^{k_{\rho-1}}Y|\,|S[W]_{h_\rho}| = |{}^{k_{\rho-1}}Y|\,|S[P]|$$

and

$$\sum_{<{}^{k_\rho}Y>} |{}^{k_\rho}Y|^{-z_\rho} = |{}^{k_{\rho-1}}Y|^{-z_\rho} \sum_{\{P\}} |S[P]|^{-z_\rho}$$

$$= \{Z(s_\rho, s_{\rho+1}, h_\rho, h_{\rho+1})\}^{-1} |{}^{k_{\rho-1}}Y|^{-z_\rho} \phi_{h_\rho}(S, z_\rho)$$

$$= \left\{ \prod_{\nu=0}^{h_\rho-1} \zeta(2z_\rho - \nu) \right\}^{-1} |{}^{k_{\rho-1}}Y|^{-z_\rho} \phi_{h_\rho}(S, z_\rho).$$

This proves that

$$\prod_{\nu=0}^{h_\rho-1} \zeta(2z_\rho - \nu) \cdot \zeta^*(Y; s, h) =$$

$$= \sum_{p} \left\{ \prod_{\substack{\nu=1 \\ \nu\neq\rho,\rho-1}}^{r} |{}^{k_\nu}Y|^{-z_\nu} \right\} |{}^{k_{\rho-1}}Y|^{-z_{\rho-1}-z_\rho} \phi_{h_\rho}(S, z_\rho).$$

In case of $\rho = 1$ we have to omit the power of $|{}^{k_{\rho-1}}Y|$ and to replace $S$ by ${}^{k_2}Y$. By the result of §10 we can make the following assertion:

(I) $$\begin{cases} \displaystyle\sum_{p} \left\{ \prod_{\substack{\nu=1 \\ \nu\neq\rho,\rho-1}}^{r} |{}^{k_\nu}Y|^{-z_\nu} \right\} |{}^{k_{\rho-1}}Y|^{-z_{\rho-1}-z_\rho} \phi_{h_\rho}(S, z_\rho) \\ \text{converges for } \mathfrak{Re}\, z_\nu > \tfrac{1}{2}(h_{\nu+1} + h_\nu) \quad (1 \leqq \nu \leqq r). \end{cases}$$

The substitution $h_\rho \mapsto h_{\rho+1}$, $h_{\rho+1} \mapsto h_\rho$, $h_\nu \mapsto h_\nu$ $(\nu \neq \rho, \rho+1)$, $s \mapsto s$ has the effect

$$z_\nu \mapsto \begin{cases} z_\nu & \text{for } \nu \neq \rho\pm 1, \\ z_{\rho+1} + \frac{1}{4}(h_\rho - h_{\rho+1}) & \text{for } \nu = \rho+1, \quad \rho < r, \\ z_{\rho-1} + \frac{1}{4}(h_{\rho+1} - h_\rho) & \text{for } \nu = \rho-1, \quad \rho > 1, \end{cases}$$

$$k_\nu \mapsto \begin{cases} k_\nu & \text{for } \nu \neq \rho, \\ k_\rho + h_{\rho+1} - h_\rho & \text{for } \nu = \rho, \end{cases}$$

hence $S \mapsto S$ because of $\overset{k_\nu}{Y} \mapsto \overset{k_\nu}{Y}$ for $\nu \neq \rho$ and so (I) transforms into the assertion

$$\text{(II)} \quad \begin{cases} \sum_p \Big\{ \prod_{\substack{\nu=1 \\ \nu\neq\rho,\rho\pm1}}^{r} |\overset{k_\nu}{Y}|^{-z_\nu} \Big\} |\overset{k_{\rho-1}}{Y}|^{-z_{\rho-1}-z_\rho-\frac{1}{4}(h_{\rho+1}-h_\rho)} \cdot \\ \qquad \cdot |\overset{k_{\rho+1}}{Y}|^{-z_{\rho+1}-\frac{1}{4}(h_\rho-h_{\rho+1})} \phi_{h_{\rho+1}}(S, z_\rho) \\ \text{converges for } \quad \mathcal{R}e\, z_\nu > \frac{1}{2}(h_{\nu+1}+h_\nu) \quad (1 \leqq \nu \leqq r, \quad \nu \neq \rho\pm1), \\ \mathcal{R}e\, z_{\rho-1} > \frac{1}{4}(h_{\rho+1}+h_\rho+2h_{\rho-1}) \quad \text{if } \rho > 1, \\ \qquad \mathcal{R}e\, z_{\rho+1} > \frac{1}{4}(2h_{\rho+2}+h_{\rho+1}+h_\rho) \quad \text{if } \rho < r, \end{cases}$$

provided that possible powers of $|\overset{k_\nu}{Y}|$ for $\nu = 0$ and $r+1$ in $\sum_p$ are omitted.

Let us choose $z_\rho \in \vartheta(h_\rho + h_{\rho+1})$ and apply the estimate for

$\phi_n(S^{(m)},z)$ proved just before the statement of Theorem 3 with $h_\rho,\ h_\rho + h_{\rho+1},\ z_\rho$ instead of $n,\ m,\ z$. Observing that

$$|S| = |{}^{k_{\rho+1}}Y|\,|{}^{k_{\rho-1}}Y|^{-1}$$

we obtain

$$|\phi_{h_\rho}(S,z_\rho)| < c_4\Big\{\phi_{h_\rho}\Big(S\,,\,\frac{h_{\rho+1}+h_\rho+1}{2}\Big) +$$

$$+\,|{}^{k_{\rho+1}}Y|^{\frac{1}{2}(h_{\rho+1}+1)}\,|{}^{k_{\rho-1}}Y|^{-\frac{1}{2}(h_{\rho+1}+1)}\,\phi_{h_{\rho+1}}\Big(S\,,\,\frac{h_{\rho+1}+h_\rho+1}{2}\Big)\Big\}$$

and conclude that an upper bound of

$$\Big|\prod_{\nu=0}^{h_\rho-1}\zeta(2z_\rho-\nu)\cdot\zeta^*(Y;s,h)\Big| \qquad \text{for} \quad z_\rho\in\vartheta(h_\rho+h_{\rho+1})$$

is given by the sum of the two series

$$c_4\sum_{p}\Big\{\prod_{\substack{\nu=1\\ \nu\neq\rho,\rho-1}}^{r}|{}^{k_\nu}Y|^{-x_\nu}\Big\}\,|{}^{k_{\rho-1}}Y|^{-x_{\rho-1}-x_\rho}\,\phi_{h_\rho}\Big(S\,,\,\frac{h_{\rho+1}+h_\rho+1}{2}\Big)$$

and

$$c_4\sum_{p}\Big\{\prod_{\substack{\nu=1\\ \nu\neq\rho,\rho\pm1}}^{r}|{}^{k_\nu}Y|^{-x_\nu}\Big\}\,|{}^{k_{\rho+1}}Y|^{-x_{\rho+1}+\frac{1}{2}(h_{\rho+1}+1)}\,.$$

$$\cdot\,|{}^{k_{\rho-1}}Y|^{-x_{\rho-1}-x_\rho-\frac{1}{2}(h_{\rho+1}+1)}\,\phi_{h_{\rho+1}}\Big(S\,,\,\frac{h_{\rho+1}+h_\rho+1}{2}\Big)$$

with $x_\nu = \mathcal{R}e\, z_\nu$ $(1 \leqq \nu \leqq r)$ and $x_{\rho+1} = 0$ in case of $\rho = r$. By virtue of (I) and (II) the first of these series converges for

$$x_\nu > \tfrac{1}{2}(h_{\nu+1} + h_\nu) \qquad \text{for} \quad 1 \leqq \nu \leqq r, \quad \nu \neq \rho, \rho-1,$$

$$x_{\rho-1} > \tfrac{1}{2}(h_{\rho+1} + 2h_\rho + h_{\rho-1} + 1) - x_\rho \qquad \text{for} \quad \rho > 1$$

and the second for

$$x_\nu > \tfrac{1}{2}(h_{\nu+1} + h_\nu) \qquad \text{for} \quad 1 \leqq \nu \leqq r, \quad \nu \neq \rho, \rho\pm1,$$

$$x_{\rho-1} > \tfrac{1}{2}(h_{\rho+1} + h_\rho + h_{\rho-1}) - x_\rho \qquad \text{for} \quad \rho > 1,$$

$$x_{\rho+1} > \tfrac{1}{2}(h_{\rho+2} + h_{\rho+1} + h_\rho + 1) \qquad \text{for} \quad \rho < r.$$

Now $s_\rho \in \vartheta(h_\rho + h_{\rho+1})$ implies $x_\rho \geqq -\frac{1}{2}$ and so all inequalities will be simultaneously satisfied if

$$x_\nu > \tfrac{1}{2}(h_{\nu+1} + h_\nu) \qquad \text{for} \quad 1 \leqq \nu \leqq r, \quad \nu \neq \rho, \rho\pm1,$$

$$x_{\rho-1} > \tfrac{1}{2}(h_{\rho+1} + 2h_\rho + h_{\rho-1} + 2) \qquad \text{for} \quad \rho > 1,$$

$$x_{\rho+1} > \tfrac{1}{2}(h_{\rho+2} + h_{\rho+1} + h_\rho + 1) \qquad \text{for} \quad \rho < r$$

and the more if $x_\nu > n$ for $1 \leqq \nu \leqq r$, $\nu \neq \rho$. Denote the interior of $\vartheta$ by $\mathring{\vartheta}$ ; then we can state that

$$\left\{ \begin{array}{l} \displaystyle\prod_{\nu=0}^{h_\rho - 1} \zeta(2s_\rho - \nu) \cdot \zeta^*(Y; s, h) \quad \text{is holomorphic in the domain} \\[2ex] \text{described by} \quad s_\rho \in \mathring{\vartheta}(h_\rho + h_{\rho+1}), \quad \mathfrak{Re}\, s_\nu > n \quad (1 \leqq \nu \leqq r, \quad \nu \neq \rho) \end{array} \right.$$

because in any compact subdomain of this domain the function is represented as a uniformly convergent series of holomorphic functions. Since $\vartheta(h_\rho + h_{\rho+1})$ contains the annulus

$$\frac{1}{5} \leqq |z_\rho - \tfrac{1}{2}(h_{\rho+1} + h_\rho)| \leqq \frac{1}{4}$$

we can state that

$$\psi(z_\rho) = \prod_{\nu=0}^{h_\rho - 1} \zeta(2z_\rho - \nu)\cdot(z_\rho - \tfrac{1}{2}(h_{\rho+1} + h_\rho))\zeta^*(Y;s,h)$$

$$= \sum_{P} \left\{ \prod_{\substack{\nu=1 \\ \nu\neq\rho,\rho-1}}^{r} |\overset{k_\nu}{Y}|^{-z_\nu} \right\} |\overset{k_{\rho-1}}{Y}|^{-z_{\rho-1}-z_\rho} (z_\rho - \tfrac{1}{2}(h_{\rho+1} + h_\rho))\phi_{h_\rho}(S,z_\rho)$$

is holomorphic for $\frac{1}{5} < |z_\rho - \frac{1}{2}(h_{\rho+1} + h_\rho)| < \frac{1}{4}$, $\mathfrak{Re}\, z_\nu > n$ $(\nu \neq \rho)$. The general term of this series is holomorphic for $|z_\rho - \frac{1}{2}(h_{\rho+1}+h_\rho)| < \frac{1}{2}$ because

$$(z_\rho - \tfrac{1}{2}(h_{\rho+1} + h_\rho))\phi_{h_\rho}(S,z_\rho)$$

is holomorphic for $|z_\rho - \frac{1}{2}(h_{\rho+1} + h_\rho)| < \frac{1}{2}$ as we have proved in §16. There we have shown also that the residue of $\phi_{h_\rho}(S,z_\rho)$ at the point $z_\rho = \frac{1}{2}(h_{\rho+1} + h_\rho)$ is given by

$$\operatorname{Res} \phi_{h_\rho}(S,z_\rho) = c_\rho |S|^{-\frac{1}{2}h_\rho} = c_\rho |\overset{k_{\rho+1}}{Y}|^{-\frac{1}{2}h_\rho} |\overset{k_{\rho-1}}{Y}|^{\frac{1}{2}h_\rho}$$

with a certain positive constant $c_\rho$. By virtue of these properties we can conclude that $\psi(z_\rho)$ is holomorphic for $|z_\rho - \frac{1}{2}(h_{\rho+1} + h_\rho)| < \frac{1}{4}$, $\mathfrak{Re}\, z_\nu > n$ $(\nu \neq \rho)$ and that

$$\psi\left(\frac{h_{\rho+1} + h_\rho}{2}\right) =$$

$$= c_\rho \sum_p \left\{ \prod_{\substack{\nu=1 \\ \nu \neq \rho, \rho\pm 1}}^{r} |\overset{k_\nu}{Y}|^{-z_\nu} \right\} |\overset{k_{\rho-1}}{Y}|^{-z_{\rho-1} - \frac{1}{2}h_{\rho+1}} |\overset{k_{\rho+1}}{Y}|^{-z_{\rho+1} - \frac{1}{2}h_\rho}$$

$$= c_\rho \zeta^*(Y; \hat{s}, \hat{h}) |Y|^{-\frac{1}{2}h_r^*} \qquad (z_{\rho+1} = 0 \text{ if } \rho = r)$$

with $h_r^*$, $\hat{s}$, $\hat{h}$ as in the statement of Theorem 3. Since $\prod_{\nu=0}^{h_\rho - 1} \zeta(2z_\rho - \nu)$ does not vanish at $z_\rho = \frac{1}{2}(h_{\rho+1} + h_\rho)$ the first part of this theorem is proved.

We prove that $\zeta^*(Y; s, h)$ is meromorphic by induction on $n-r$. We have treated already the case $r = n-1$. Thus we can assume that $\zeta^*(Y; s, h)$ is meromorphic in $\mathbb{C}^{r+1}$ for a given $r \leqq n-1$. Then also

$$\prod_{\nu=0}^{h_\rho - 1} \zeta\left(2(s_{\rho+1} - s_\rho) + \frac{h_{\rho+1} + h_\rho}{2} - \nu\right) \cdot \left(s_{\rho+1} - s_\rho - \frac{h_{\rho+1} + h_\rho}{4}\right) \zeta^*(Y; s, h)$$

is meromorphic in $\mathbb{C}^{r+1}$. This implies that the value of this function on $s_{\rho+1} - s_\rho - \frac{1}{4}(h_{\rho+1} + h_\rho) = 0$ and thus $\zeta^*(Y; \hat{s}, \hat{h})$ is a meromorphic function of $\hat{s} \in \mathbb{C}^r$. Every zeta function $\zeta^*(Y; \hat{s}, \hat{h})$ with given $\hat{h}$ can be obtained in this way. The proof by induction is now complete.

The closing part of this paragraph concerns the analytic relations between, and in particular the functional equations of, the zeta functions $\zeta^*(Y; s, h)$ in terms of

$$\zeta_0^*(Y; s, h) = |Y|^{s_{r+1} - \frac{1}{2}h_{r+1}} \zeta^*(Y; s, h)$$

$$= |Y|^{s_{r+1} - \frac{1}{2}h_{r+1}} \sum_p \left\{ \prod_{\substack{\nu=1 \\ \nu \neq \rho, \rho-1}}^{r} |\overset{k_\nu}{Y}|^{-z_\nu} \right\} |\overset{k_{\rho-1}}{Y}|^{-z_{\rho-1} - z_\rho} \phi_{h_\rho}^*(S, z_\rho)$$

with $\phi_{h_\rho}^*$ defined by

$$\phi_{h_\rho}(S,z_\rho) = \prod_{\nu=0}^{h_\rho-1} \zeta(2z_\rho - \nu)\phi^*_{h_\rho}(S,z_\rho)$$

so that the relation

$$\phi^*_{h_\rho}(S,z_\rho) = \zeta^*(S;s_\rho,s_{\rho+1},h_\rho,h_{\rho+1})$$

holds. In §16 we proved a functional equation of $\phi_{h_\rho}(S,z_\rho)$ which can be rewritten as

$$\phi^*_{h_\rho}(S,z_\rho) = \prod_{\nu=0}^{h_\rho-1} \frac{\eta(h_\rho + h_{\rho+1} - \nu - 2z_\rho)}{\eta(2z_\rho - \nu)} |S|^{-\frac{1}{2}h_\rho} \phi^*_{h_\rho}(S^{-1}, \frac{h_{\rho+1} + h_\rho}{2} - z_\rho)$$

with

$$\eta(z) = \pi^{-\frac{1}{2}z}\Gamma(\tfrac{z}{2})\zeta(z).$$

By means of

$$\phi^*_{h_\rho}(S^{-1},z_\rho) = |S|^{z_\rho} \phi^*_{h_{\rho+1}}(S,z_\rho)$$

$$|S| = |\overset{k_{\rho+1}}{Y}||\overset{k_{\rho-1}}{Y}|^{-1}$$

we obtain the formal relation

$$\zeta^*_0(Y;s,h) = \prod_{\nu=0}^{h_\rho-1} \frac{\eta(h_\rho + h_{\rho+1} - \nu - 2z_\rho)}{\eta(2z_\rho - \nu)} \zeta^*_0(Y;s',h')$$

where

$$\zeta^*_0(Y;s',h') = |Y|^{s_{r+1}-\frac{1}{2}h_{r+1}} \sum_{p} \left\{ \prod_{\substack{\nu=1\\ \nu\neq\rho,\rho\pm1}}^{r} |\overset{k_\nu}{Y}|^{-z_\nu} \right\} |\overset{k_{\rho-1}}{Y}|^{-z_{\rho-1}-\frac{1}{2}h_{\rho+1}} \cdot$$

$$\cdot |\overset{k_{\rho+1}}{Y}|^{-z_{\rho+1}-z_\rho+\frac{1}{2}h_{\rho+1}} \phi^*_{h_{\rho+1}}(S, \frac{h_{\rho+1} + h_\rho}{2} - z_\rho)$$

and $z_{\rho+1} = 0$ in case of $\rho = r$. The sequences

$$s' = (s'_1, s'_2, \ldots, s'_{r+1}), \quad h' = (h'_1, h'_2, \ldots, h'_{r+1})$$

and

$$z'_\nu = s'_{\nu+1} - s_\nu + \frac{1}{4}(h'_{\nu+1} + h'_\nu) \qquad (1 \leqq \nu \leqq r),$$

$$k'_\nu = h'_1 + h'_2 + \ldots + h'_\nu \qquad (1 \leqq \nu \leqq r+1)$$

are determined by

$$k'_\nu = k_\nu \qquad \text{for } \nu \neq \rho,$$

$$h'_\rho = h_{\rho+1},$$

$$z'_\nu = z_\nu \qquad \text{for } \nu \neq \rho, \rho\pm 1,$$

$$z'_{\rho-1} + z'_\rho = z_{\rho-1} + \frac{1}{2}h_{\rho+1} \qquad \text{for } \rho > 1,$$

$$z'_\rho = \frac{1}{2}(h_{\rho+1} + h_\rho) - z_\rho,$$

$$z'_{\rho+1} = z_{\rho+1} + z_\rho - \frac{1}{2}h_{\rho+1} \qquad \text{for } \rho < r,$$

$$s'_{r+1} - \frac{1}{4}h'_{r+1} = \begin{cases} s_{r+1} - \frac{1}{4}h_{r+1} & \text{for } \rho < r, \\ s_{r+1} + \frac{1}{4}h_{r+1} - z_r & \text{for } \rho = r. \end{cases}$$

It is not hard to check that the transformation $\tau_\rho : s,h \mapsto s',h'$ represents a simultaneous transposition of the sequences $s$ and $h$ with the effect

$$\tau_\rho : s_\rho \leftrightarrow s_{\rho+1},\ h_\rho \leftrightarrow h_{\rho+1}, \text{ whereas } s_\nu,\ h_\nu\ (\nu \neq \rho, \rho+1) \text{ remain fixed.}$$

The above formal relation can be considered as an analytic identity as soon as we know that there exists an open domain in which both series for $\zeta_0^*(Y;s,h)$ and $\zeta_0^*(Y;s',h')$ after multiplication with

$$\prod_{\nu=0}^{h_\rho-1} \zeta(2s_\rho-\nu) \quad \text{and} \quad \prod_{\nu=0}^{h'_\rho-1} \zeta(2s'_\rho-\nu),$$

respectively, converge. Such a domain is actually defined by

$$s_\rho \in \mathring{\vartheta}(h_\rho+h_{\rho+1}), \qquad \mathfrak{Re}\, s_\nu > \tfrac{3}{2}n \qquad (\nu \neq \rho)$$

because these conditions imply obviously

$$s'_\rho \in \mathring{\vartheta}(h_\rho+h_{\rho+1}), \qquad \mathfrak{Re}\, s'_\nu > n \qquad (\nu \neq \rho)$$

since $\mathfrak{Re}\, s_\rho > -\frac{1}{2}$ and

$$\begin{aligned} s'_\nu &= s_\nu && \text{for } \nu \neq \rho, \rho\pm 1, \\ s'_\rho &= \tfrac{1}{2}(h_{\rho+1}+h_\rho) - s_\rho, && \\ s'_{\rho-1} &= s_{\rho-1}+s_\rho-\tfrac{1}{2}h_\rho && \text{for } \rho > 1, \\ s'_{\rho+1} &= s_{\rho+1}+s_\rho-\tfrac{1}{2}h_{\rho+1} && \text{for } \rho < r. \end{aligned}$$

Thus the modified series are both convergent. Observing that every permutation $\tau$ can be composed by the transpositions $\tau_\rho$ $(1 \leqq \rho \leqq r)$ we obtain finally the following result.

*Theorem* 4. *The meromorphic functions*

$$\zeta_0^*(Y;s,h) = |Y|^{s_{r+1}-\frac{1}{2}h_{r+1}}\, \zeta^*(Y;s,h)$$

*satisfy the analytic relations*

$$\frac{\zeta_0^*(Y;\tau_\rho s,\tau_\rho h)}{\zeta_0^*(Y;s,h)} = \prod_{\nu=0}^{h_\rho-1} \frac{\eta(2s_\rho-\nu)}{\eta(h_\rho+h_{\rho+1}-\nu-2s_\rho)} \qquad (1 \leqq \rho \leqq r),$$

*and for any permutation* $\tau$ *the quotient*

$$\zeta_0^*(Y;\tau s,\tau h):\ \zeta_0^*(Y;s,h)$$

*is an elementary function composed only by values of the function*

$$\eta(z) = \pi^{-\frac{1}{2}z}\Gamma(\tfrac{z}{2})\zeta(z).$$

There is only an apparent asymmetry in the relation with $\tau_\rho$, because the product on the right-hand side does not change if $\nu$ ranges from 0 to $h_{\rho+1}-1$ instead of $h_\rho - 1$. This is a simple consequence of the functional equation $\eta(1-z) = \eta(z)$ of Riemann's zeta function.

## §18. *Non-analytic Eisenstein series*

Let $\alpha, \beta$ be a pair of complex numbers, set $r = \alpha - \beta$, $q = \alpha + \beta$, and assume that $r$ is an even integer and $\mathcal{R}e\, q > n+1$. We introduce the non-analytic Eisenstein series

$$G(Z,\bar{Z}) = G(Z,\bar{Z};\alpha,\beta) = \sum_{\{C,D\}} |CZ+D|^{-\alpha}|C\bar{Z}+D|^{-\beta}$$

$$= \sum_{\{C,D\}} |CZ+D|^{-r}\|CZ+D\|^{-2\beta}$$

where $C,D$ runs over a complete set of non-associated coprime symmetric pairs of $n$-rowed square matrices (§11), $Z = Z^{(n)} = X+iY$, $Y > 0$ and we assume that

$$\arg|CZ+D| + \arg|C\bar{Z}+D| = 0,$$

so that

$$|CZ+D|^{-\beta}|C\bar{Z}+D|^{-\beta} = \|CZ+D\|^{-2\beta}.$$

For $\beta = 0$ this series is identical with the analytic Eisenstein series $E^{r}_{n,0}(Z,1)$ introduced in §14. The proof of Theorem 1 of §14 shows that for any given $\varepsilon > 0$ the series $G(Z,\bar{Z})$ converges absolutely and uniformly in the strip

$$\mathcal{V}_n(\varepsilon) = \{Z \in \mathfrak{H}_n \mid \sigma(X^2) \leq \tfrac{1}{\varepsilon},\ \ Y \geq \varepsilon E\}$$

provided that $\mathcal{R}e\, q > n+1$, which we have assumed.

In §8 we have seen that, with the notation introduced there, the series $G$ satisfies the differential equations $\Omega_{\alpha\beta}G = 0$ and the transformation formula $G|M_{\alpha,\beta} = G$, where

$$G|M_{\alpha,\beta}(Z,\bar{Z}) = G(M\langle Z\rangle, M\langle \bar{Z}\rangle)|CZ+D|^{-\alpha}|C\bar{Z}+D|^{-\beta}$$

and $M$ is an arbitrary modular matrix (§11) of degree $n$. In particular $G$ is invariant under $X \mapsto X+S$, where $S = S'$ is an arbitrary integral matrix. We shall compute the Fourier-expansion of $G$.

The one-to-one correspondence

$$\{C,D\} \leftrightarrow \{C_o^{(j)}, D_o^{(j)}\}, \{Q^{(n,j)}\}, \quad |C_o| \neq 0, \quad j = \text{rank } C,$$

described in the Lemma of §11, yields (cf. the beginning of §12)

$$G(Z,\bar{Z}) = 1 + \sum_{j=1}^{n} \sum_{\{Q\}} \sum_{\{C_o,D_o\}} |C_o Z[Q] + D_o|^{-\alpha} |C_o \bar{Z}[Q] + D_o|^{-\beta}.$$

The summation over $\{C_o, D_o\}$ can be replaced by a summation over all symmetric rational matrices $R^{(j)} = C_o^{-1} D_o$ (§12). By elementary divisor theory we can determine unimodular matrices $U, V$ such that $URV^{-1} = (\delta_{\mu\nu} e_\nu)$, where

$$e_\nu = \frac{a_\nu}{b_\nu} \geqq 0, \quad b_\nu > 0, \quad (a_\nu, b_\nu) = 1, \quad e_\nu | e_{\nu+1} \quad (1 \leqq \nu < j).$$

Then

$$\{C_o, D_o\} = \{(\delta_{\mu\nu} b_\nu) U, (\delta_{\mu\nu} a_\nu) V\},$$

so that $\|C_o\| = \prod_{\mu=1}^{j} b_\mu = \nu(R)$ is the product of the denominators of of the elementary divisors of $R$. We obtain

$$G(Z,\bar{Z}) = 1 + \sum_{j=1}^{n} \sum_{\{Q\}} \sum_{R} (\nu(R))^{-q} |Z[Q] + R|^{-\alpha} |\bar{Z}[Q] + R|^{-\beta}$$

$$= 1 + \sum_{j=1}^{n} \sum_{\{Q\}} \sum_{R \bmod 1} (\nu(R))^{-q} \psi(Z[Q] + R\,,\, \bar{Z}[Q] + R),$$

where

$$\psi = \psi(Z,\bar{Z}) = \sum_{\substack{S=S' \\ \text{integral}}} |Z + S|^{-\alpha} |\bar{Z} + S|^{-\beta} \qquad (\mathcal{R}\!e\, q > n+1).$$

The function $\psi$ is invariant under $X \mapsto X + S$ and has therefore a Fourier expansion

$$\psi(Z,\bar{Z}) = \sum_{\substack{T=T' \\ \text{semi-integral}}} c(Y,T) e^{2\pi i \sigma(TX)}.$$

Denote by $\mathcal{X}$ the unit cube in $X$-space. Then

$$c(Y,T) = \int_{\mathcal{X}} \cdots \int \psi(Z,\bar{Z}) e^{-2\pi i \sigma(TX)} [dX]$$

$$= \sum_{S} \int_{\mathcal{X}} \cdots \int |Z + S|^{-\alpha} |\bar{Z} + S|^{-\beta} e^{-2\pi i \sigma(T(X+S))} [dX]$$

$$= \int_{-\infty}^{\infty} \cdots \int |X + iY|^{-\alpha} |X - iY|^{-\beta} e^{-2\pi i \sigma(TX)} [dX].$$

If $\mathcal{R}\!e\,\alpha$, $\mathcal{R}\!e\,\beta > \frac{n-1}{2}$ then the formulae

$$\int_{P>0} e^{i\sigma(ZP)} |P|^{\alpha - \frac{1}{2}(n+1)} [dP] = \Gamma_n(\alpha) |-iZ|^{-\alpha},$$

$$\int_{Q>0} e^{-i\sigma(\bar{Z}Q)} |Q|^{\beta - \frac{1}{2}(n+1)} [dQ] = \Gamma_n(\beta) |i\bar{Z}|^{-\beta}$$

hold with

$$\Gamma_n(s) = \pi^{\frac{1}{4}n(n-1)} \prod_{\nu=0}^{n-1} \Gamma(s - \frac{\nu}{2})$$

(§6). We multiply together these integrals and substitute $V = P - Q$, $W = P + Q$, so that $W + V > 0$, $W - V > 0$ and furthermore

$$[dP][dQ] = 2^{-\frac{1}{2}n(n+1)}[dW][dV].$$

We get

$$\Gamma_n(\alpha)\Gamma_n(\beta)|-iZ|^{-\alpha}|i\bar{Z}|^{-\beta} =$$

$$= 2^{-n(q-\frac{1}{2}(n+1))} \int_{W\pm V>0} \cdots \int e^{-\sigma(YW)+i\sigma(XV)} \cdot$$

$$\cdot\ |W + V|^{\alpha-\frac{1}{2}(n+1)}|W - V|^{\beta-\frac{1}{2}(n+1)}[dW][dV]$$

or

$$\int_{-\infty}^{\infty} \cdots \int e^{i\sigma(XV)} \cdot$$

$$\cdot \left\{ \int_{W\pm V>0} \cdots \int e^{-\sigma(YW)}|W + V|^{\alpha-\frac{1}{2}(n+1)}|W - V|^{\beta-\frac{1}{2}(n+1)}[dW] \right\} [dV]$$

$$= e^{\frac{1}{2}\pi i r n} 2^{n(q-\frac{1}{2}(n+1))}\Gamma_n(\alpha)\Gamma_n(\beta)|Z|^{-\alpha}|\bar{Z}|^{-\beta}.$$

The Fourier inversion formula yields

$$\int_{W\pm V>0} \cdots \int e^{-\sigma(YW)}|W + V|^{\alpha-\frac{1}{2}(n+1)}|W - V|^{\beta-\frac{1}{2}(n+1)}[dW] =$$

$$= e^{\frac{1}{2}\pi i r n_2 n(q-1)}(2\pi)^{-\frac{1}{2}n(n+1)}\Gamma_n(\alpha)\Gamma_n(\beta)\ \cdot$$

$$\cdot \int_{-\infty}^{\infty}\!\!\cdots\!\int |X+iY|^{-\alpha}|X-iY|^{-\beta}e^{-i\sigma(XV)}[dX].$$

Substituting $W = 2\pi H$, $V = 2\pi T$ we get

$$c(Y,T) = \frac{e^{-\frac{1}{2}\pi i r n_2 n_\pi nq}}{\Gamma_n(\alpha)\Gamma_n(\beta)} h_{\alpha\beta}(Y,T)$$

where

$$h_{\alpha\beta}(Y,T) = \int_{H\pm T>0}\!\!\cdots\!\int e^{-2\pi\sigma(YH)}|H+T|^{\alpha-\frac{1}{2}(n+1)}|H-T|^{\beta-\frac{1}{2}(n+1)}[dH]$$

so that

$$\psi(Z,\overline{Z}) = \frac{e^{-\frac{1}{2}\pi i r n_2 n_\pi nq}}{\Gamma_n(\alpha)\Gamma_n(\beta)} \sum_T h_{\alpha\beta}(Y,T)e^{2\pi i\sigma(XT)}$$

provided that

$$\mathcal{R}\!\mathit{e}\,\alpha > \frac{n-1}{2}, \qquad \mathcal{R}\!\mathit{e}\,\beta > \frac{n-1}{2}.$$

In the analytic case $\beta = 0$ we proceed as follows. We define

$$\phi(H) = \begin{cases} |H|^{\alpha-\frac{1}{2}(n+1)}e^{2\pi i\sigma(HZ)} & \text{if } H > 0, \\ 0 & \text{if } H \not> 0, \end{cases}$$

and consider the periodic function

$$\chi(H) = \sum_{\substack{T=T' \\ \text{semi-integral}}} \phi(H+T)$$

which has a Fourier-expansion of the type

$$\chi(H) = \sum_{\substack{S=S' \\ \text{integral}}} \gamma(S) e^{2\pi i\sigma(HS)} .$$

A fundamental domain for the group of translations $H \mapsto H+T$ is given by

$$\mathfrak{F} = \{H = (h_{\mu\nu}) \mid 0 \leqq h_{\nu\nu} \leqq 1, \quad 0 \leqq h_{\mu\nu} \leqq \tfrac{1}{2} \quad \text{for} \quad \mu \neq \nu\}.$$

Thus we have

$$\gamma(S) = 2^{\frac{1}{2}n(n-1)} \int_{\mathfrak{F}} \cdots \int \chi(H) e^{-2\pi i\sigma(HS)} [dH]$$

$$= 2^{\frac{1}{2}n(n-1)} \sum_{T} \int_{\mathfrak{F}} \cdots \int \phi(H+T) e^{-2\pi i\sigma((H+T)S)} [dH]$$

$$= 2^{\frac{1}{2}n(n-1)} \int_{-\infty}^{\infty} \cdots \int \phi(H) e^{-2\pi i\sigma(HS)} [dH]$$

$$= 2^{\frac{1}{2}n(n-1)} \int_{H>0} \cdots \int e^{2\pi i\sigma(H(Z-S))} |H|^{\alpha-\frac{1}{2}(n+1)} [dH]$$

$$= e^{\frac{1}{2}\pi i\alpha n} \, 2^{-n(\alpha-\frac{1}{2}(n-1))} \, \pi^{-\alpha n} \Gamma_n(\alpha) |Z-S|^{-\alpha}$$

and

$$\chi(0) = \sum_{\substack{T>0 \\ \text{semi-integral}}} |T|^{\alpha-\frac{1}{2}(n+1)} e^{2\pi i\sigma(TZ)} = \sum_{\substack{S=S' \\ \text{integral}}} \gamma(S)$$

$$= e^{\frac{1}{2}\pi i\alpha n} \, 2^{-n(\alpha-\frac{1}{2}(n-1))} \, \pi^{-\alpha n} \Gamma_n(\alpha) \sum_{\substack{S=S' \\ \text{integral}}} |Z+S|^{-\alpha}$$

or

$$\psi = \psi(Z) = \frac{e^{-\frac{1}{2}\pi i \alpha n} 2^{n(\alpha-\frac{1}{2}(n-1))} \pi^{\alpha n}}{\Gamma_n(\alpha)} \sum_{T>0} |T|^{\alpha-\frac{1}{2}(n+1)} e^{2\pi i \sigma(TZ)}.$$

Substituting these expressions for $\psi$ into the above formula for $G(Z,\bar{Z})$, we obtain the Fourier-expansion of the Eisenstein-series:

$$G(Z,\bar{Z}) = 1 + \sum_{j=1}^{n} \sum_{T^{(j)}} \sum_{\{Q^{(n,j)}\}} a_{\alpha\beta}(T) h_{\alpha\beta}(Y[Q],T) e^{2\pi i \sigma(X[Q]T)},$$

where

$$a_{\alpha\beta}(T) = \frac{e^{-\frac{1}{2}\pi i r j} 2^{j} \pi^{jq}}{\Gamma_j(\alpha)\Gamma_j(\beta)} S(q,T),$$

$$S(q,T) = \sum_{R^{(j)} \bmod 1} \nu(R)^{-q} e^{2\pi i \sigma(RT)}$$

provided that

$$\mathcal{Re}\,\alpha > \frac{n-1}{2}, \qquad \mathcal{Re}\,\beta > \frac{n-1}{2}.$$

In the analytic case $\beta = 0$ we get

$$G(Z) = 1 + \sum_{j=1}^{n} \sum_{T^{(j)}>0} \sum_{\{Q^{(n,j)}\}} a_\alpha(T) e^{2\pi i \sigma(ZT[Q'])},$$

where

$$a_\alpha(T) = \frac{e^{-\frac{1}{2}\pi i \alpha j} 2^{j(\alpha-\frac{1}{2}(j-1))} \pi^{j\alpha}}{\Gamma_j(\alpha)} |T|^{\alpha-\frac{1}{2}(j+1)} S(\alpha,T).$$

Siegel [26] proved that for $T^{(j)} > 0$, $1 \leqq j < n+1 < k$, $k \equiv 0 \pmod 2$ the coefficients $a_k(T)$ are rational numbers. He established furthermore the following facts [31]: Denote by $B_j$ the $j^{\text{th}}$ Bernoulli number, by $d_k$ the product of the denominators of the $k$ rational numbers $B_{2j}/j$ where $j = 1,2,\ldots,k-1$ and $\frac{1}{2}k$. Let $2z_k$ be the highest power of 2 which is less than $k$. Then if $k \equiv 0 \pmod 4$ a common denominator of all the Fourier coefficients $a_k(T)$ with $T = T^{(j)} > 0$, $1 \leqq j < n+1 < k$, is equal to $d_k$, and if $k \equiv 2 \pmod 4$ a common denominator is equal to $z_k d_k$. It is well-known that the only prime numbers $p$ which divide $d_k$ are those which are irregular in the sense of Kummer, i.e., such that the class number of the field $\mathbb{Q}(e^{2\pi i/p})$ of the $p^{\text{th}}$ roots of unity is divisible by $p$. In the case $n = 2$ the coefficients $a_k(T)$, $T^{(2)} > 0$, have been determined explicity [20].

For singular matrices $T$ very little is known about the "singular series" $S(q,T)$. It can be shown that

$$S(q,0^{(n)}) = \frac{\zeta(q-n)}{\zeta(q)} \prod_{\nu=1}^{n} \frac{\zeta(2q-n-\nu)}{\zeta(2q-2\nu)},$$

where $\zeta(q)$ denotes Riemann's zeta function.

Now the question arises whether it is possible to attach Dirichlet series by means of integral transforms to the non-anayltic Eisenstein series $G(Z,\bar{Z})$ or, more generally, to automorphic forms of the same type, which might be described by the transformation property and the Fourier series. Without any restriction this is possible in the case $n = 1$. But already in the case $n = 2$ difficulties come up which show that one can not proceed in the usual way. They are related to the fact that in the case of $T = T^{(2)}$, $\sqrt{-|T|} =$ rational $\neq 0$, the group $\Gamma(T) = \{U \mid T[U] = T\}$ of units of $T$ is finite, so that the fundamental domain of $\Gamma(T)$ in $\mathcal{P}(T) = \{P \mid P > 0,\ PT^{-1}P = T\}$ has infinite volume (cf. end of §2). Therefore certain integrals do not exist.

It seems to be remarkable that the quotient

$$\frac{a_{\alpha+1,\beta-1}(T^{(j)})}{a_{\alpha\beta}(T^{(j)})} = (-1)^j \frac{\varepsilon_j(\beta-1)}{\varepsilon_j(\alpha)}$$

is on the whole independent of $T$. So one may guess that there is a close relation between the Eisenstein series $G(Z,\bar{Z};\alpha,\beta)$ and $G(Z,\bar{Z};\alpha+1,\beta-1)$ so that the proportionality of the Fourier coefficients is not accidental. This is actually true: we shall prove in §19 the existence of a differential operator $M_\alpha$ such that

$$M_\alpha G(Z,\bar{Z};\alpha,\beta) = \varepsilon_n(\alpha)G(Z,\bar{Z};\alpha+1,\beta-1),$$

and, more generally, a certain kind of non-analytic automorphic form of "type $\{\alpha,\beta\}$" is mapped by $M_\alpha$ in an analogous way. If $\beta$ is a natural number, then $G(Z,\bar{Z};\alpha,\beta)$, and also the non-analytic automorphic forms of type $\{\alpha,\beta\}$, can be transformed into analytic forms by the product $M_{\alpha+\beta-1} \cdots M_{\alpha+1}M_\alpha$. By this detour it is possible to attach Dirichlet series having reasonable properties also to non-analytic forms. The procedure is to take the Dirichlet series attached to the analytic form into which the non-analytic form is transformed. Of course this is no solution to the general problem because of the restriction $\beta \in \mathbb{N}$. But already in this case there are interesting applications to the theory of indefinite quadratic forms, defined by rational symmetric matrices with positive determinant. The general part of this theory will be treated in the next section.

## §19. *The differential operator* $M_\alpha$

Using the notation of §6 we define

$$M_\alpha = \sum_{h=0}^{n} \frac{\varepsilon_n(\alpha)}{\varepsilon_h(\alpha)} s_h(Z-\bar{Z}, \frac{\partial}{\partial Z}),$$

where $\varepsilon_h(\alpha) = \alpha(\alpha - \frac{1}{2}) \dots (\alpha - \frac{h-1}{2})$ for $h > 0$, $\varepsilon_0(\alpha) = 1$, and

$$s_h(Z-\bar{Z}, \frac{\partial}{\partial Z}) = \sum_{\substack{i_1<\dots<i_h \\ k_1<\dots<k_h}} \begin{pmatrix} i_1 & \dots & i_h \\ k_1 & \dots & k_h \end{pmatrix}_{Z-\bar{Z}} \begin{pmatrix} k_1 & \dots & k_h \\ i_1 & \dots & i_h \end{pmatrix}_{\frac{\partial}{\partial Z}} \qquad \text{for } h > 0,$$

$$s_0(Z-\bar{Z}, \frac{\partial}{\partial Z}) = 1.$$

In particular $s_n(Z-\bar{Z}, \frac{\partial}{\partial Z}) = |Z-\bar{Z}|\,|\frac{\partial}{\partial Z}|$. Let $f$ be an arbitrary function of $Z$ and $\bar{Z}$ and

$$S = \begin{pmatrix} A & B \\ C & D \end{pmatrix}$$

an arbitrary symplectic matrix of degree $n$. With the notation

$$f|S_{\alpha,\beta}(Z,\bar{Z}) = f(S\langle Z\rangle, S\langle\bar{Z}\rangle)\,|CZ+D|^{-\alpha}\,|C\bar{Z}+D|^{-\beta}$$

we then have

$$M_\alpha(f|S_{\alpha,\beta}) = (M_\alpha f)|S_{\alpha+1,\beta-1}.$$

It is easy to see that this transformation formula is equivalent to

$$|CZ+D|^{-1-\alpha}|C\bar{Z}+D|^{1-\beta}\hat{M}_\alpha = M_\alpha|CZ+D|^{-\alpha}|C\bar{Z}+D|^{-\beta},$$

where $\hat{M}_\alpha$ is the image of $M_\alpha$ under $Z \mapsto S\langle Z\rangle$, $\bar{Z} \mapsto S\langle\bar{Z}\rangle$. If the transformation formula holds for two symplectic matrices $S_1$ and $S_2$, then it also holds for their product $S_1S_2$, hence it is sufficient to prove it for generators of the symplectic group, e.g., for the matrices

$$\begin{pmatrix} E & T \\ 0 & E \end{pmatrix}, \quad \begin{pmatrix} V' & 0 \\ 0 & V^{-1} \end{pmatrix}, \quad \begin{pmatrix} 0 & -E \\ E & 0 \end{pmatrix},$$

where $T = \bar{T} = T'$ and $|V| \neq 0$, $V$ real. The assertion is obvious for the first type of matrices and very easy to prove for the second type of matrices [21, p. 47] but much more difficult for the third matrix. In this last case the transformation formula reduces to

$$|Z|^{-1-\alpha}|\bar{Z}|\hat{M}_\alpha|Z|^\alpha = M_\alpha$$

since $\hat{M}_\alpha$ commutes with $|\bar{Z}|$. Up to now it is only possible to prove this formula by means of complicated determinant relations [21]. The proof is cumbersome, so we confine the consideration to some remarks. The main idea of the proof is to express $s_h(Z-\bar{Z}, \frac{\partial}{\partial Z})$ in terms of the entries of $Z\frac{\partial}{\partial Z} = (D_\mu^\nu)$ ($\mu$ = row index, $\nu$ = column index), because on the basis of

$$Z\frac{\partial}{\partial Z}|Z|^\alpha = |Z|^\alpha(Z\frac{\partial}{\partial Z} + \alpha E)$$

it is then easy to see how $s_h(Z-\bar{Z}, \frac{\partial}{\partial Z})$ commutes with $|Z|^{-\alpha}$. This requires a generalization concerning symmetric matrices of the Capelli identity which is known for arbitrary square matrices [35, pp. 39-42]. The outcome of this generalization is the formula

$$\begin{pmatrix} \mu_1 & \cdots & \mu_h \\ \nu_1 & \cdots & \nu_h \end{pmatrix}_{\frac{\partial}{\partial Z}} = \sum_{\rho_1<\cdots<\rho_h} \begin{pmatrix} \mu_1 & \cdots & \mu_h \\ \rho_1 & \cdots & \rho_h \end{pmatrix}_{Z^{-1}} |D_{\rho_i}^{\nu_k} + \delta_{\rho_i\nu_k}\frac{h-k}{2}|,$$

where $1 \leqq i,k \leqq h$ ($i$ = row index, $k$ = column index). Since the entries of the last determinant do not commute, a special instruction is necessary how to compute it. In general we define a determinant $|\omega_{ik}|$ with $i$ as row index, $k$ as column index, $i,k = 1,\ldots,h$, and with elements $\omega_{ik}$ from a non-commutative ring, by

$$|\omega_{ik}| = \sum_{\mu_1,\ldots,\mu_h} \varepsilon_{\mu_1\mu_2\ldots\mu_h}\, \omega_{\mu_1 1}\, \omega_{\mu_2 2} \cdots \omega_{\mu_h h},$$

where $\mu_1,\ldots,\mu_h$ runs through all the $h!$ permutations of $1,2,\ldots,h$, and $\varepsilon_{\mu_1\mu_2\ldots\mu_h}$ denotes the sign of the permutation

$$\begin{pmatrix} 1 & 2 & & h \\ \mu_1 & \mu_2 & \ldots & \mu_h \end{pmatrix}.$$

Since

$$D_{\rho_i}^{\nu_k} |z|^{-\alpha} = |z|^{-\alpha}(D_{\rho_i}^{\nu_k} - \alpha\delta_{\rho_i\nu_k}),$$

the generalized Capelli identity now yields at once

$$\begin{pmatrix} \mu_1 & \ldots & \mu_h \\ \nu_1 & \ldots & \nu_h \end{pmatrix}_{\frac{\partial}{\partial z}} |z|^{-\alpha} = (-1)^h \varepsilon_h(\alpha) \begin{pmatrix} \mu_1 & \ldots & \mu_h \\ \nu_1 & \ldots & \nu_h \end{pmatrix}_{z^{-1}} |z|^{-\alpha}.$$

Denote by

$$\overline{\begin{pmatrix} \mu_1 & \ldots & \mu_h \\ \nu_1 & \ldots & \nu_h \end{pmatrix}_A}$$

the algebraic complement of

$$\begin{pmatrix} \mu_1 & \ldots & \mu_h \\ \nu_1 & \ldots & \nu_h \end{pmatrix}_A$$

in $|A|$ for any square matrix $A$, where we always assume that $\mu_1 < \dots < \mu_h$, $\nu_1 < \dots < \nu_h$. Then

$$\overline{\begin{pmatrix} \mu_1 & \dots & \mu_h \\ \nu_1 & \dots & \nu_h \end{pmatrix}_A} = (-1)^{\mu_1+\dots+\mu_h+\nu_1+\dots+\nu_h} \begin{pmatrix} \kappa_1 & \dots & \kappa_{n-h} \\ \lambda_1 & \dots & \lambda_{n-h} \end{pmatrix}_A$$

$$= |A| \begin{pmatrix} \nu_1 & \dots & \nu_h \\ \mu_1 & \dots & \mu_h \end{pmatrix}_{A^{-1}} ,$$

where we have denoted by $\kappa_1 \dots \kappa_{n-h}$ and $\lambda_1 \dots \lambda_{n-k}$ the terms left in $1,2,\dots,n$ after omitting $\mu_1,\dots,\mu_h$ or $\nu_1,\dots,\nu_h$, respectively. Thus we can rewrite the above relation in the form

$$\begin{pmatrix} \mu_1 & \dots & \mu_h \\ \nu_1 & \dots & \nu_h \end{pmatrix}_{\frac{\partial}{\partial Z}} |Z|^{-\alpha} = (-1)^h \varepsilon_h(\alpha) \overline{\begin{pmatrix} \nu_1 & \dots & \nu_h \\ \mu_1 & \dots & \mu_h \end{pmatrix}_Z} |Z|^{-\alpha-1}$$

[21, formula (45)]. In [13] this identity was proved in a complicated way. The asserted transformation formula follows from this identity.

Now we get easily

$$M_\alpha |CZ + D|^{-\alpha} |C\bar{Z} + D|^{-\beta} = \varepsilon_n(\alpha) |CZ + D|^{-\alpha-1} |C\bar{Z} + D|^{-\beta+1},$$

where $C,D$ is the second matrix row of an arbitrary symplectic matrix. The transformation formula of $M_\alpha$ applied to the function 1 yields indeed

$$M_\alpha |CZ + D|^{-\alpha} |C\bar{Z} + D|^{-\beta} = |CZ + D|^{-1-\alpha} |C\bar{Z} + D|^{1-\beta} \hat{M}_\alpha 1$$

$$= \varepsilon_n(\alpha) |CZ + D|^{-1-\alpha} |C\bar{Z} + D|^{-\beta+1},$$

since obviously $M_\alpha 1 = \hat{M}_\alpha 1 = \varepsilon_n(\alpha)$. Consequently we get the relation announced in §18:

$$M_\alpha G(Z,\overline{Z};\alpha,\beta) = \varepsilon_n(\alpha)G(Z,\overline{Z};\alpha+1,\beta-1).$$

We introduce now the generalized confluent hypergeometric function

$$I_{\alpha\beta}(Z,\overline{Z},T) = \int_{-\infty}^{\infty}\!\!\cdots\!\int |Z+V|^{-\alpha}|\overline{Z}+V|^{-\beta}\, e^{-2\pi i\sigma(VT)}[dV]$$

$$= \int_{-\infty}^{\infty}\!\!\cdots\!\int |V+iY|^{-\alpha}|V-iY|^{-\beta}\, e^{-2\pi i\sigma(VT)}[dV]e^{2\pi i\sigma(XT)},$$

where we take $\mathcal{R}\mathit{n}\, q > n+1$, $q = \alpha+\beta$, $T = \overline{T} = T'$, $V = V'$. For $I_{\alpha 0}(Z,\overline{Z},T)$ we shall simply write $I_\alpha(Z,T)$. In the case

$$\mathcal{R}\mathit{n}\,\alpha > \frac{n-1}{2}, \qquad \mathcal{R}\mathit{n}\,\beta > \frac{n-1}{2}$$

we have

$$I_{\alpha\beta}(Z,\overline{Z},T) = c(Y,T)e^{2\pi i\sigma(XT)}$$

$$= \frac{e^{-\frac{1}{2}\pi i rn}2^n\pi^{nq}}{\Gamma_n(\alpha)\Gamma_n(\beta)}\, h_{\alpha\beta}(Y,T)e^{2\pi i\sigma(XT)}$$

with the notation of §18.

In order to compute $I_q(Z,T)$, we determine the Fourier transform

$$\Phi(V) = \int_{-\infty}^{\infty}\!\!\cdots\!\int \phi(T)e^{2\pi i\sigma(TV)}[dT] \qquad (V' = V)$$

of the function

$$\phi(T) = \begin{cases} |T|^{q-\frac{1}{2}(n+1)} e^{2\pi i \sigma(TZ)} & \text{for } T > 0 \\ 0 & \text{for } T \not> 0. \end{cases}$$

One has immediately

$$\Phi(V) = e^{\frac{1}{2}\pi i n q} (2\pi)^{-nq} \Gamma_n(q) |Z + V|^{-q},$$

hence by the Fourier inversion formula

$$I_q(Z,T) = \frac{e^{-\frac{1}{2}\pi i n q}(2\pi)^{nq}}{\Gamma_n(q)} \int_{-\infty}^{\infty}\!\!\cdots\!\!\int \Phi(V) e^{-2\pi i \sigma(VT)} [dV]$$

$$= \frac{e^{-\frac{1}{2}\pi i n q}(2\pi)^{nq}}{2^{\frac{1}{2}n(n-1)}\Gamma_n(q)} \phi(T)$$

$$= \frac{e^{-\frac{1}{2}\pi i n q}(2\pi)^{nq}}{2^{\frac{1}{2}n(n-1)}\Gamma_n(q)} \cdot \begin{cases} |T|^{q-\frac{1}{2}(n+1)} e^{2\pi i \sigma(TZ)} & \text{for } T > 0, \\ 0 & \text{for } T \not> 0. \end{cases}$$

We replace in our formulae $Z$, $\bar{Z}$, $T = T^{(n)}$ by $Z[Q]$, $\bar{Z}[Q]$, $T^{(j)}$, respectively, where $Q = Q^{(n,j)}$ has rank $j$. Then we get

$$I_{\alpha\beta}(Z[Q],\bar{Z}[Q],T) = \int_{-\infty}^{\infty}\!\!\cdots\!\!\int |Z[Q] + V|^{-\alpha} |\bar{Z}[Q] + V|^{-\beta} e^{-2\pi i \sigma(VT)} [dV],$$

$$I_q(Z[Q],T) = \frac{e^{-\frac{1}{2}\pi i j q}(2\pi)^{jq}}{2^{\frac{1}{2}j(j-1)}\Gamma_j(q)} \cdot \begin{cases} |T|^{q-\frac{1}{2}(j+1)} e^{2\pi i \sigma(Z[Q]T)} & \text{for } T > 0, \\ 0 & \text{for } T \not> 0. \end{cases}$$

Complete $Q$ so that $U^{(n)} = (Q\ *)$, $|U| = 1$, then

$$(C,D) = \left(\begin{pmatrix} E^{(j)} & 0 \\ 0 & 0 \end{pmatrix} U', \begin{pmatrix} V^{(j)} & 0 \\ 0 & E \end{pmatrix} U^{-1}\right)$$

is a symmetric pair, i.e., the second matrix row of a symplectic matrix. We have

$$|CZ + D| = |Z[Q] + V|,$$

so by our above relation

$$M_\alpha |Z[Q] + V|^{-\alpha} |\bar{Z}[Q] + V|^{-\beta} = \varepsilon_n(\alpha) |Z[Q] + V|^{-\alpha-1} |\bar{Z}[Q] + V|^{-\beta+1}$$

and this yields

$$M_\alpha I_{\alpha\beta}(Z[Q],\bar{Z}[Q],T) = \frac{\Gamma_n(\alpha+1)}{\Gamma_n(\alpha)} I_{\alpha+1,\beta-1}(Z[Q],\bar{Z}[Q],T).$$

In the case $\beta \in \mathbb{N}$ it is reasonable to consider the product

$$M = M_{\alpha+\beta-1} \cdots M_{\alpha+1} M_\alpha$$

and to apply this operator to $I_{\alpha\beta}$. The result is

$$MI_{\alpha\beta}(Z[Q],\bar{Z}[Q],T) = \frac{\Gamma_n(q)}{\Gamma_n(\alpha)} I_q(Z[Q],T)$$

$$= \begin{cases} \dfrac{e^{-\frac{1}{2}\pi i j q}(2\pi)^{jq}\Gamma_n(q)}{\Gamma_j(q)\Gamma_n(\alpha)} |T|^{q-\frac{1}{2}(j+1)} e^{2\pi i \sigma(Z[Q]T)} & \text{for } T > 0, \\ 0 & \text{for } T \not> 0. \end{cases}$$

We add the formula

$$M1 = \frac{\Gamma_n(q)}{\Gamma_n(\alpha)} .$$

Our method of transforming $G(Z,\bar{Z};\alpha,\beta)$ in the case $\beta\in\mathbb{N}$ into an analytic modular form also works in the following general situation, when so-called non-analytic automorphic forms of type $\{\alpha,\beta\}$ enter: Assume that $f_k(Z,\bar{Z})$ $(1\leqq k\leqq h)$ are real analytic in $\mathfrak{H}$, have expansions

$$f_k(Z,\bar{Z}) = c_k + \sum_{j=1}^{n}\sum_{T^{(j)}}\sum_{Q^{(n,j)}} c_{kj}(T,Q)I_{\alpha\beta}(Z[Q],\bar{Z}[Q],T),$$

and that

$$f_k|S_{\alpha\beta} = \sum_{l=1}^{h}\gamma_{kl}(S)f_l$$

for every matrix $S$ belonging to a discrete subgroup $\Gamma$ of the symplectic group. If $\beta\in\mathbb{N}$, the functions $g_k = Mf_k$ $(1\leqq k\leqq h)$ are holomorphic in $\mathfrak{H}$ and satisfy the transformation formula

$$g_k|S = \sum_{l=1}^{h}\gamma_{kl}(S)g_l$$

for

$$S = \begin{pmatrix} A & B\\ C & D\end{pmatrix}\in\Gamma,$$

where

$$g|S(Z) = g(S\langle Z\rangle)|CZ+D|^{-q}.$$

This is obvious because $M(f_k|S_{\alpha\beta}) = (Mf_k)|S_{q0}$.

Atle Selberg mentioned in a personal conversation that the operator

$$|Z-\bar{Z}|^{\frac{1}{2}(n+1)-\alpha}\left|\frac{\partial}{\partial \bar{Z}}\right||Z-\bar{Z}|^{\alpha-\frac{1}{2}(n-1)}$$

has the same characteristic transformation property as $M_\alpha$. Actually the two operators are identical, so it is justified to write

$$M_\alpha = |Z-\bar{Z}|^{\frac{1}{2}(n+1)-\alpha}\left|\frac{\partial}{\partial \bar{Z}}\right||Z-\bar{Z}|^{\alpha-\frac{1}{2}(n-1)}.$$

To prove the transformation formula we first assert that

$$\left|\frac{\partial}{\partial \hat{\bar{Z}}}\right| = |CZ+D|^{\frac{1}{2}(n+3)}\left|\frac{\partial}{\partial \bar{Z}}\right||CZ+D|^{\frac{1}{2}(1-n)}$$

for $\hat{Z} = S\langle Z\rangle$ and an arbitrary symplectic matrix $S = \begin{pmatrix} A & B \\ C & D \end{pmatrix}$.

It suffices to prove this for a system of generators of the symplectic group. The relation is obvious for matrices of the type

$$\begin{pmatrix} E & T \\ 0 & E \end{pmatrix}, \quad \begin{pmatrix} U' & 0 \\ 0 & U^{-1} \end{pmatrix},$$

and for

$$S = \begin{pmatrix} 0 & E \\ -E & 0 \end{pmatrix}$$

it follows from

$$\left|\frac{\partial}{\partial \hat{Y}}\right| = (-1)^n |Y|^{\frac{1}{2}(n+3)}\left|\frac{\partial}{\partial Y}\right||Y|^{\frac{1}{2}(1-n)} \qquad \hat{Y} = Y^{-1}$$

(§6). For $\hat{Z} = \hat{X} + i\hat{Y} = S\langle Z\rangle$ we have

$$|\hat{Y}| = |Y||CZ+D|^{-1}|C\bar{Z}+D|^{-1},$$

i.e.,

$$|\hat{Z}-\hat{\bar{Z}}| = |Z-\bar{Z}||CZ+D|^{-1}|C\bar{Z}+D|^{-1},$$

so that one gets finally for the image $\hat{M}_\alpha$ of $M_\alpha$ under $Z \mapsto \hat{Z}$, $\bar{Z} \mapsto \hat{\bar{Z}}$ the desired relation

$$\hat{M}_\alpha = |CZ + D|^{\alpha+1} |C\bar{Z} + D|^{\beta-1} M_\alpha |CZ + D|^{-\alpha} |C\bar{Z} + D|^{-\beta}.$$

## §20. *Final aspects*

The operator $M_n = |Y|\,|\frac{\partial}{\partial Y}|$ defined in §6 can be defined with analogous properties also in more general domains, namely in the domains of positivity of M. Koecher. We refer to [22] for details, and give here only a brief report on those domains which are of special interest for the treatment of the zeta functions of semi-simple algebras over the field $\mathbb{Q}$ of rational numbers. Such a domain is contained in the algebra of all $n$-rowed square matrices with entries in a certain field $K$. We use the notation $K^{(n\times n)}$ for this algebra, and, more generally, $K^{(m\times n)}$ for the module of all matrices $A = A^{(m,n)}$ with entries in $K$. For $K$ we have to take into consideration the fields $\mathbb{R}$, $\mathbb{C}$ and the field of real quaternions

$$\mathbb{H} = \{z + wj \mid z,w \in \mathbb{C}\},$$

where $wj = j\bar{w}$ and $j^2 = -1$. We introduce the canonical anti-automorphism $y \mapsto y^*$ of $\mathbb{H}$ by

$$(z + wj)^* = \bar{z} - wj.$$

This mapping is an involution, it leaves $\mathbb{R}$ fixed and maps $\mathbb{C}$ onto itself, hence $K^* = K$. Clearly $y = y^*$ if and only if $y \in \mathbb{R}$. For any matrix $A = (a_{\mu\nu})$ with entries in $K$ we define $A^* = (a^*_{\nu\mu})$; then

$$(A + B)^* = A^* + B^*, \quad (AB)^* = B^*A^*$$

provided that $A + B$ and $AB$ are defined. On $K^{(n\times n)}$ the mapping $A \to A^*$ defines an involutory anti-automorphism.

A matrix $Y \in K^{(n\times n)}$ will be called *Hermitian* if $Y^* = Y$.

Together with $Y$ also $Y[C] = C^*YC$ is Hermitian. If $Y$ is Hermitian, then for any column vector $\varphi \in K^{(n\times 1)}$ the number $Y[\varphi]$ will be real. A Hermitian matrix $Y$ is said to be *positive*, in symbol $Y > 0$, if $Y[\varphi] > 0$ for every $\varphi \neq 0$. Now we can introduce the domain of positivity

$$\mathcal{Y}_n(K) = \{Y \mid Y \in K^{(n\times n)},\ Y = Y^* > 0\}.$$

As we shall explain, $\mathcal{Y}_n(K)$ is a weakly symmetric Riemannian space for $K = \mathbb{C}$ and $\mathbb{H}$, similarly to the case $K = \mathbb{R}$ (§6).

First of all we define an isomorphism of $\mathbb{H}$ into $\mathbb{C}^{(2\times 2)}$ by

$$z + wj \mapsto \begin{pmatrix} z & w \\ -\bar{w} & \bar{z} \end{pmatrix}.$$

Identifying $\mathbb{H}$ with its image in $\mathbb{C}^{(2\times 2)}$, we obtain an imbedding of the subfield $K$ of $\mathbb{H}$ into $\mathbb{C}^{(2\times 2)}$, and more generally of $K^{(n\times n)}$ into $\mathbb{C}^{(2n\times 2n)}$. In $\mathbb{C}^{(2n\times 2n)}$ the mapping $A \mapsto \bar{A}'$ is well-defined, and it is easy to see that $A^* = \bar{A}'$ for $A \in K^{(n\times n)} \subset \mathbb{C}^{(2n\times 2n)}$. For $Z = (z_{\mu\nu}) \in \mathbb{C}^{(2n\times 2n)}$ we set $\sigma(Z) = \sum_{\nu=1}^{2n} z_{\nu\nu}$ and for $A \in K^{(n\times n)}$ we define the *reduced trace* of $A$ by $\mathrm{tr}(A) = \frac{1}{2}\sigma(A)$. Then $\mathrm{tr}(AB) = \mathrm{tr}(BA)$ and $\mathrm{tr}(A) \in \mathbb{R}$ if $A = A^*$. Similarly, if $Y = Y^* > 0$, and if $|Y|$ denotes the determinant of $Y$ as an element of $\mathbb{C}^{(2n\times 2n)}$, then we define the *reduced determinant* by $\det(Y) = +\sqrt{|Y|}$.

Now we can assert that the metric fundamental form

$$ds^2 = \mathrm{tr}(Y^{-1}dY\,Y^{-1}dY)$$

on $\mathcal{Y}_n(K)$ is positive for $dY \neq 0$ and invariant under the maps

$$Y \mapsto Y[C], \qquad C \in GL(n,K),$$

$$\mu\colon Y \mapsto \mu(Y) = Y^{-1}.$$

For a given matrix $Y \in \mathcal{Y}_n(K)$ there exists a unitary matrix $U \in K^{(n\times n)}$, i.e., satisfying $UU^* = E$, such that $Y = U^*DU$, where $D$ is a real positive diagonal matrix. This shows that the square root $\sqrt{Y}$ exists in $\mathcal{Y}_n(K)$, namely $\sqrt{Y} = U^*\sqrt{D}\,U$. The proof that $\mathcal{Y}_n(K)$ is a weakly symmetric Riemannian space with respect to $GL(n,K)$ and $\mu$ is now the same as for $K = \mathbb{R}$.

Using the relation

$$\det(Y[C]) = \sqrt{|\bar{C}'Y C|} = \det(Y)\cdot\sqrt{|C\bar{C}'|} = \det(Y)\det(CC^*),$$

we get from $Y = U^*DU$, $U^*U = E$, $D = (\delta_{\mu\nu}d_\nu)$ that

$$\det(Y) = \det(D) = d_1 d_2 \dots d_n,$$

since the matrix in $\mathbb{C}^{(2n\times 2n)}$ corresponding to $(\delta_{\mu\nu}d_\nu) \in K^{(n\times n)}$ is

$$\begin{pmatrix} d_1 & & & & & \\ & d_1 & & & & \\ & & \cdot & & & \\ & & & \cdot & & \\ & & & & d_n & \\ & & & & & d_n \end{pmatrix}.$$

Also $\operatorname{tr}(Y^h) = \operatorname{tr}(D^h) = \sum_{\nu=1}^{n} d_\nu^h$, hence the function $\det(Y)$ is a polynomial of $\operatorname{tr}(Y^h)$, $h = 1,2,\dots,n$, thus a rational function of the real coordinates of $Y$.

It is not hard to prove that the invariant volume element on $\mathcal{Y}_n(K)$ has the form

$$dv = (\det Y)^{-q}[dY],$$

where

$$qn = \text{real dimension of } \mathcal{Y}_n(K) = \begin{cases} \frac{n(n+1)}{2} & \text{for } K = \mathbb{R}, \\ n^2 & \text{for } K = \mathbb{C}, \\ n(2n-1) & \text{for } K = \mathbb{H}, \end{cases}$$

and

$$[dY] = \begin{cases} \prod_{\mu \leq \nu} dy_{\mu\nu} & \text{for } K = \mathbb{R}, \\ \prod_{\nu} dy_{\nu\nu} \prod_{\mu<\nu} (\frac{i}{2}\, dy_{\mu\nu} d\bar{y}_{\mu\nu}) & \text{for } K = \mathbb{C}, \\ \prod_{\nu} dy_{\nu\nu} \prod_{\mu<\nu} (\frac{i}{2}\, dz_{\mu\nu} d\bar{z}_{\mu\nu} \frac{i}{2}\, dw_{\mu\nu} d\bar{w}_{\mu\nu}) & \text{for } K = \mathbb{H}, \end{cases}$$

with $Y = (y_{\mu\nu})$ and $y_{\mu\nu} = z_{\mu\nu} + w_{\mu\nu} j$ for $\mu \neq \nu$ in the third case.

We introduce the gamma-function of $\mathcal{Y}_n(K)$ by setting

$$\Gamma_n(s,K) = \int_{Y>0} e^{-\mathrm{tr}(Y)} (\det Y)^s dv.$$

There are two methods for evaluating this function, one due to Selberg which for $K = \mathbb{R}$ was explained in §6, the other due to Siegel which uses induction on $n$. The result of the calculation is

$$\Gamma_n(s,K) = \pi^{\frac{1}{4}n(n-1)h} \Gamma(s) \Gamma(s - \tfrac{h}{2}) \ldots \Gamma(s - \tfrac{h}{2}(n-1)),$$

where $h = (K:\mathbb{R})$ for $K = \mathbb{R}, \mathbb{C}, \mathbb{H}$.

There exists exactly one operator matrix $\frac{\partial}{\partial Y}$ such that

$$df = \mathrm{tr}(dY \tfrac{\partial}{\partial Y}) f \quad \text{and} \quad (\tfrac{\partial}{\partial Y} f)^* = \tfrac{\partial}{\partial Y} f$$

for an arbitrary real function $f = f(Y)$. Thus $Y \mapsto Y[C]$ with $C \in GL(n,K)$ implies the transformation $\frac{\partial}{\partial Y} \mapsto \frac{\partial}{\partial Y}[C^{*-1}]$ and consequently

$$M_n = \det(Y)\det(\tfrac{\partial}{\partial Y})$$

is an invariant differential operator.

For $T = T^*$ we have

$$d\,\mathrm{tr}(YT) = \mathrm{tr}(dY \tfrac{\partial}{\partial Y})\mathrm{tr}(YT) = \mathrm{tr}(dYT),$$

hence

$$\frac{\partial}{\partial Y}\,\mathrm{tr}(YT) = T$$

and

$$\det(\tfrac{\partial}{\partial Y})e^{-\mathrm{tr}(YT)} = (-1)^n \det(T)e^{-\mathrm{tr}(YT)}.$$

Applying $\det(\frac{\partial}{\partial T})$ to

$$\int_{Y>0} e^{-\mathrm{tr}(YT)}(\det Y)^s dv = \Gamma_n(s,K)(\det T)^{-s} \qquad (T = T^* > 0)$$

we get

$$\det(\tfrac{\partial}{\partial T})(\det T)^{-s} = (-1)^n \frac{\Gamma_n(s+1,K)}{\Gamma_n(s,K)}(\det T)^{-s-1}$$

$$= (-1)^n \prod_{\nu=0}^{n-1} (s - \tfrac{h}{2}\nu)(\det T)^{-s-1}.$$

The image $\hat{M}_n$ of $M_n$ under the substitution $Y \mapsto Y^{-1}$ can be determined explicitly with the help of a generalization of the Cauchy formula for several complex variables due to S. Bochner [1] (see [22]). One obtains

$$\hat{M}_n = (-1)^n(\det Y)^{q-1} M_n(\det Y)^{1-q},$$

which shows in particular that $\hat{M}_n$ is adjoint to $M_n$ with respect to the invariant volume element $dv$. We introduce, as we did for $K = \mathbb{R}$, the operator

$$P_k = P_k(Y) = (\det Y)^{-k}\hat{M}_n(\det Y)^k M_n$$

and see that the image $\hat{P}_k$ of $P_k$ under $Y \mapsto Y^{-1}$ is given by

$$\hat{P}_k = (\det Y)^k P_k (\det Y)^{-k}.$$

Let $\mathfrak{A}$ be a semi-simple algebra over $\mathbb{Q}$ and assume that $\mathfrak{A} \otimes \mathbb{R}$ splits into

$$\mathfrak{A} \otimes \mathbb{R} = \bigoplus_{\nu=1}^{r} K_\nu^{(n_\nu \times n_\nu)}$$

with

$$K_\nu \in \{\mathbb{R}, \mathbb{C}, \mathbb{H}\}, \qquad n_\nu \in \mathbb{N}.$$

Then it is meaningful to consider theta-series of the type

$$\theta(Y,S;\mathfrak{M}) = \sum_{G \in \mathfrak{M}} \prod_{\nu=1}^{r} e^{-\pi \operatorname{tr}(S_\nu[G_\nu]Y_\nu)},$$

where $\mathfrak{M}$ is a lattice in $\mathfrak{A} \otimes \mathbb{R}$ of maximal dimension,

$$Y = \bigoplus_{\nu=1}^{r} Y_\nu, \qquad S = \bigoplus_{\nu=1}^{r} S_\nu, \qquad Y_\nu, S_\nu \in \mathcal{P}_{n_\nu}(K_\nu),$$

and

$$G = \bigoplus_{\nu=1}^{r} G_\nu, \qquad G_\nu \in K_\nu^{(n_\nu \times n_\nu)}.$$

One has to prove a transformation formula of the well-known type for

$\theta(Y,S;\mathfrak{m})$ with respect to $Y \mapsto Y^{-1}$, $S \mapsto S^{-1}$.

K. Hey [3] was the first who tried to get functional equations for the zeta functions of $\mathfrak{A}$ by using theta-series. But she succeeded only for division algebras. The appropriate restriction is made explicitly only once in a subordinate clause in the middle of p. 21. The difficulties in the general case arise from the zero divisors. $G \in \mathfrak{A} \otimes \mathbb{R}$ is a zero-divisor if and only if $|G_\nu| = 0$ for at least one $\nu$, and this is equivalent to $\det(S_\nu[G_\nu]) = 0$. But now it is very easy to eliminate the zero-divisors by applying $\prod_{\nu=1}^{r} P_{\frac{1}{2}n_\nu}(Y_\nu)$ to $\theta(Y,S;\mathfrak{m})$. Then the Mellin transformation gives the desired result.

## References


1. S. Bochner, *Group invariance of Cauchy's formula in several variables*, Ann. of Math. (2) 45 (1944), 686-707.

2. E. Freitag, *Zur Theorie der Modulformen zweiten Grades*, Nachr. Akad. Wiss. Göttingen, II. Math.-Phys. Kl. 1965, 151-157.

3. K. Hey, *Analytische Zahlentheorie in Systemen hyperkomplexer Zahlen* (Dissertation), Lütcke und Wulff, Hamburg, 1929.

4. L. Hörmander, *An introduction to complex analysis in several variables*, Van Nostrand, Princeton, N. J., 1966.

5. J. Igusa, *On Siegel modular forms of genus two*, Amer. J. Math. 84 (1962), 306-316.

6. ———, *On Siegel modular forms of genus two*, Amer. J. Math. 86 (1964), 392-412.

7. G. Kaufhold, *Dirichletsche Reihe mit Funktionalgleichung in der Theorie der Modulfunktion 2. Grades*, Math. Ann. 137 (1959), 454-476.

8. H. Klingen, *Charakterisierung der Siegelschen Modulgruppe durch ein endliches System definierender Relationen*, Math. Ann. 144 (1961), 64-72.

9. ———, *Zum Darstellungssatz für Siegelsche Modulformen*, Math. Z. 102 (1967), 30-43; Berichtigung, 105 (1968), 399-400.

10. M. Koecher, *Über Dirichlet-Reihen mit Funktionalgleichung*, J. Reine Angew. Math. 192 (1953), 1-23; see also the review in Zentralblatt f. Math. 53, 55.

11. ———, *Über Thetareihen indefiniter quadratischer Formen*, Math. Nachr. 9 (1953), 51-85.

12. H. Maaß, *Modulformen zweiten Grades und Dirichletreihen*, Math. Ann. 122 (1950), 90-108.

13. ———, *Die Differentialgleichungen in der Theorie der Siegelschen Modulfunktionen*, Math. Ann. 126 (1953), 44-68.

14. ———, *Die Bestimmung der Dirichletreihen mit Größencharakteren zu den Modulformen n-ten Grades*, J. Indian Math. Soc. 19 (1955), 1-23.

15. ———, *Spherical functions and quadratic forms*, J. Indian Math. Soc. 20 (1956), 117-162.

16. ———, *Zetafunktionen mit Größencharakteren und Kugelfunktionen*, Math. Ann. 134 (1957), 1-32.

17. H. Maaß, *Zur Theorie der Kugelfunktionen einer Matrixvariablen*, Math. Ann. 135 (1958), 391-416.

18. ————, *Über die räumliche Verteilung der Punkte in Gittern mit indefiniter Metrik*, Math. Ann. 138 (1959), 287-315.

19. ————, *Die Multiplikatorsysteme zur Siegelschen Modulgruppe*, Nachr. Akad. Wiss. Göttingen, II. Math.-Phys. Kl. 1964, 125-135.

20. ————, *Die Fourierkoeffizienten der Eisensteinreihen zweiten Grades*, Mat.-Fys. Medd. Danske Vid. Selsk. 34, no. 7, (1964).

21. ————, *Modulformen zu indefiniten quadratischen Formen*, Math. Scand. 17 (1965), 41-55.

22. H. L. Resnikoff, *On singular automorphic forms in several complex variables*, Amer. J. Math., to appear.

23. A. Selberg, *Harmonic analysis and discontinuous groups*, J. Indian Math. Soc. 20 (1956), 47-87.

24. ————, *A new type of zeta functions connected with quadratic forms*, Report of the Institute in the Theory of Numbers, Boulder, Colorado, 1959, 207-210.

25. ————, *Discontinuous groups and harmonic analysis*, Proc. Intern. Congr. of Mathematicians, Stockholm, 1962, 177-189.

26. C. L. Siegel, *Einführung in die Theorie der Modulfunktionen n-ten Grades*, Math. Ann. 116 (1939), 617-657; Gesammelte Abhandlungen, Springer-Verlag, Berlin-Heidelberg-New York, 1966, no. 32, vol. II, 97-137.

27. ————, *Einheiten quadratischer Formen*, Abh. Math. Sem. Univ. Hamburg 13 (1940), 209-239; Gesammelte Abhandlungen, no. 33, vol. II, 138-168.

28. ————, *Symplectic geometry*, Amer. J. Math. 65 (1943), 1-86; Gesammelte Abhandlungen, no. 41, vol. II, 274-359.

29. ————, *Some remarks on discontinuous groups*, Ann. of Math. (2) 46 (1945), 708-718; Gesammelte Abhandlungen, no. 53, vol. III, 67-77.

30. ————, *Zur Bestimmung des Volumens des Fundamentalbereichs der unimodularen Gruppe*, Math. Ann. 137 (1959), 427-432; Gesammelte Abhandlungen, no. 73, vol. III, 328-333.

31. ————, *Über die Fourierschen Koeffizienten der Eisensteinschen Reihen*, Mat.-Fys. Medd. Danske Vid. Selsk. 34, no. 6, (1964); Gesammelte Abhandlungen, no. 79, vol. III, 443-458.

32. ————, *Über die Fourierschen Koeffizienten von Eisensteinschen Reihen der Stufe T*, Math. Z. 105 (1968), 257-266.

33. A. Terras, *A generalization of Epstein's zeta function*, to appear.

34. ————, *Functional equations of generalized Epstein zeta functions in several complex variables*, to appear.

35. H. Weyl, *The classical groups*, Princeton mathematical series no. 1, Princeton University Press, Princeton, N. J., 1939.

36. E. Witt, *Eine Identität zwischen Modulformen zweiten Grades*, Abh. Math. Sem. Univ. Hamburg 14 (1941), 323-337.